Mathematics and Visualization

For further volumes:
http://www.springer.com/series/4562

Peer-Timo Bremer • Ingrid Hotz •
Valerio Pascucci • Ronald Peikert
Editors

Topological Methods in Data Analysis and Visualization III

Theory, Algorithms, and Applications

With 98 Figures, 69 in color

Springer

Editors

Peer-Timo Bremer
Center for Applied Scientific Computing
Lawrence Livermore National Laboratory
Livermore
California
USA

Ingrid Hotz
Comparative Visualization Group
Konrad-Zuse-Zentrum für Informationstechnik
Berlin
Germany

Valerio Pascucci
School of Computing and SCI Institute
University of Utah
Salt Lake City
Utah
USA

Ronald Peikert
Department of Computer Science
ETH Zürich
Zürich
Switzerland

ISSN 1612-3786 ISSN 2197-666X (electronic)
ISBN 978-3-319-04098-1 ISBN 978-3-319-04099-8 (eBook)
DOI 10.1007/978-3-319-04099-8
Springer Cham Heidelberg New York Dordrecht London

Library of Congress Control Number: 2011944972

Mathematical Subject Classification (2010): 76M24, 53A45, 62-07, 62H35, 65D18, 65U05, 68U10

Printed on acid-free paper

Springer is part of Springer Science+Business Media (www.springer.com)

Preface

As many areas in science and engineering are relying more and more heavily on computational science, the analysis of extremely large and complex data sets is becoming ever more crucial. Even though they are a comparatively recent development, topological techniques have proven highly valuable in this context as they can provide a high level, abstract view of data which is well aligned with human intuition. As a result, topological concepts often translate directly into features of interest in various applications and thus can reduce the time-to-insight.

The quick rise in popularity of topological techniques has led to an interest and exciting state of the art that spans the entire gamut of analysis of extremely large data to fundamental, theoretical work. Similar to previous events, TopoInVis 2013 was designed to provide a forum for research in the entire spectrum of techniques to be discussed with a focus on experimental solutions for open problems. The event was held in Davis, California, on the beautiful campus of the University of California with the gracious support of the Scientific Computing and Imaging Institute of the University of Utah and the Computer Science Department at UC Davis. The two and a half day event led to a number of interesting discussions on the current state of the art in topological research as well as interesting panel discussions and off-line research collaborations. The conference was followed by an open call for contributions to this volume in which all presenters as well as the community at large was invited to submit novel research contributions. This resulted in 17 interesting chapters ranging from contributions to flow field analysis to applications in medical and material science.

Preface

Acknowledgements

TopoInVis 2013 has been supported by the Scientific Computing and Imaging Institute of the University of Utah and the Computer Science Department at UC Davis. In particular, we would like to acknowledge the help of Deborah Zemek and Nathan Galli at Utah in organizing and the support that we received from our colleagues of the Institute for Data Analysis and Visualization (IDAV) and the Department of Computer Science at UC Davis.

Contents

Part I
Robust Topological Analysis

Robust Detection of Singularities in Vector Fields

Harsh Bhatia, Attila Gyulassy, Hao Wang, Peer-Timo Bremer, and Valerio Pascucci

Abstract Recent advances in computational science enable the creation of massive datasets of ever increasing resolution and complexity. Dealing effectively with such data requires new analysis techniques that are provably robust and that generate reproducible results on any machine. In this context, combinatorial methods become particularly attractive, as they are not sensitive to numerical instabilities or the details of a particular implementation. We introduce a robust method for detecting singularities in vector fields. We establish, in combinatorial terms, necessary and sufficient conditions for the existence of a critical point in a cell of a simplicial mesh for a large class of interpolation functions. These conditions are entirely local and lead to a provably consistent and practical algorithm to identify cells containing singularities.

1 Introduction

Vector fields, which are either acquired from real-world experiments or generated by computer simulations, are ubiquitous in scientific research. In recent years, the increase in computational power coupled with the ability to simulate ever more complex data has greatly increased the need for the automatic analysis of large-scale vector fields. To handle the complexity and size of modern simulations,

H. Bhatia (✉) • P.-T. Bremer
SCI Institute, University of Utah, Salt Lake City, UT, USA

Center for Applied Scientific Computing, Lawrence Livermore National Laboratory, Livermore, CA, USA
e-mail: hbhatia@sci.utah.edu; bhatia4@llnl.gov; ptbremer@sci.utah.edu; bremer5@llnl.gov

A. Gyulassy • V. Pascucci • H. Wang
SCI Institute, University of Utah, Salt Lake City, USA
e-mail: jediati@sci.utah.edu; pascucci@sci.utah.edu; haow@cs.utah.edu

P.-T. Bremer et al. (eds.), *Topological Methods in Data Analysis and Visualization III*, Mathematics and Visualization, DOI 10.1007/978-3-319-04099-8_1,

flexible and multi-scale methods are needed. These techniques must be able to define features at different levels of resolution, remove noise, and most importantly be computationally robust and consistent with the mathematical theory of vector fields. State-of-the-art topological techniques fulfill many of these requirements by defining features based on the global behavior of *streamlines*, which are the curves parallel to the direction of flow at a given instant, and by providing feature-based simplification. For example, topological techniques for 2D vector fields use *singularities* – points where the field is zero, such as sources, sinks, saddles, etc., together with *separatrices* – the streamlines of saddles, and *limit cycles* – the streamlines which wrap back onto themselves, to decompose the vector field into regions of similar flow behavior. Such decompositions are then used to analyze, visualize [21,24], simplify [29,31], and compress [23] vector fields.

However, the majority of analysis and visualization techniques are based on the direct application of the theory of smooth vector fields to sampled flows, ignoring the fact that real numbers are replaced by finite-precision floating-point arithmetic. As a result, such numerical approaches may lead to the lack of *consistency*, meaning that the fundamental laws of smooth mathematical theory may not be preserved in practical applications. One type of inconsistency is the topological inconsistency, e.g., intersection of streamlines, violation of the Poincaré-Hopf theorem, etc. In order to obtain consistent results, the analysis must have numerical *robustness*, meaning that it is independent of the underlying machine and/or floating-point standards. Numerical instabilities may produce topological structures which are incorrect or inconsistent, hence questioning the fidelity of any subsequent analysis.

In several application areas, singularities of scalar fields have been shown to correspond to features of interest. For example, Laney et al. [9] use the gravitational potential on an envelope surface to indicate mixing structures in a Rayleigh-Taylor simulation. Mascarenhas et al. [15] and Bremer et al. [2] use extrema in fields derived from different combustion simulations to count regions of flame extinction and strong burning, respectively. Robust singularity detection in vector fields promises similar results for new applications such as turbulent flow analysis. As a result, the focus of the current work is a robust technique for the detection of singularities, which allows for consistent analysis.

Real-world data is most often available as discrete samples at the vertices of a mesh, and a continuous function is recovered through interpolation of these values. The particular case of piecewise-linear (PL) interpolants has been studied extensively in the context of identification and representation of singularities [8,27, 29,31]. The simplicity of linear interpolation makes PL vector fields preferred in many applications. However, even with this simple interpolant, robust identification of singularities remains a challenge, as in the case of many non-combinatorial geometric algorithms [22, 32]. For example, linear interpolation on a simplicial domain may admit singularities on the boundaries of simplices – an unstable configuration highly sensitive to numerical perturbation. Figure 1 shows how non-combinatorial methods may yield false positives or false negatives in these configurations, leading to inconsistent analysis. Furthermore, a large number of fluid simulations impose *no-slip* boundary conditions, meaning that the vectors at the

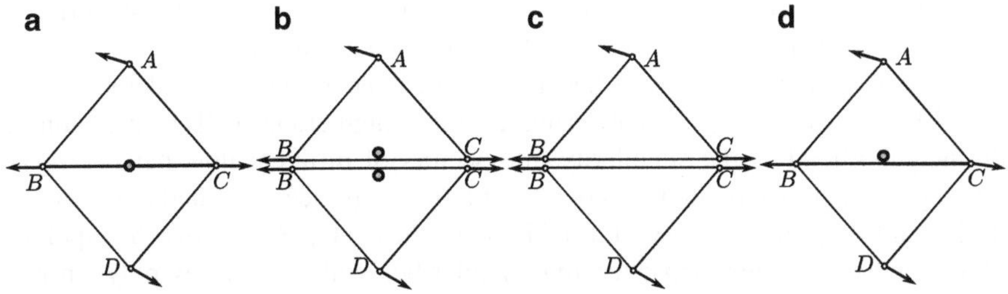

Fig. 1 (**a**) When $\mathbf{v}_B$ and $\mathbf{v}_C$ are anti-parallel, a singularity exists on the edge BC. Numerical techniques for identifying triangles containing critical points might test positive for (**b**) both ABC and BDC, or (**c**) neither of them. (**d**) A practical implementation of our combinatorial technique uses *Simulation of Simplicity* [5] to consistently determine the triangle containing the critical point

boundary are zero. These zero-vectors are the artifacts of the boundary/simulation rather than singularities of the field. While it is possible to filter them out of the critical point identification, it requires an explicit manual intervention, on top of a priori knowledge of the boundary conditions.

In contrast to numerical approaches, some combinatorial alternatives [18, 19] have been proposed to extract the topological structure from vector fields. Such techniques typically convert the input data into a discrete form to obtain numerical robustness, however, they are severely limited since it is not known how well do these discrete techniques approximate the field.

To address these problems, new combinatorial techniques are required which are guaranteed to detect all topological structures in a manner consistent with the fundamental principles of vector fields. To this end, our contributions are:

- We prove the necessary and sufficient conditions for the existence of singularities within cells of a simplicial mesh for a broad class of interpolation functions.
- We show how to turn these necessary and sufficient conditions for the existence of singularities into a combinatorial algorithmic approach.

2 Related Work

The topological skeleton of vector fields introduced by Helman and Hesselink [8] is of special interest to many researchers. It consists of important features of the field such as singularities, saddle separatrices, and limit cycles. Thus, the identification and classification of these features is an integral part of the analysis and visualization of vector fields. However, most of the early attempts at identification of singularities were based on numerical analysis. For example, isolated non-degenerate singularities were identified using numerically integrated tangent curves (streamlines) and classified, based on eigenvalue analysis [8]. Lavin et al. [10] and Batra and

Hesselink [1] extract singularities in a PL vector field $\mathbf{V} = \mathbf{Ax} + \mathbf{o}$ by numerically solving the system $\mathbf{Ax} = \mathbf{0}$ for each cell in a triangulated domain.

The detection of singularities has also been extended to higher-order singularities [20]. Tricoche et al. [27] analyze higher-order singularities in 2D by partitioning the neighborhood of the singularity into sectors of different flow behavior. The topological analysis of higher-order singularities provides a foundation for the design and simplification of vector fields. Tricoche et al. [28] simplify the topology of vector fields by merging clustered singularities within a convex polygon into higher-order singularities. These ideas have been extended to 3D by Weinkauf et al. [30, 31]. It is more challenging to identify singularities in nonlinear vector fields. Li et al. [11] subdivide the simplicial mesh and compute the vector field by side-vertex interpolation in polar coordinates. Singularities are then ensured to be located at the vertices.

In general, detection of singularities can be reformulated as solving nonlinear systems of equations. The Newton-Raphson method and Broyden's method can be used to solve such systems. However, techniques aimed at solving generic nonlinear systems are sensitive to perturbation and not guaranteed to find all the solutions. For multivariate rational splines, Elber and Kim [6] apply the bisection method to localize the potential regions containing roots. However, computational complexity is a major concern of their method.

Consistency and robustness are particularly desired when computing the topological skeleton of a vector field. A number of techniques have been proposed to extract it in a stable and efficient manner. Such techniques range from deriving some properties from the original vector field, and basing the extraction on those properties; to converting the vector field into a simpler combinatorial form which makes the extraction more robust.

For example, Polthier and Preuß [17] detect singularities as the extrema of the rotation-free and divergence-free potentials obtained from the discrete Helmholtz-Hodge decomposition of the vector field. This method, however, only works for piecewise-constant (PC) vector fields. Chen et al. propose the Entity Connection Graph (ECG) [3] and the Morse Connection Graph (MCG) [4] as the topological representation of PL vector fields. However, both ECG and MCG do not represent higher-order features of the field. On the other hand, Reininghaus et al. [18, 19] construct a combinatorial vector field. While using their combinatorial field enables the extraction of a consistent topological structure, it is unclear how close the resulting combinatorial field is to the original field. By comparison, this work proposes a robust and consistent combinatorial identification of singularities by working directly on the input vector field.

In addition to the techniques discussed above, the notion of Poincaré index also inspires combinatorial approaches to detect critical points. In this context, Garth et al. [7] propose a method to detect and track singularities in time-dependent vector fields by ensuring the Poincaré index is always preserved. Other techniques ensuring the validity of Poincaré index include the works of Mann and Raywood [14] and Trioche et al. [25, 26].

3 Foundations

Let D be a bounded, open subset of $\mathbb{R}^n$. The closure and the boundary of D are denoted by $\bar{D}$ and ∂D respectively. A point $\mathbf{x} \in \mathbb{R}^n$ is denoted as $\mathbf{x} = (\mathbf{x}_0, \ldots, \mathbf{x}_{n-1})$, and has an L-infinity norm $|\mathbf{x}| = \max\{|\mathbf{x}_i|; i = 0, \cdots, n-1\}$. $C(\bar{D})$ denotes the class of continuous functions $\phi \in C(\bar{D})$, such that $\phi(\mathbf{x}) : \bar{D} \to \mathbb{R}^n$ with the norm $||\phi|| = \sup_{\mathbf{x} \in D} |\phi(\mathbf{x})|$. $C^1(\bar{D})$ is a subset of $C(\bar{D})$ such that $\phi \in C^1(\bar{D})$ has continuous first-order partial derivatives. Let $\mathbf{p} = \phi(\mathbf{x}) = (\phi_0(\mathbf{x}), \cdots, \phi_{n-1}(\mathbf{x}))$, then, the Jacobian matrix $\mathbf{J}$ of ϕ is given as $\mathbf{J}\phi(\mathbf{x}) = [\nabla\phi_0, \cdots, \nabla\phi_{n-1}]^T$. For $\phi \in C^1(\bar{D})$, $\mathbf{p} = \phi(\mathbf{x})$ is called a *degenerate value* of ϕ if there exists $\mathbf{x} \in \bar{D}$ such that $\det(\mathbf{J}\phi(\mathbf{x})) = \mathbf{0}$, otherwise $\mathbf{p}$ is a *regular value* of ϕ.

3.1 Degree Theory

The proposed critical point detection technique requires results relating the existence of certain values in the image of a function to the span of the image. We are interested in a particular class of functions $\phi : \mathbb{R}^n \to \mathbb{R}^n$ for which the pre-image of a non-degenerate value $\mathbf{p}$ is a finite set of points. Therefore, we utilize the Brouwer degree as a tool for determining whether or not a particular value exists inside the image of a simplex. The following definitions have been made by Lloyd [12].

For a bounded, open subset $D \subset \mathbb{R}^n$, and a continuous function $\phi \in C(\bar{D})$, the Brouwer degree of ϕ in D with respect to the value $\mathbf{p}$, where $\mathbf{p} \notin \phi(\partial D)$ is defined as follows:

Definition 1 (Brouwer degree for $C^1(\bar{D})$ and regular value p). If $\phi \in C^1(\bar{D})$ and $\mathbf{p}$ is a regular value of ϕ, then

$$\deg(\phi, D, \mathbf{p}) = \sum_{\mathbf{x} \in \phi^{-1}(\mathbf{p})} \operatorname{sign}(\det(\mathbf{J}\phi(\mathbf{x}))).$$

An intuition behind the concept of Brouwer degree is illustrated in Fig. 2(a). It is essentially the count of the net crossings of $\mathbf{p}$ by the image of $\bar{D}$ under ϕ. The above definition is limited to regular values $\mathbf{p}$ only. Not all values in a sampled function reconstructed through interpolation are regular, so our definition must encompass degenerate values as well.

Definition 2 (Brouwer degree for $C^1(\bar{D})$ and degenerate value p). If $\phi \in C^1(\bar{D})$ and $\mathbf{p}$ is a degenerate value of ϕ, then

$$\deg(\phi, D, \mathbf{p}) = \deg(\phi, D, \mathbf{p}_1)$$

where $\mathbf{p}_1$ is any regular value such that $|\mathbf{p} - \mathbf{p}_1| < \operatorname{dist}(\mathbf{p}, \phi(\partial D))$.

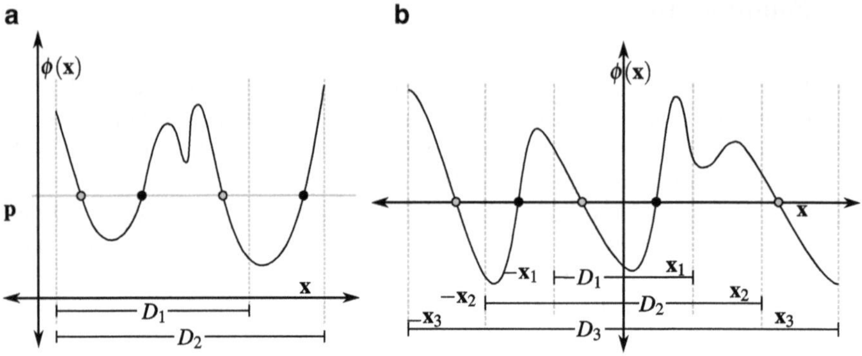

Fig. 2 The Brouwer degree counts the net number of times **p** is crossed by the image of $\bar{D}$ under ϕ. The "positive" ($\det(\mathbf{J}\phi(\mathbf{x})) > 0$) and "negative" ($\det(\mathbf{J}\phi(\mathbf{x})) < 0$) crossings are shown as *solid* and *hollow dots* respectively. (**a**) For the open and bounded sets D_1 and D_2, the Brouwer degree of $\phi(\mathbf{x})$ with respect to the value **p** is -1 and 0 respectively. (**b**) When the sets D_i are symmetric, Theorem 2 guarantees that if the values on the boundary have different signs, then the Brouwer degree with respect to **0** is odd, as is the case for D_2 and D_3

The existence of a non-degenerate value in every neighborhood of **p** is guaranteed by Sard's Theorem [12]. Similarly, we would like to extend this definition to functions that are continuous, but not necessarily in $C^1(\bar{D})$.

Definition 3 (Brouwer degree for $C(\bar{D})$). If $\phi \in C(\bar{D})$, then

$$\deg(\phi, D, \mathbf{p}) = \deg(\phi_1, D, \mathbf{p})$$

where ϕ_1 is any function in $C^1(\bar{D})$ such that for any $\mathbf{x} \in \bar{D}$, $|\phi(\mathbf{x}) - \phi_1(\mathbf{x})| < \operatorname{dist}(\mathbf{p}, \phi(\partial D))$.

Basically, we can find a function ϕ_1 that is "close" to ϕ and has continuous derivatives, and define the Brouwer degree with respect to this function.

Using the Brouwer degree, the net number of crossings of **p** by the image of $\bar{D}$ can be counted. If this number is nonzero, then there exists at least one **x** such that $\phi(\mathbf{x}) = \mathbf{p}$. This leads to the following theorems:

Theorem 1 (Kronecker's existence theorem). *If* $\deg(\phi, D, \mathbf{p}) \neq 0$, *then the equation* $\phi(\mathbf{x}) = \mathbf{p}$ *has at least one solution in* D.

Theorem 2. *Let* D *be a bounded, open, symmetric subset of* $\mathbb{R}^n$ *containing the origin. If* $\phi : \bar{D} \to \mathbb{R}^n$ *is continuous,* $\mathbf{0} \notin \phi(\partial D)$, *and for all* $\mathbf{x} \in \partial D$

$$\frac{\phi(\mathbf{x})}{|\phi(\mathbf{x})|} \neq \frac{\phi(-\mathbf{x})}{|\phi(-\mathbf{x})|}$$

then $\deg(\phi, D, \mathbf{0})$ *is an odd number [12].*

Intuitively, Theorem 2 ensures that ϕ crosses $\phi(\mathbf{x}) = 0$ at least once if no antipodal vectors of ϕ are parallel, as can be seen in Fig. 2b.

3.2 Sampled Vector Fields

The proposed technique addresses the detection of singularities for interpolated vector fields, where vectors are defined on the vertices of a simplicial complex and then interpolated on the interior of simplices.

A *k-simplex*, $\mathbb{S}^k$, is the convex hull of $k + 1$ affinely-independent vertices, such that $\mathbb{S}^k = \{\mathbf{x}_i\}, \mathbf{x}_i \in \mathbb{R}^n$, $0 \leqslant i \leqslant k \leqslant n$. A simplex $\mathbb{S}^l$ is called a *l-face* of $\mathbb{S}^k$ for $l \leqslant k$ if $\mathbb{S}^l \subseteq \mathbb{S}^k$, and a *proper l-face* of $\mathbb{S}^k$ for $l < k$ if $\mathbb{S}^l \subset \mathbb{S}^k$. A proper $k-1$-face $\mathbb{S}_j^{k-1}$ of $\mathbb{S}^k$ is called its *facet* if $\mathbb{S}_j^{k-1} = \{\mathbf{x}_i\}, \mathbf{x}_i \subset \mathbb{S}^k$, $0 \leqslant i \leqslant k \leqslant n$, $i \neq j$, i.e., it is constructed by removing vertex $\mathbf{x}_j$ from $\mathbb{S}^k$. If $\mathbb{S}^l$ is a (proper) face of $\mathbb{S}^k$, then $\mathbb{S}^k$ is called a *(proper) coface* of $\mathbb{S}^l$. $\mathring{\mathbb{S}}$ denotes the interior of a simplex, and is given by removing all the proper faces from a simplex. A *simplicial complex*, denoted as $\mathcal{M}$, is a collection of simplices such that $\mathbb{S}_i \in \mathcal{M}$ implies that all the faces of $\mathbb{S}_i$ are in $\mathcal{M}$, and the intersection of any two simplices is a face of both or empty. The local neighborhood of a vertex $\mathbf{x}_j$ can be defined in terms of its *star* $\mathrm{S}(\mathbf{x}_j)$, which consists of all cofaces of $\mathbf{x}_j$. The star is not closed under taking faces.

A sampled vector field is given as a simplicial complex $\mathcal{M}$ with $m + 1$ vertices $\{\mathbf{x}_0, \ldots, \mathbf{x}_m\}$, $\mathbf{x}_i \in \mathbb{R}^n$ and vector values $\{\mathbf{v}_0, \ldots, \mathbf{v}_m\}, \mathbf{v}_i \in \mathbb{R}^n$ defined at the vertices. The vector field is called *generic* when

1. $||\mathbf{v}_i|| > 0, \forall i \in \{0, \ldots, m\}$,
2. The vectors $\{\mathbf{v}_{i_0}, \ldots, \mathbf{v}_{i_d}\}$ at vertices $\{\mathbf{x}_{i_0}, \ldots, \mathbf{x}_{i_d}\}$ of the d-simplex $\mathbb{S}_i \in \mathcal{M}$ are affinely independent.

Note that not every vector field sampled from observations or simulations is generic, since the sampled vector magnitudes can be zero violating condition 1. While we assume a generic vector field in the following discussion, Sect. 4.2 discusses how this assumption can be relaxed.

In this chapter, we focus on a class of interpolating vector valued functions which can be expressed as

$$\mathbf{V}(\mathbf{x}) = \sum_{\mathring{\mathbb{S}}_i \in \mathcal{M}} \mathbf{V}_i(\mathbf{x}), \qquad \text{such that,} \qquad \mathbf{V}_i(\mathbf{x}) = \sum_{\mathbf{x}_j \in \mathbb{S}_i} w_j(\mathbf{x})\mathbf{v}_j$$

where, the weight functions $w_j(\mathbf{x})$ defined for vertices $\mathbf{x}_j$ are continuous, non-negative, and local, meaning $w_j(\mathbf{x}) > 0, \forall \mathbf{x} \in \mathrm{S}(\mathbf{x}_j)$, and $w_j(\mathbf{x}) = 0, \forall \mathbf{x} \notin \mathrm{S}(\mathbf{x}_j)$.

Following the definition of the weight functions $w_j(\mathbf{x})$, it is clear that for the simplex $\mathbb{S}_i = \{\mathbf{x}_{i_0}, \ldots, \mathbf{x}_{i_d}\}$, $w_j(\mathbf{x}) > 0$ only for $j \in \{i_0, \ldots, i_d\}$. Furthermore, $w_j(\mathbf{x}) \to 0$ as $\mathbf{x} \to \mathbb{S}_j$, where $\mathbb{S}_j$ is a facet of $\mathbb{S}_i$. Since $\mathbf{V}_i$ are defined only on the interior of simplices, $\mathbf{V}$ is C^0 continuous across the faces of the simplices. It is

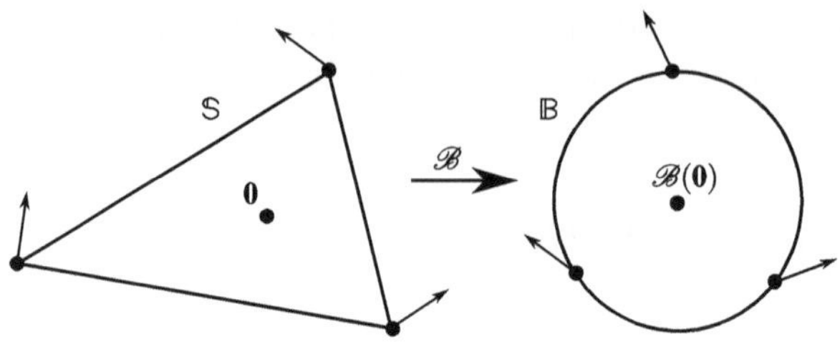

Fig. 3 The mapping $\mathscr{B} : \mathbb{S} \rightarrow \mathbb{B}$ in $\mathbb{R}^2$. The origin $\mathbf{0}$ is contained in the interior of $\mathbb{S}$

simple to confirm that PL interpolation falls into this class of functions. Also, a variant of Radial Basis Function (RBF) interpolation falls into this class where the weights smoothly fall to zero at the boundary of the vertex stars.

4 Critical Point Detection

This section discusses how Brouwer degree theory can be used to robustly detect singularities in the class of interpolated vector fields defined in Sect. 3.2.

4.1 Main Result

Using the concepts introduced above we will show that a simplex $\mathbb{S}$ contains a critical point if and only if the origin, $\mathbf{0}$, lies in the convex hull of the vectors at the simplex's vertices. To connect the results of Sect. 3 with vector fields on simplices, we define a one-to-one mapping from a simplex to an enclosing ball.

Let $\mathbb{S} = \{\mathbf{x}_0, \ldots, \mathbf{x}_m\}, m \leqslant n, \mathbf{x}_i \in \mathbb{R}^n$ be a m-simplex of $\mathscr{M}$. We can assume, without loss of generality, that the origin $\mathbf{0}$ is in the interior of $\mathbb{S}$. Let $\mathbb{B}$ be the unit ball. Let $\mathbf{x}$ ($\neq \mathbf{0}$) be a point in the interior of $\mathbb{S}$, and $\mathbf{x}'$ be the intersection of the ray from the origin through $\mathbf{x}$ with the boundary of $\mathbb{S}$. Then, we can define a mapping $\mathscr{B} : \mathbb{S} \rightarrow \mathbb{B}$ as $\mathscr{B}(\mathbf{x}) = \mathbf{x}/||\mathbf{x}'||$ for $\mathbf{x} \neq \mathbf{0}$, and $\mathscr{B}(\mathbf{0}) = \mathbf{0}$. (see Fig. 3)

Assuming non-degenerate simplices, $\mathscr{B}$ is continuous and invertible, with $\mathbf{0}$ as its fixed point, and we can use it to map $\mathbf{V}$ from any simplex in $\mathscr{M}$ onto an enclosing ball $\mathbb{B}$ with center in the interior. We now show that if $\mathbf{0}$ is contained in the convex hull of $\mathbf{V}$, the Brouwer degree of this simplex with respect to $\mathbf{0}$ is odd.

Lemma 1. *Let* $\mathbb{S} = \{\mathbf{x}_0, \ldots, \mathbf{x}_n\}, \mathbf{x}_i \in \mathbb{R}^n$ *be a simplex of* $\mathscr{M}$ *containing the origin* $\mathbf{0}$*, and* $\mathbf{V}$ *be a vector field as defined above. If* $\mathbf{0}$ *lies in the convex hull of* $\{\mathbf{v}_0, \ldots, \mathbf{v}_n\}$*, then* $\deg(\mathbf{V}, \mathbb{S}, \mathbf{0})$ *is odd.*

Proof. Let $\mathbb{B}$ be the unit ball of $\mathbb{S}$ as defined above. Furthermore, define $\bar{\mathbf{V}} : \mathbb{B} \rightarrow \mathbb{R}^n$, as $\bar{\mathbf{V}}(\mathbf{x}) = \mathbf{V}(\mathscr{B}^{-1}(\mathbf{x}))$. Assume, there exists an $\mathbf{x}_b \in \partial\mathbb{B}$ such that

$$\bar{\mathbf{V}}(\mathbf{x}_b) = a \cdot \bar{\mathbf{V}}(-\mathbf{x}_b), \text{ with } a > 0.$$

The following argumentation proves that this assumption leads to a contradiction.

Due to continuity of $\mathscr{B}$, it can not map adjacent points of $\mathbb{S}$ to antipodal points of $\mathbb{B}$, therefore it follows that there exist two different facets of $\mathbb{S}$, namely $\mathbb{S}_j$ and $\mathbb{S}_k$, containing parallel vectors. Let $\mathbf{x} \in \mathbb{S}_j$ and $\mathbf{y} \in \mathbb{S}_k$, such that for some $a > 0$, $\sum_{i \neq j} w_i(\mathbf{x})\mathbf{v}_i = a \cdot \sum_{i \neq k} w_i(\mathbf{y})\mathbf{v}_i$, $j \neq k$.

$$\sum_{i \neq j} w_i(\mathbf{x})\mathbf{v}_i - a \cdot \sum_{i \neq k} w_i(\mathbf{y})\mathbf{v}_i = 0$$

$$\sum_{i \neq j,k} (w_i(\mathbf{x}) - a w_i(\mathbf{y}))\, \mathbf{v}_i - a w_j(\mathbf{y})\mathbf{v}_j + w_k(\mathbf{x})\mathbf{v}_k = 0$$

Also, since $\mathbf{0}$ lies in the convex hull of $\{\mathbf{v}_0, \ldots, \mathbf{v}_n\}$, we can find c_i such that $\sum_{i=0}^{n} c_i \mathbf{v}_i = \mathbf{0}$, and $c_i \geqslant 0$, $\forall i$, and $\sum_{i=0}^{n} c_i = 1$. Now, since $\{\mathbf{v}_0, \ldots, \mathbf{v}_n\}$ forms an affine combination, $\sum \lambda_i \mathbf{v}_i = \sum \mu_i \mathbf{v}_i \Leftrightarrow \lambda_i = \mu_i \ \forall i$. This implies that

$$\begin{aligned} c_i &= w_i(\mathbf{x}) - a w_i(\mathbf{y}) && \forall i \neq j,k \\ c_k &= w_k(\mathbf{x}) \\ c_j &= -a w_j(\mathbf{y}). \end{aligned}$$

However, since all c_i's, w_i's, and a are positive, the third condition gives a contradiction. Hence, a point $\mathbf{x}_b$ with $\bar{\mathbf{V}}(\mathbf{x}_b) = a \cdot \bar{\mathbf{V}}(-\mathbf{x}_b)$ does not exist. By Theorem 2, $\deg(\bar{\mathbf{V}}, \mathbb{B}, \mathbf{0})$, and hence $\deg(\mathbf{V}, \mathbb{S}, \mathbf{0})$ is odd. □

Combining all the results presented above allows us to prove the main result:

Theorem 3. *Let* $\mathbb{S} = \{\mathbf{x}_0, \ldots, \mathbf{x}_n\}, \mathbf{x}_i \in \mathbb{R}^n$ *be a simplex of* $\mathscr{M}$ *and* $\mathbf{V}$ *be a vector field on* $\mathscr{M}$ *as defined above. Then* $\mathbb{S}$ *contains a critical point if and only if* $\mathbf{0}$ *is in the interior of the convex hull of* $\{\mathbf{v}_0, \ldots, \mathbf{v}_n\}$.

Proof. If $\mathbf{0}$ is in the interior of the convex hull of $\{\mathbf{v}_0, \ldots, \mathbf{v}_n\}$, then the Brouwer degree of the origin inside $\mathbb{S}$ is odd (Lemma 1). By Theorem 1, there exists a critical point in $\mathbb{S}$.

If a critical point is located at $\mathbf{x}$ in $\mathbb{S}$, then

$$\mathbf{v}(\mathbf{x}) = \sum_{i=0}^{n} w_i(\mathbf{x})\mathbf{v}_i = \mathbf{0}$$

$$\Rightarrow \sum_{i=0}^{n} \frac{w_i(\mathbf{x})}{\sum_{j=0}^{n} w_j(\mathbf{x})} \mathbf{v}_i = \sum_{i=0}^{n} c_i(\mathbf{x})\mathbf{v}_i = \mathbf{0}.$$

where, $c_i(\mathbf{x}) = \frac{w_i(\mathbf{x})}{\sum_{j=0}^{n} w_j(\mathbf{x})}$. Since all the weights w_i's are non-negative, we have $0 \leqslant c_i \leqslant 1$, $\forall i$, and $\sum_{i=0}^{n} c_i = 1$. Hence, $\mathbf{0}$ is in the interior of the convex hull of $\mathbf{v}_i$'s. □

The practical implication of this result is that, to identify a simplex containing a critical point, one does not need to know the actual interpolation function as long as it satisfies the properties discussed in Sect. 3.2. Furthermore, it suffices to check the convex hull of the vectors at the vertices to find singularities. Thus, detection of singularities for any interpolation scheme is reduced to a simpler PL test which can be performed using the Simulation of Simplicity (SoS) [5] as will be discussed later.

Notice that if the space is sampled too sparsely, or a more complex interpolation scheme is used, multiple critical points may appear within a single simplex. In this case, our technique will assume the simplest possible interpretation of the vectors at the vertices. In particular, if the singularities in the same simplex cancel each other, then no singularities will be reported, since the vector field on the boundary can be completed with an entirely regular vector field in its interior.

The topological consistency of our technique can be proved as below.

Corollary 1. *Using Theorem 3 always leads to topologically consistent critical point detection for* $\mathbf{V}$ *defined on* $\mathscr{M}$ *as above.*

Proof. Given a closed orientable n-dimensional manifold $\mathbb{M}$, we consider its corresponding simplicial complex $\mathscr{M}$. We know that any smooth vector field on $\mathbb{M}$ must have an even number of critical points. Therefore, to demonstrate consistency, we show that our technique detects an even number of critical points for the sampled vector field $\mathbf{V}$ on $\mathscr{M}$.

We study $\mathscr{M}$ in function (vector) space by considering the simplicial complex defined by the vectors of $\mathbf{V}$, and denote it as $\mathscr{M}_{\mathbf{v}}$. Note that an orientable and closed $\mathscr{M}$ implies an orientable and closed $\mathscr{M}_{\mathbf{v}}$. There exists an embedding of $\mathscr{M}_{\mathbf{v}}$ in $\mathbb{R}^{n+1}$ where the $n+1^{\text{th}}$ component is chosen at random. We trace a line ℓ through the origin and orthogonal to the $\mathbb{R}^n$ subspace. Note that ℓ corresponds to zero vector $\mathbf{0}$. By Theorem 3, singularities are not allowed to exist on the faces of simplices of $\mathscr{M}$. Since $\mathscr{M}_{\mathbf{v}}$ is closed, there exist an even number of transveral intersections between $\mathscr{M}_{\mathbf{v}}$ and ℓ. Consequently, each intersection corresponds to a single simplex containing a critical point at its interior, leading to an even number of simplices with critical points; therefore, even number of critical points. □

4.2 Robust Computation Using the Simulation of Simplicity

Detection of critical point in a simplex $\mathbb{S}$ requires a robust way of determining whether the zero vector is contained in the convex combination of the vectors of $\mathbb{S}$. Let $\mathbb{S}_{\mathbf{v}}$ denote the simplex created by the vectors of $\mathbf{V}$ at the vertices of $\mathbb{S}$, then detection of critical point simply translates into testing whether the origin $\mathbf{0}$

Algorithm 1 Positive$_n(\mathbf{x}_{j_0}, \ldots, \mathbf{x}_{j_n})$

$s \leftarrow$ *Sort* $(j_0, \ldots, j_n)$: *s is the number of swaps needed to sort*
$d \leftarrow$ *Sign* $\det(\mathbf{x}_{j_0}, \ldots, \mathbf{x}_{j_n})$
if $odd(s)$ **then**
  $d \leftarrow -d$
end if
return d

lies inside $\mathbb{S}_\mathbf{v}$. This point-in-simplex test can be further reduced to computing the orientation of the test point with respect to the facets of the simplex. This section describes how this procedure can be performed in a combinatorial manner.

The orientation of $n + 1$ points in n-dimensional space can be computed as the sign of the following determinant

$$\det(\mathbf{x}_0, \ldots, \mathbf{x}_n) = \begin{pmatrix} \mathbf{x}_{0,0} & \ldots & \mathbf{x}_{0,n-1} & 1 \\ \vdots & \ddots & \vdots & 1 \\ \mathbf{x}_{n,0} & \ldots & \mathbf{x}_{n,n-1} & 1 \end{pmatrix}.$$

The determinant is zero if all the points lie on a common hyperplane. For example, in 2D, the orientation of three points is counter-clockwise (positive) if the sign of the corresponding determinant is positive, or clockwise (negative) if the sign is negative. The degenerate case where the three points are collinear leads to the determinant being zero. Recall that the sign of a determinant switches if an odd number of row-exchanges are carried out, therefore, care must be taken with respect to the order of the points. Algorithm 1 [5] calculates the orientation of a set of points, by first assigning them a consistent order (sorting), and then computing the sign of the determinant.

As discussed in Sect. 3.2, the determinant is always non-zero for generic vectors. To impose genericity, we use the Simulation of Simplicity (SoS) [5], which is a general-purpose programming technique to handle degeneracies in the data. The SoS applies a symbolic perturbation to the data preventing any of the determinants from becoming 0, thus providing a non-degenerate point-in-simplex test. Intuitively, if a point lies on a shared coface of two or more simplices, the SoS makes a consistent choice by enforcing it to be contained inside one of them, while it lies outside the rest of them.

Numerical robustness is achieved by first converting the data to a fixed precision, and then representing the values as long integers in computing the determinant (as used in the Geomdir library [16]), hence removing the need for floating-point arithmetic. As a result, the determinant computation is both robust and combinatorial. We point out that Algorithm 1 is the fundamental step in the process of critical point detection, and helps achieve robustness by replacing numerical data (vector components) with combinatorial information (orientation).

Algorithm 2 Point $\mathbf{x}$ in simplex $\mathbb{S} = \{\mathbf{x}_0, \ldots, \mathbf{x}_n\}, \mathbf{x}_i \in \mathbb{R}^n$

```
s ← Positive_n(x_0, ..., x_n)
for i = 0 to n do
    (x'_0, ..., x'_n) ← (x_0, ..., x_n)
    x'_i ← x
    s_i ← Positive_n(x'_0, ..., x'_n)
    if s ≠ s_i then
        return false
    end if
end for
return true
```

Using Algorithm 1 for orientation computation, Algorithm 2 [5] performs the point-in-simplex test. Given a simplex $\mathbb{S} = \{\mathbf{x}_0, \ldots, \mathbf{x}_n\}, \mathbf{x}_i \in \mathbb{R}^n$, the first step is to determine the orientation of the facets of the simplex. Then, one by one each facet is replaced by the given test point and the corresponding orientation is computed. If all such combinations have the same orientation, then the test point lies inside the simplex, otherwise not.

5 Experimental Results

We apply our technique to identify critical points in PL vector fields on a triangulated domain. For each simplex in the domain, we test for origin to lie inside the simplex created by the vectors at its vertices using the SoS, and use the Geomdir library [16] to compute the determinants robustly. We compare our method with the numerical method to detect critical points [10], which solves a linear system for every simplex in the domain.

To evaluate the two techniques, we create a synthetic 2D vector field with known critical points, two of which are carefully placed on an edge and a vertex. In Fig. 4, we see that for both of these critical points, the numerical technique determined that all the triangles sharing the corresponding edge and vertex respectively, contain critical points – a topologically inconsistent result. Our proposed method using SoS, however, makes a choice by representing each critical point as one triangle only, which matches the known ground truth.

To demonstrate dimension-independence of our technique, we test it on the Lorenz system (A 3D vector field $\mathbf{V}(x, y, z) = (\sigma(y - x),\ x(\rho - z) - y,\ xy - \beta z))$ with parameters $\sigma = 10$, $\beta = 8/3$, and $\rho = 1/2$. For $\rho < 1$, we expect the system to contain a singularity at the origin. Figure 5 shows the comparison where our technique selects a single tetrahedron touching the origin to represent the singularity. The numerical technique, on the other hand, detects all the 20 tetrahedra touching the origin as critical.

To compare the running times of the two methods, we test them on the top slice of a 3D simulation of global ocean eddies [13], shown in Fig. 6. The ocean

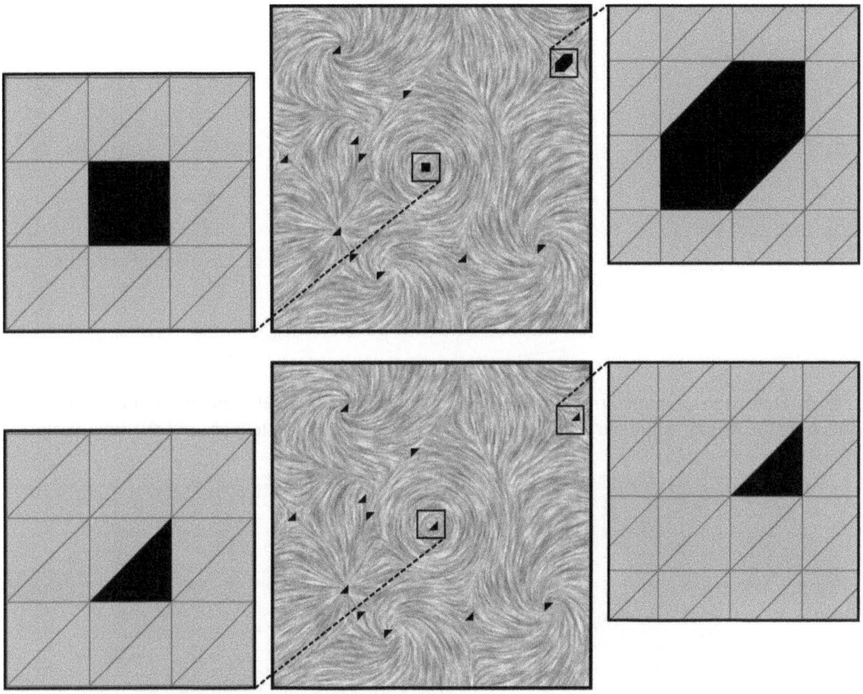

Fig. 4 Detection of singularities in 2D synthetic vector field. The *top row* shows the triangles where critical points were detected, numerically. *Zoom-in views* show that multiple triangles test positive for two critical points lying on the boundaries of triangles. *Bottom row* shows the consistent results detected using our algorithm where only one triangle per critical point tested positive

surface is represented by 10,633,760 triangles. The numerical method [10] takes $\approx$ 89.22 seconds to detect 24,552 critical points. On the other hand, our method takes only $\approx$ 6.85 seconds to detect the same critical points, and thus, is significantly faster.

6 Conclusions and Future Work

In this chapter, we provide a necessary and sufficient condition for the existence of critical points in a simplex for a broad class of interpolated vector fields. Our existence condition for critical points allows us to develop a robust method to detect critical points in n-dimensions. Furthermore, when given finite-precision values, the technique is guaranteed to use finite-precision, and therefore the result can be computed exactly in a combinatorial manner. In the future, we wish to investigate further the class of interpolation functions we showed this result for, identifying which other interpolation techniques fall into this class. We also wish to extend this work to include classification of critical points.

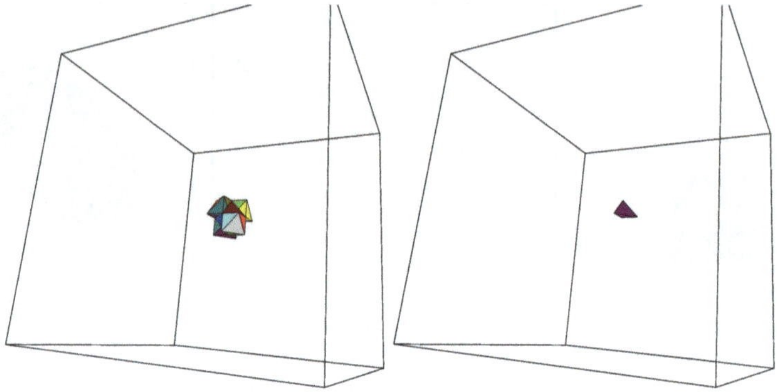

Fig. 5 Detection of singularities in the Lorenz system. Numerical method (*left*) detects 20 tets touching the origin, while our technique (*right*) detects a single tet to represent the singularity

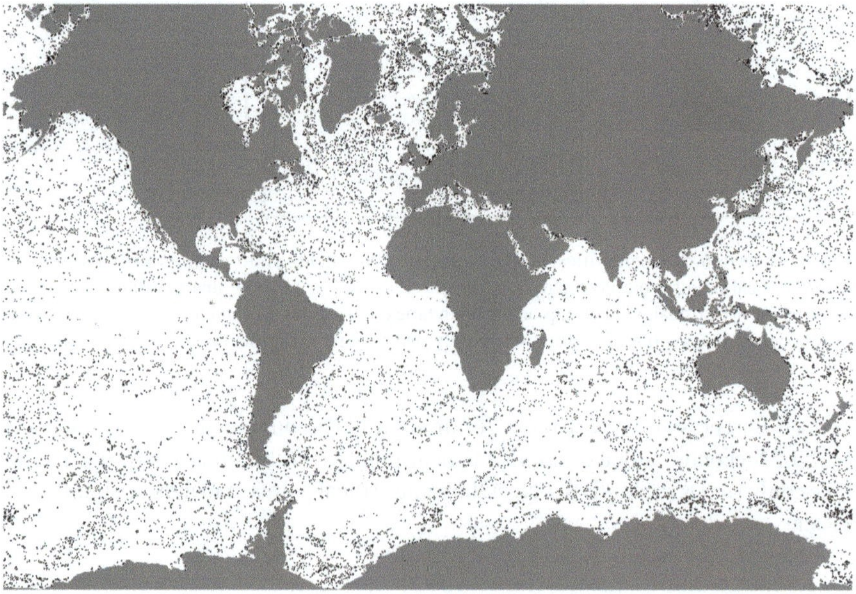

Fig. 6 Detection of singularities in a 2D slice of simulation of global oceanic eddies [13]. The data contains 24,552 triangles with critical points, each represented by a *black dot*

Acknowledgements We thank Mathew Maltude from the Climate, Ocean and Sea Ice Modelling program at Los Alamos National Laboratory and the BER Office of Science UV-CDAT team for providing us access to the ocean data from Fig. 6. This work was supported in part by NSF OCI-0906379, NSF OCI-0904631, DOE/NEUP 120341, DOE/MAPD DESC000192, DOE/LLNL B597476, DOE/Codesign P01180734, and DOE/SciDAC DESC0007446. This work was also performed under the auspices of the US Department of Energy (DOE) by Lawrence Livermore National Laboratory (LLNL) under contract DE-AC52-07NA27344. LLNL-PROC-644955.

References

1. R. Batra, L. Hesselink, Feature comparisons of 3-D vector fields using earth mover's distance, in *Proceedings of IEEE Visualization*, San Francisco, 1999, pp. 105–114
2. P.-T. Bremer, G. Weber, V. Pascucci, M. Day, J. Bell, Analyzing and tracking burning structures in lean premixed hydrogen flames. IEEE Trans. Vis. Comput. Graph. **16**(2), 248–260 (2010)
3. G. Chen, K. Mischaikow, R.S. Laramee, P. Pilarczyk, E. Zhang, Vector field editing and periodic orbit extraction using morse decomposition. IEEE Trans. Vis. Comput. Graph. **13**(4), 769–785 (2007)
4. G. Chen, K. Mischaikow, R.S. Laramee, E. Zhang, Efficient Morse decompositions of vector fields. IEEE Trans. Vis. Comput. Graph. **14**(4), 848–862 (2008)
5. H. Edelsbrunner, E.P. Mücke, Simulation of simplicity: a technique to cope with degenerate cases in geometric algorithms. ACM Trans. Graph. **9**, 66–104 (1990)
6. G. Elber, M.-S. Kim, Geometric constraint solver using multivariate rational spline functions, in *Proceedings of ACM Symposium on Solid Modeling and Applications (SMA '01)*, Ann Arbor, 2001, pp. 1–10
7. C. Garth, X. Tricoche, G. Scheuermann, Tracking of vector field singularities in unstructured 3D time-dependent datasets, in *Proceedings of IEEE Visualization*, Austin, 2004, pp. 329–336
8. J.L. Helman, L. Hesselink, Representation and display of vector field topology in fluid flow data sets. IEEE Comput. **22**(8), 27–36 (1989)
9. D. Laney, P.-T. Bremer, A. Mascarenhas, P. Miller, V. Pascucci, Understanding the structure of the turbulent mixing layer in hydrodynamic instabilities. IEEE Trans. Vis. Comput. Graph. **12**(5), 1052–1060 (2006)
10. Y. Lavin, R. Batra, L. Hesselink, Feature comparisons of vector fields using earth mover's distance, in *Proceedings of IEEE Visualization*, Research Triangle Park, 1998, pp. 103–109
11. W.-C. Li, B. Vallet, N. Ray, B. Levy, Representing higher-order singularities in vector fields on piecewise linear surfaces. IEEE Trans. Vis. Comput. Graph. **12**(5), 1315–1322 (2006)
12. N.G. Lloyd, Degree theory, Cambridge University Press, 1978
13. M. Maltrud, F. Bryan, S. Peacock, Boundary impulse response functions in a century-long eddying global ocean simulation. Environ. Fluid Mech. **10**, 275–295 (2010)
14. S. Mann, A. Rockwood, Computing singularities of 3D vector fields with geometric algebra, in *Proceedings of IEEE Visualization*, Boston, 2002, pp. 283–290
15. A. Mascarenhas, R.W. Grout, P.-T. Bremer, E.R. Hawkes, V. Pascucci, J.H. Chen, Topological feature extraction for comparison of terascale combustion simulation data, in *Topological Methods in Data Analysis and Visualization*, ed. by V. Pascucci, X. Tricoche, H. Hagen, J. Tierny. Mathematics and Visualization (Springer, Berlin/Heidelberg, 2011), pp. 229–240
16. E.P. Mucke, Geomdir. http://www.geom.uiuc.edu/software/cglist/GeomDir/
17. K. Polthier, E. Preuß, Identifying vector field singularities using a discrete Hodge decomposition, in *Mathematical Visualization III*, ed. by H. Hege, K. Polthier (Springer, Berlin/New York, 2003) pp. 112–134
18. J. Reininghaus, I. Hotz, Combinatorial 2D vector field topology extraction and simplification, in *Topological Methods in Data Analysis and Visualization*, ed. by V. Pascucci, X. Tricoche, H. Hagen, J. Tierny. Mathematics and Visualization (Springer, Berlin/Heidelberg, 2011), pp. 103–114
19. J. Reininghaus, C. Löwen, I. Hotz, Fast combinatorial vector field topology. IEEE Trans. Vis. Comput. Graph. **17**, 1433–1443 (2011)
20. G. Scheuermann, H. Krüger, M. Menzel, A.P. Rockwood, Visualizing nonlinear vector field topology. IEEE Trans. Vis. Comput. Graph. **4**(2), 109–116 (1998)
21. G. Scheuermann, X. Tricoche, Topological methods for flow visualization, in *The Visualization Handbook*, ed. by C.D. Hansen, C.R. Johnson (Elsevier, Oxford, 2005), pp. 341–356
22. S. Schirra, Robustness and precision issues in geometric computation, in *Handbook of Computational Geometry*, chapter 14, ed. by J.-R. Sack, J. Urrutia, (Elsevier, Amsterdam/New York, 2000)

23. H. Theisel, C. Rössl, H.-P. Seidel, Compression of 2D vector fields under guaranteed topology preservation. Comput. Graph. Forum (Proc. Eurographics), **22**(3), 333–342 (2003)
24. H. Theisel, T. Weinkauf, H.-C. Hege, H.-P. Seidel, Saddle connectors – an approach to visualizing the topological skeleton of complex 3D vector fields, in *Proceedings of IEEE Visualization*, Seattle, 2003
25. X. Tricoche, C. Garth, A. Sanderson, Visualization of topological structures in area preserving maps. IEEE Trans. Vis. Comput. Graph. **17**(12), 1765–1774 (2011)
26. X. Tricoche, C. Garth, A. Sanderson, K. Joy, Visualizing invariant manifolds in area-preserving maps, in *Topological Methods in Data Analysis and Visualization II*, ed. by R. Peikert, H. Hauser, H. Carr, R. Fuchs. Mathematics and Visualization (Springer/Berlin Heidelberg, 2012), pp. 109–124
27. X. Tricoche, G. Scheuermann, H. Hagen, Higher order singularities in piecewise linear vector fields, in *The Mathematics of Surfaces IX*, ed. by R. Cipolla, R. Martin (Springer, London), pp. 99–113
28. X. Tricoche, G. Scheuermann, H. Hagen, A topology simplification method for 2D vector fields, in *Proceedings of IEEE Visualization*, Salt Lake City, 2000, pp. 359–366
29. X. Tricoche, G. Scheuermann, H. Hagen, Continuous topology simplification of planar vector fields, in *Proceedings of IEEE Visualization*, San Diego, 2001, pp. 159–166
30. T. Weinkauf, H. Theisel, H.-C. Hege, H.-P. Seidel, Toplogical construction and visualization of higher order 3D vector fields. Comput. Graph. Forum (Proc. Eurographics), **23**(3), 469–478 (2004)
31. T. Weinkauf, H. Theisel, K. Shi, H.-C. Hege, H.-P. Seidel, Extracting higher order critical points and topological simplification of 3D vector fields, in *Proceedings of IEEE Visualization*, Minneapolis, 2005
32. C.K. Yap, Robust geometric computation, in *Handbook of Discrete and Computational Geometry*, chapter 41, ed. by J.E. Goodman, J. O'Rourke (Chapmen & Hall, Boca Raton, 2004), pp. 927–952

Interpreting Feature Tracking Through the Lens of Robustness

Primoz Skraba and Bei Wang

Abstract A key challenge in the study of a time-varying vector fields is to resolve the correspondences between features in successive time steps and to analyze the dynamic behaviors of such features, so-called feature tracking. Commonly tracked features, such as volumes, areas, contours, boundaries, vortices, shock waves and critical points, represent interesting properties or structures of the data. Recently, the topological notion of robustness, a relative of persistent homology, has been introduced to quantify the stability of critical points. Intuitively, the robustness of a critical point is the minimum amount of perturbation necessary to cancel it. In this chapter, we offer a fresh interpretation of the notion of feature tracking, in particular, critical point tracking, through the lens of robustness. We infer correspondences between critical points based on their closeness in stability, measured by robustness, instead of just distance proximities within the domain. We prove formally that robustness helps us understand the sampling conditions under which we can resolve the correspondence problem based on region overlap techniques, and the uniqueness and uncertainty associated with such techniques. These conditions also give a theoretical basis for visualizing the piecewise linear realizations of critical point trajectories over time.

P. Skraba
Jozef Stefan Institute, Ljubljana, Slovenia
e-mail: primoz.skraba@ijs.si

B. Wang (✉)
Scientific Computing and Imaging Institute, University of Utah, Salt Lake City, UT, USA
e-mail: beiwang@sci.utah.edu; wang.bei@gmail.com

P.-T. Bremer et al. (eds.), *Topological Methods in Data Analysis and Visualization III*, Mathematics and Visualization, DOI 10.1007/978-3-319-04099-8_2,

1 Introduction

A number of techniques have been proposed to define, extract and track features from vector field data [22, 26]. There are many different types of tracked features including volumes, contours, and vortices, which represent interesting properties or structures of the vector fields. In this chapter, we restrict ourselves to critical points [12, 13, 19, 26, 35–37] and sublevel sets. Feature tracking, which is initially inspired by object tracking in computer vision [42], has been intensively researched where most techniques could be classified into three categories [26]. The first approach does not rely on temporal interpolation but focuses on feature extractions at individual time slices and subsequently feature matching via region correspondences or attribute correspondences [22]. Correspondences could be found based on distance proximity [16, 28], attribute similarity [28], or spatial overlap of features [30–32], or alternatively, using prediction and verification [23–25, 29]. Level sets components volume overlap has been proposed in tracking contour tree evolution [34] and contour tree matching [20]. The second approach is based on temporal interpolation and considers time as an additional dimension of the space-time domain. Iso-surfaces are extracted and tracked in 4D space-time in scalar field [17, 40], and for vortex tracking in scale space [1]. Topological structures, such as Reeb graphs [8, 39], and Jacobi sets [7], could be employed in feature tracking (and specifically critical point tracking [8]). Temporal linear interpolation in combination with critical points tracking in 2D and 3D flow fields have been developed in [12, 37]. The third approach represents the dynamic behavior of features as streamlines of a higher-dimensional vector field, called feature flow fields [35, 41], with combinatorial extensions developed in [18, 26]. Critical points are tracked by computing streamlines using combinatorial feature flow fields [26], whose importance measure, referred to as integrated persistence, combines spatial persistence of a critical point along its temporal dimension.

Among these various feature tracking approaches, tracking the temporal evolution of critical points (and their corresponding sublevel sets) plays an important role in understanding the behavior of time-varying vector fields. Recently, the topological notion of robustness [5, 6, 11], a relative of persistent homology, has been introduced to quantify the stability of critical points [5, 38]. Intuitively, the robustness of a critical point is the minimum amount of perturbation necessary to cancel it. It has been shown to be useful for vector field analysis, visualization [38] and simplification [33]. The work in [27] also strongly advocated the need for importance measures for critical points and proposed such a measure closely related to persistence. Although robustness is also closely related to persistence, in the sense that the robustness of features in level and interlevel sets, quantified through well groups, can be read off the persistence diagram of the function [2]; however in more general settings the reduction from robustness to persistence is not known and the authors in [11] have conjectured that robustness may sit somewhere between the 1-parameter notion of persistence and its multi-parameter generalization [3].

In this chapter, we offer a fresh interpretation of the notion of feature tracking, in particular, critical point tracking, through the lens of robustness. We obtain our theoretical results by relating critical points tracking with their stability. That is, in a nutshell, stable critical points could be tracked more easily and more accurately. We prove formally that robustness can help us understand the sampling conditions under which we can resolve the correspondence problem based on commonly used region correspondence techniques. (e.g. [20,30–32,34]). It also gives a theoretical basis for visualizing the piecewise-linear (PL) realizations of critical point trajectories.

2 Background

We provide the relevant background for well groups [5, 6] and degree theory. In particular, we review the notion of static robustness and its properties explicitly stated in [38]. The main components for proving our results rely on these properties, as well as the Stability Theorem of Well Diagrams.

Degrees. In a 2D vector field, the degree of a critical point x, denoted as $\deg(x)$, equals its Poincaré index. Sources and sinks have a degree of $+1$ while saddles have a degree of -1. A path-connected component C in the domain that encloses a set of critical points $\{x_i\}$ has a degree that sums the degrees of the individual critical points: $\deg(C) = \sum_i \deg(x_i)$. For the formal definition of the *degree* of a continuous mapping see [14] (page 134) and [6].

Merge tree. Given a continuous 2D vector field $f : \mathbb{R}^2 \to \mathbb{R}^2$, we define a scalar function $f_0 : \mathbb{R}^2 \to \mathbb{R}$ such that the value at each point $x \in \mathbb{R}^2$ is the Euclidean norm of the vector field at x, $f_0(x) = ||f(x)||_2$. Let $\mathbb{F}_r = f_0^{-1}(-\infty, r]$ be the *sublevel set* of f_0 for some $r \geqslant 0$. A value $r > 0$ is a *regular value* of f_0 if $\mathbb{F}_r$ is a 2-manifold, and for all sufficiently small $\epsilon > 0$, $f_0^{-1}[r-\epsilon, r+\epsilon]$ deformation retracts to $f_0^{-1}(r)$; otherwise it is a *critical value*. We further assume f_0 has a finite number of critical values and f has finite number of critical points (that is, the number of components in $\mathbb{F}_0$ is finite).

We construct a *merge tree* (or a *join tree* [4]), which tracks the (connected) components of $\mathbb{F}_r$ as they appear and merge, as we increase r from 0 (or $-\infty$). This corresponds to the 0-dimensional persistent homology [9] of the sublevel set filtration of f_0. The leaves of the tree represent the creation of a component while the root represents the entire domain of f_0. An internal node represents the merging of two or more components. We then assign an integer to each node in the tree that record the degree of the corresponding component in the sublevel set. The degree of any such component is determined by the sum of degrees of the critical points lying in it [5].

Well groups and well diagrams. To understand the concepts of well groups and well diagrams first introduced in [10], we need to introduce our particular notion of vector field perturbation. Let $f, h : \mathbb{R}^2 \to \mathbb{R}^2$ be two continuous 2D vector fields. A continuous mapping h is an *r-perturbation* of f, if the distance between the two mappings $\mathrm{d}(f, h) := \sup_{x \in \mathbb{R}^2} ||f(x) - h(x)||_2 \leqslant r$.

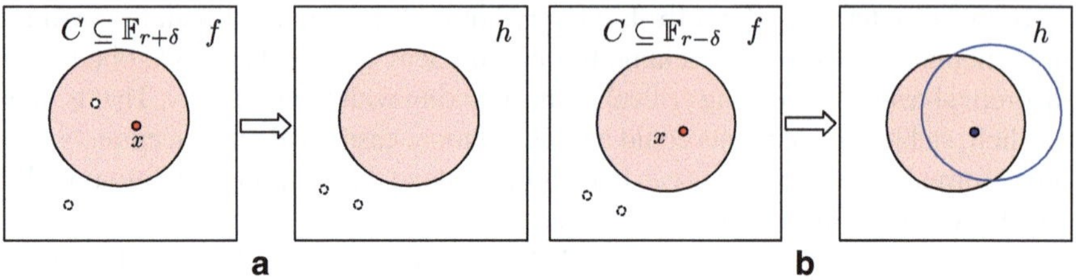

Fig. 1 Illustrations for (**a**) Lemma 1 and (**b**) Lemma 2

As we track the connected components over a filtration (the sublevel set of f_0) at each value of r, we are computing the *0-dimensional homology* groups over a field. These groups are vector spaces whose ranks equal the number of components presented in the associated sublevel sets. Furthermore, if h is an r-perturbation of f, then $\mathbb{H}_0 = h^{-1}(0)$ is a subspace of $\mathbb{F}_r$. The 0-dimensional homology groups are denoted as $\mathsf{H}(\mathbb{H}_0)$ and $\mathsf{H}(\mathbb{F}_r)$. The subspace relation induces a linear map $j_h : \mathsf{H}(\mathbb{H}_0) \rightarrow \mathsf{H}(\mathbb{F}_r)$ between the two vector spaces. The *well group*, $\mathsf{U}(r)$, is the subgroup of $\mathsf{H}(\mathbb{F}_r)$, whose elements belong to the image of each j_h, for all r-perturbation h of f [5]. That is, $\mathsf{U}(r) = \bigcap_h \text{im}\, j_h$. Intuitively, an element in $\mathsf{U}(r)$ is considered a stable element in $\mathsf{H}(\mathbb{F}_r)$ if it does not disappear with respect to any perturbation. The rank of $\mathsf{U}(0)$ is the number of critical points of f. A point $r \in (0, \infty)$ belongs to the *well diagram* of f_0, $\text{Dgm}(f_0)$, with multiplicity k if the rank of the well group drops by k at r [5]. For reasons of stability, the point 0 is counted with infinite multiplicity. The point ∞ is counted with multiplicity k if for all sufficiently large values of r, the rank of $\mathsf{U}(r)$ is k. An algorithm to compute the well diagram is suggested by the Equivalence Theorem [5]. It states that, if r is a regular value of f_0, then the rank of the well group $\mathsf{U}(r)$ is the number of components C of $\mathbb{F}_r$ such that $\deg(C) \neq 0$. We demonstrate by an example below that the constructed augmented merge tree is sufficient to derive its corresponding well diagram.

Stability of well diagrams. We now introduce the notion of stability for the well diagrams. Let $f, g : \mathbb{R}^2 \rightarrow \mathbb{R}^2$ be two vector fields. Construct a bijection $\mu : \text{Dgm}(f_0) \rightarrow \text{Dgm}(g_0)$ that maps the kth highest point in $\text{Dgm}(f_0)$ to the kth highest point in $\text{Dgm}(g_0)$. Since the point 0 in each well diagram has an arbitrary multiplicity, by choosing the appropriate multiplicities for 0, μ becomes a bijection. The *bottleneck distance* between $\text{Dgm}(f_0)$ and $\text{Dgm}(g_0)$ is $W_\infty(\text{Dgm}(f_0), \text{Dgm}(g_0)) = \sup_{a \in \text{Dgm}(f_0)} |a - \mu(a)|$. The Stability Theorem of Well Diagrams [11] states that the bottleneck distance between two well diagrams is bounded by the distance between the mappings, that is, $W_\infty(\text{Dgm}(f_0), \text{Dgm}(g_0)) \leqslant \text{d}(f, g)$.

Static robustness and its properties. The *static robustness* of a critical point is the height of its lowest degree zero ancestor in the merge tree [5, 38]. The static robustness quantifies the stability of a critical point with respect to perturbations of the vector fields through the following lemmas first introduced in [38]. Both lemmas are illustrated in Fig. 1.

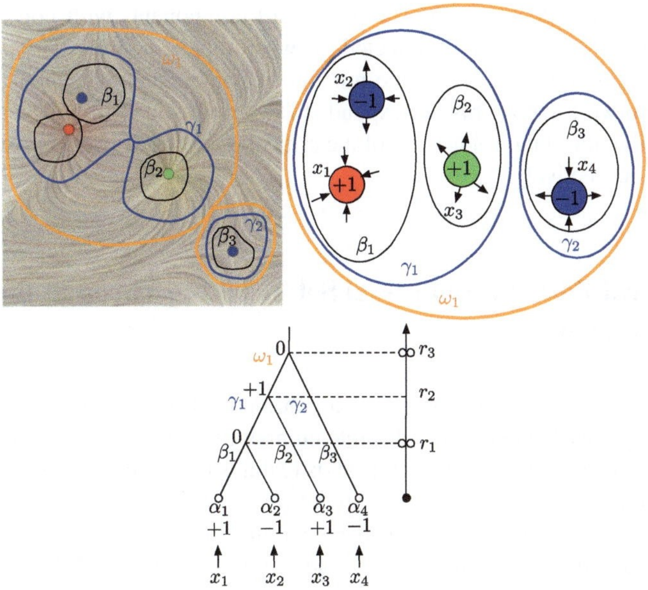

Fig. 2 *Top*, vector field f (*left*) and relations among components of $\mathbb{F}_r$ (*right*). *Bottom*, augmented merge tree (*left*) and its well diagram (*right*) (Figure adapted from [38])

Lemma 1 (Critical Point Cancellation [38]). *Suppose a critical point x of f has static robustness r. Let C be the connected component of $\mathbb{F}_{r+\delta}$ containing x, for an arbitrarily small $\delta > 0$. Then, there exists an $(r+\delta)$-perturbation h of f, such that $h^{-1}(0) \cap C = \emptyset$ and $h = f$ except possibly within the interior of C.*

Lemma 2 (Degree and Critical Point Preservation [38]). *Suppose a critical point x of f has static robustness r. Let C be the connected component of $\mathbb{F}_{r-\delta}$ containing x, for some $0 < \delta < r$ and $r - \delta$ being a regular value. Then for any ϵ-perturbation h of f where $\epsilon \leqslant r-\delta$, the sum of the degrees of the critical points in $h^{-1}(0) \cap C$ is* $\deg(C)$. *Furthermore, if C contains only one critical point x, we have* $\deg(h^{-1}(0) \cap C) = \deg(x)$. *In other words, there is no ϵ-perturbation ($\epsilon \leqslant r - \delta$) that could cancel the critical point in C; that is, x is preserved.*

Example. We illustrate the above concepts through an example shown Fig. 2 adapted from [38]. A 2D vector field f (on the left) contains four critical points, a sink x_1 (red), a source x_3 (green), and two saddles x_2 and x_4 (blue). Its corresponding mapping f_0 has three critical values, denoted as r_1, r_2 and r_3, respectively. The merge tree (on the right) tracks the components of the sublevel sets $\mathbb{F}_r$ as they appear and merge, as r increases from 0. We use α, β, γ etc. to represent components of certain sublevel sets at the critical values. At $r = 0$, four components α_1 to α_4 appear that correspond to the four critical points. At $r = r_1$, components represented by α_1 and α_2 merge into a single component represented by β_1, which has degree zero. The number of components with non-zero degree drops from four to two, this is reflected by two points in the well diagram $\mathrm{Dgm}(f_0)$ with value r_1. Then at $r = r_3$ the number of components with non-zero degree drops from two to zero, this corresponds to two points in $\mathrm{Dgm}(f_0)$ with value r_3. By definition, the static robustness of the critical points x_1, x_2, x_3, and x_4 are r_1, r_1, r_3, and r_3, respectively.

3 Critical Point and Sublevel Set Tracking Through the Lens of Robustness

In practice, we do not have access to a continuous time-varying 2D vector field, but rather a dataset consists of a discrete number of snapshots at different points in time. This means that to track vector fields features (i.e. critical points and sublevel sets) we must first resolve the *correspondence problem*. That is, determining the correspondences between the critical points and sublevel sets in successive time steps, that actually represent the same object at different times [22]. In this section, we prove formally that robustness helps us understand the sampling conditions under which we can resolve the correspondence problem based on region overlap techniques, and the uniqueness and uncertainty associated with such techniques.

The stability of well diagrams and the properties associated with (static) robustness allow us to give a theoretical underpinning to this approach by requiring that the vector field changes slowly enough and then treating adjacent time steps as small perturbations of each other. We assume that the underlying time-varying vector field is c-Lipschitz and that we have an ϵ-sampling in space and time. It follows that the vector fields at each time steps are $c\epsilon$-perturbations of each other. Formally, suppose $f : \mathbb{X} \subseteq \mathbb{R}^2 \to \mathbb{R}^2$ is a c-Lipschitz function. That is, $\forall x, x' \in \mathbb{X}$, $||f(x) - f(x')||_2 \leqslant c||x - x'||_2$. Given a triangulation K of $\mathbb{X}$ and f valued at its vertices, we could linearly interpolate over its simplexes (that is, edges and triangles) resulting in a continuous function $\hat{f} : |K| \to \mathbb{R}^2$ [5]. If vertices P in K are ϵ-sampling of $\mathbb{X}$ (namely, $\forall x \in \mathbb{X}$, $d(x, P) := \inf_{y \in P} ||x - y||_2 \leqslant \epsilon$), then we have $\forall x \in \mathbb{X}$, $||f(x) - \hat{f}(x)||_2 \leqslant c\epsilon$. This observation allows us to move from continuous to the piecewise-linear (PL) setting by noting that for a c-Lipschitz function, a linear interpolation between samples results in an error of at most $c\epsilon$ from the true underlying function. As a consequence of the Stability Theorem of Well Diagrams, we have,

Lemma 3 (Triangulation Lemma [5]). *The bottleneck distance between the well diagrams of f and the PL interpolation $\hat{f}$ is bounded by $W_\infty(\mathrm{Dgm} f, \mathrm{Dgm} \hat{f}) \leqslant c\epsilon$.*

In the time varying setting, to accommodate the additional dimension of time we make a small change of notation by referring to $f : \mathbb{X} \times \mathbb{R} \to \mathbb{R}^2$ as a time-varying 2D vector field over domain $\mathbb{X} \subseteq \mathbb{R}^2$, where $f_t(x) = f(x,t) : \mathbb{X} \to \mathbb{R}^2$ represents a 2D vector field at time $t \in \mathbb{R}$. For notational simplicity we assume we have an

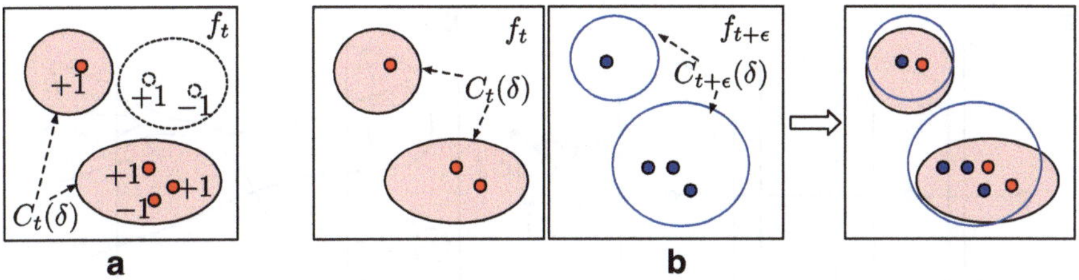

Fig. 3 (**a**) Definition of $C_t(\delta)$: sublevel set with non-zero degree. (**b**) Illustration for Lemma 4

ϵ-sampling and that the Lipschitz constant is c in the time domain as well. That is, $\forall x \in \mathbb{X}, ||f_t(x) - f_{t+\epsilon}(x)||_2 \leqslant c\epsilon$.

We now give guarantees on the correspondence of critical points across time slices with respect to robustness. We assume that we have a PL-interpolation of the c-Lipschitz time-varying vector field built on certain ϵ-sample. The proofs do not depend on this interpolation but rather require a bound on the error from approximating the underlying function. Better interpolation methods will lead to better approximations and therefore better constants in the theoretical guarantees.

We introduce some additional notation. Recall $f : X \times \mathbb{R} \to \mathbb{R}^2$ is a time-varying 2D vector field over domain $\mathbb{X}$, and $f_t(x) = f(x,t) : \mathbb{X} \to \mathbb{R}^2$ represents a 2D vector field at time t. A crucial concept is the sublevel set of the Euclidean norm of f_t, $||f_t(x)||_2$. Let $C_t(\delta) = \{x \in \mathbb{X} \mid ||f_t(x)||_2 \leqslant \delta\}$ denote its sublevel set for any $\delta > 0$ whose degree is non-zero, see Fig. 3a. If we consider a specific connected component of $C_t(\delta)$, we denoted it by $C_t^i(\delta)$. Furthermore, let $z_t = C_t(0) = \{x \in \mathbb{X} \mid ||f_t(x)||_2 = 0\}$ represent the set of critical points. When considering a single critical point in the set we will add an index to the notation (e.g. z_t^i). We begin our discussion of correspondence by making the following observation: the critical points with high robustness in two adjacent time steps must be contained in the interior of the intersection of the corresponding sublevel sets. Formally,

Lemma 4 (Critical Points Containment). *For two adjacent time steps of the vector field $f_t, f_{t+\epsilon} : \mathbb{X} \to \mathbb{R}^2$, the critical points in both time steps belong to* $\text{int}\,(C_t(\delta) \cap C_{t+\epsilon}(\delta))$ *for all $\delta > c\epsilon$.*

Proof. The lemma is illustrated in Fig. 3b. Consider $C_t(\delta)$, the sublevel set of f_t, where $\delta > 0$. By the Lipschitz assumption, $\forall x \in \mathbb{X}, ||f_t(x) - f_{t+\epsilon}(x)||_2 \leqslant c\epsilon$. It follows that $\forall \delta > c\epsilon$, the critical points in f_t, $z_t = C_t(0) \subseteq C_{t+\epsilon}(c\epsilon) \subset C_{t+\epsilon}(\delta)$, and the critical point in $f_{t+\epsilon}$, $z_{t+\epsilon} = C_{t+\epsilon}(0) \subseteq C_t(c\epsilon) \subset C_t(\delta)$. On the other hand, $z_t = C_t(0) \subset C_t(\delta)$ and $z_{t+\epsilon} = C_{t+\epsilon}(0) \subset C_{t+\epsilon}(\delta)$. Hence, $z_t, z_{t+\epsilon} \subseteq \text{int}\,(C_t(\delta) \cap C_{t+\epsilon}(\delta))$. □

Lemma 4 states that critical points are contained in intersections across time steps. While this is an important observation, it does not imply correspondence. The argument we would ultimately like to make is that we can find correspondences between two adjacent time slices represented by a (bounded)-homotopy. Recall a

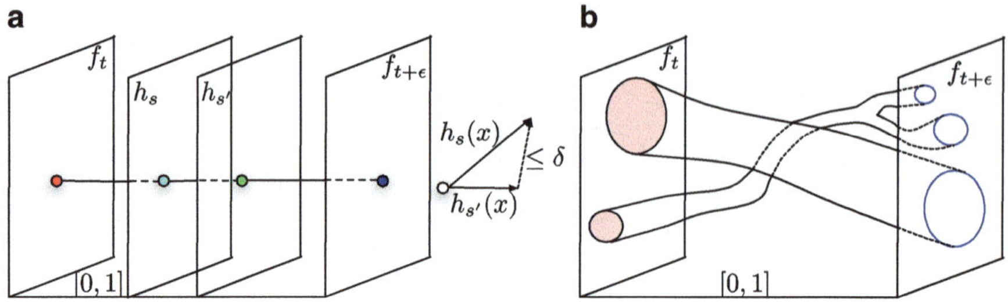

Fig. 4 Illustrations of (**a**) bounded-homotopy and (**b**) δ-tube

homotopy between two continuous functions that map between topological spaces, $f_t, f_{t+\epsilon} : \mathbb{X} \to \mathbb{R}^2$, is defined to be a continuous function $H : \mathbb{X} \times [0,1] \to \mathbb{R}^2$, such that $\forall x \in \mathbb{X}$, $H(x,0) = f_t(x)$ and $H(x,1) = f_{t+\epsilon}(x)$. We then use $h_s(x) = H(x,s) : \mathbb{X} \to \mathbb{R}^2$ ($s \in [0,1]$) to represent intermediate time slices. Such a homotopy is *δ-bounded* (or has a maximum deformation of at most δ), if $\forall x \in \mathbb{X}$ and $\forall s, s' \in [0,1]$, $||h_s(x) - h_{s'}(x)||_2 \leqslant \delta$. Figure 4a gives an illustration.

In order to obtain a correspondence, we will need to impose some further conditions. First we formally define a correspondence. A *δ-correspondence* between a pair of critical points is defined such that there exists a δ-bounded homotopy which maps the points to each other. Formally, there is a δ-correspondence between critical points $p \in z_t$ and $q \in z_{t+\epsilon}$ if there exists a δ-bounded homotopy H between f_t and $f_{t+\epsilon}$, such that $H^{-1}(0)$ contains a continuous path embedded in $\mathbb{X} \times [0,1] \subset \mathbb{R}^3$ that connects p with q. We refer to such a path as the *critical path*. A construction we will make use of is the *straight-line homotopy*. For two functions, $f_t, f_{t+\epsilon} : \mathbb{X} \to \mathbb{R}^2$, we define their straight-line homotopy as

$$f_{t+s\epsilon}(x) = h_s(x) = (1-s)f_t(x) + sf_{t+\epsilon}(x) \qquad 0 \leqslant s \leqslant 1, \quad \forall x \in \mathbb{X} \tag{1}$$

First, we examine a simple situation, which we refer to as the *unique intersection*. We assume a component at time t intersects with only one component at time $t+\epsilon$ and vice versa. Formally we say that, for a single component $C_t^i(\delta)$ in $C_t(\delta)$, there exists only a single component $C_{t+\epsilon}^j(\delta)$ in $C_{t+\epsilon}(\delta)$ such that they intersect; and for $C_{t+\epsilon}^j(\delta)$, the only component it intersects in $C_t(\delta)$ is $C_t^i(\delta)$.

Lemma 5 (Critical Points Correspondence Under Unique Intersection). *For $\delta > c\epsilon$, let $C_t^i(\delta)$ and $C_{t+\epsilon}^j(\delta)$ be components of the δ-sublevel sets with a unique intersection. If there exists a unique δ-robust critical point in each component, denoted as x and y respectively, then they are in correspondence.*

The lemma is illustrated in Fig. 5a. The main idea behind its proof is to construct a $(\delta + c\epsilon)$-bounded homotopy that maps x to y by considering tubular neighborhood surrounding the parametrized curve connecting x and y as shown in Fig. 5b.

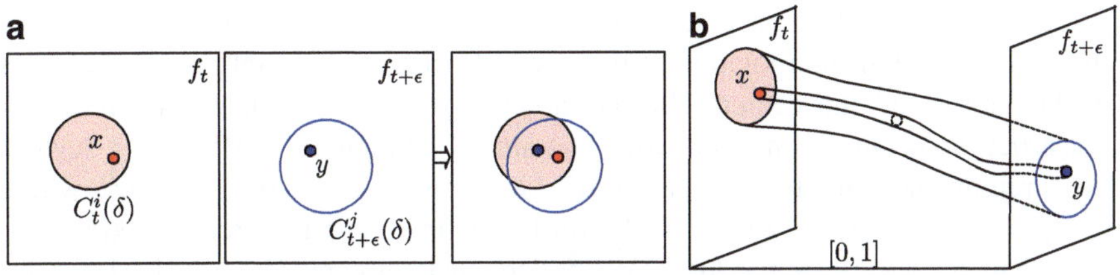

Fig. 5 (**a**) Illustration of Lemma 5 and (**b**) the main idea behind its proof

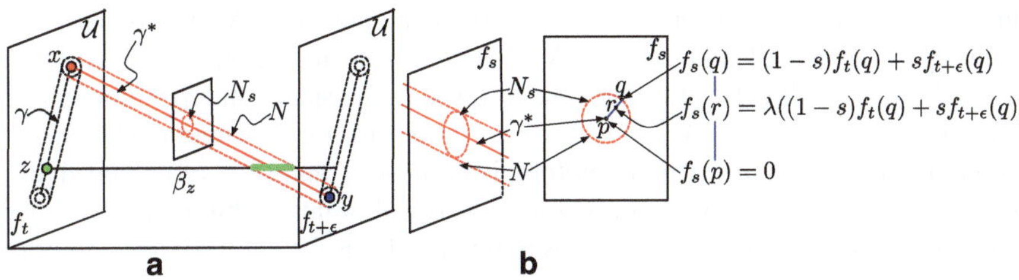

Fig. 6 (**a**) Constructing the homotopy for correspondence $H : \mathbb{X} \times [0, 1] \to \mathbb{R}^2$. Here we show two time slices at t and $t + \epsilon$. $\mathscr{U} \subset \mathbb{X}$ is represented as a *rectangle*. The path between critical points x and y is $\gamma \subset \mathscr{U}$. It is "lifted" to a parametrized curve $\gamma^* \subset \mathbb{X} \times [0, 1] \subset \mathbb{R}^3$, with N as its tubular neighborhood. (**b**) Spatial interpolation within N_s that linearly interpolate between its boundary and its center. λ is the scaling parameter used in the interpolation, where $0 \leqslant \lambda \leqslant 1$

Proof. Let $\mathscr{I} = C_t^i(\delta) \cap C_{t+\epsilon}^j(\delta)$ and $\mathscr{U} = C_t^i(\delta) \cup C_{t+\epsilon}^j(\delta)$. Suppose $\mathscr{I} \neq \emptyset$. Suppose both points x and y are δ-robust, that is, they have static robustness greater or equal to δ. First, based on the unique intersection condition, by Lemma 4, both x and y are contained in $\mathscr{I}$. Second, we claim that $\mathscr{U} \backslash C_t^i(\delta)$ (and $\mathscr{U} \backslash C_{t+\epsilon}^j(\delta)$ symmetrically) contains no critical points. Suppose there is a critical point $x' \in \mathscr{U} \backslash C_t^i(\delta)$, since $x' \notin C_t^i(\delta)$, then either (a) $x' \in C_t^{i'}(\delta)$ for some $i' \neq i$, or (b) $||f_t(x')||_2 > \delta$. (a) is impossible as it violates the unique intersection condition. For (b), based on Lipschitz condition and the reverse triangle inequality, we have $||f_{t+\epsilon}(x')||_2 \geqslant ||f_t(x')||_2 - ||f_t(x') - f_{t+\epsilon}(x')||_2 > \delta - c\epsilon > 0$. Hence x' cannot be a critical point.

For the rest of the proof, we need to show that x and y are in correspondence, by constructing the desired homotopy. We also claim that such a homotopy is $(\delta + c\epsilon)$-bounded. First, we construct a critical path. Since both $x, y \in \mathscr{U}$, there exists a parametrized continuous curve γ in $\mathscr{U}$ that connects these two points. That is, $\gamma : [0, 1] \to \mathbb{X}$ where $\gamma(0) = x$ and $\gamma(1) = y$. Such a curve could be "lifted" to $\mathbb{X} \times [0, 1]$ by defining a parametrized curve $\gamma^* : [0, 1] \to \mathbb{X} \times [0, 1]$ where $\gamma^*(t) = \gamma(t) \times t$, for $0 \leqslant t \leqslant 1$. This constitutes the desired critical path between x and y in $H^{-1}(0) \subset \mathbb{R}^3$. Our goal now is to define a continuous homotopy based on such a critical path. Such a process is shown in Fig. 6a.

Second, we consider a tubular neighborhood $N \subset \mathbb{R}^3$ of γ^*. Such a neighborhood intersects each time slice $\mathbb{X} \times s$ at N_s, $\forall s \in [0, 1]$, where $N_s \subset \mathbb{X}$

contains a zero center that is the intersection between γ^* and the time slice. We introduce a spatial interpolation within each N_s that linearly interpolate between its boundary and the center. This is shown in Fig. 6b. Third, we are ready to construct the desired homotopy with guaranteed continuity. To do so, we rewrite the straight-line homotopy for a fixed $z \in \mathbb{X}$ as $\alpha_z(s) : [0,1] \to \mathbb{R}^2$, where $\alpha_z(s) = (1-s)f_t(z) + sf_{t+\epsilon}(z)$. We further define a curve β_z for a fixed $z \in \mathbb{X}$, such that, $\beta_z : [0,1] \to z \times [0,1]$. We define our homotopy $H : \mathbb{X} \times [0,1] \to \mathbb{R}^2$ as follows. $\forall z \in \mathbb{X}$: (a) If β_z does not intersect N or it only intersects N on its boundary point (non-transversal intersection), we use the straight-line homotopy, that is, $H(z,s) = \alpha_z(s)$ for $0 \leqslant s \leqslant 1$; (b) If β_z intersects N transversally (by entering N at time s' and existing N at time s''), then $H(z,s)$ is defined to be $\alpha_z(s)$ for $s \in [0,s'] \cup [s'',1]$, otherwise, it respects the spatial interpolation within the interior of N for $s \in (s',s'')$. In case (a) and (b), the maximum deformation at any point $z \in \mathbb{X}$ during the homotopy is at most $c\epsilon$ and $(\delta + c\epsilon)$ respectively. Finally, since x and y are the only δ-robust critical points within their respective components, the uniqueness conditions imply that this is the only possible choice of correspondence. □

Remark 1. Much of the complication in constructing the above homotopy is dealing with the tubular neighborhoods to ensure the mapping is continuous. and therefore a homotopy. It is important to note that although the maximum deformation of such a homotopy is bounded, its Lipschitz constant is not. Controlling the Lipschitz constant of such a homotopy is a far more difficult problem. We merely impose a Lipschitz constant on the time-varying vector field to ensure validity of the approximation, whereas here we demonstrate the existence of a possible correspondence.

Now we extend the above lemma to cases without the uniqueness intersection assumptions. We make the following claim regarding many-to-many correspondences among critical points with large robustness. We consider the following statement a major contribution in rethinking and treating correspondence problem under the robustness framework. The key point is that we relax the uniqueness condition on the intersections. With this we lose the guarantee on uniqueness of the map, but we show that for any choice among possible correspondences, there exists a homotopy.

Lemma 6 (Robust Critical Points Correspondence). *There exists a $(\delta + c\epsilon)$-bounded homotopy between δ-robust critical points between time slices, from which a correspondence could be obtained.*

The idea is the behind the proof is illustrated in Fig. 7. Here, we consider case (a) where the sublevel sets have unique intersection, however there may be multiple δ-robust critical points in each component; and case (b) where the sublevel sets do not have unique intersection. In case (a), we choose the desired correspondence by constructing the critical paths which are no longer unique. In case (b), we require that a path exists between critical points. There are many choices of correspondences

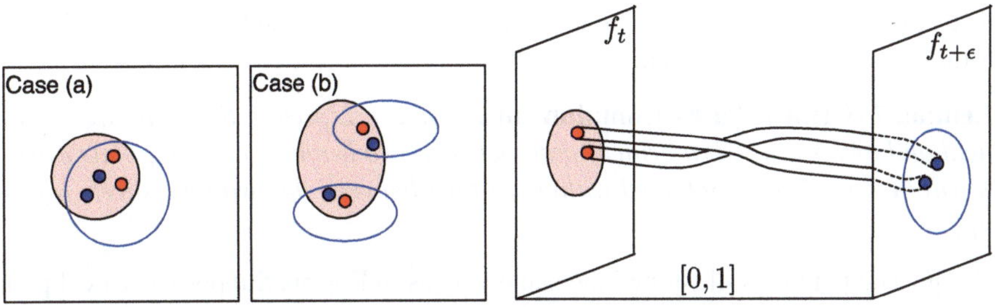

Fig. 7 Illustration of the main proof idea behind Lemma 6

(under the restriction that the corresponding points are distinct), our proposed homotopy construction works for any of them.

Proof. In case (a), suppose we still have unique intersection condition, however there exist multiple δ-robust critical points in each component. All the arguments from Lemma 5 hold except now we must first choose the desired correspondences by constructing critical paths (which are no longer unique). We also need to add some discussions, since for a fixed choice of correspondences, we have multiple critical paths and their corresponding tubular neighborhoods. In generic situations, we suppose all such critical paths with arbitrarily small tubular neighborhoods do not intersect during the construction of the homotopy. Therefore although the constructed homotopy might be more complicated (where β_z for a fixed z may intersect multiple neighborhoods), it remains continuous and bounded.

In case (b), suppose the unique intersection condition no longer holds. In the construction of the above homotopy, we require only that a path γ exists (in the union of components, i.e. $\mathscr{U}$) between critical points. Without the uniqueness assumption, we have many choices of correspondences, and the homotopy construction works for any of them (under the restriction that the corresponding points are distinct). To complete the proof, we need only to check that a path γ exists, such that, $\forall x \in \gamma$, $||f_t(x)||_2 \leqslant (\delta + c\epsilon)$ and $||f_{t+\epsilon}(x)||_2 \leqslant (\delta + c\epsilon)$. This follows from the Lipschitz assumptions. □

Robustness also provides guarantees on the *critical paths*, that is, trajectories of critical points trace over time. To make a precise statements, we define the notion of δ-*tube* and its PL counterpart. A δ-*tube* is a (connected) component in the collection of δ-sublevel sets between two adjacent discrete time steps based on straight-line homotopy. Let $C_s(\delta)$ be connected components of the δ-sublevel set at time s. Suppose f_t and $f_{t+\epsilon}$ are two adjacent time steps. A δ-tube is defined as a component in $\bigcup_{s\in[t,t+\epsilon]} C_s(\delta)$. This is illustrated in Fig. 4b. Given the i-th δ-tube between times t and $t+\delta$, without loss of generality, we denote each of its time slice as $C_s^i(\delta)$. Note that splitting and merging is possible with a given δ-tube.

A PL δ-tube is similarly defined but is based on the straightline homotopy of PL interpolations at each time step. Correspondingly, each of its time slice is denoted as $\hat{C}_s^i(\delta)$, for $t \leqslant s \leqslant t + \epsilon$. The following lemma states conditions under which a δ-tube contains a critical path.

Lemma 7 (Critical Paths Containment). *For a c-Lipschitz time-varying vector field and any $\delta > c\epsilon$, if a critical path between two δ-robust critical points exists, it will be completely contained within a δ-tube between the two time slices f_t and $f_{t+\epsilon}$.*

Proof. The lemma is illustrated the same way as in Fig. 5b. Suppose a critical path γ leaves the i-th δ-tube at some time s ($t \leqslant s \leqslant t+\epsilon$). This implies that there exists a critical point $p \in z_s$ that continuously moves outside of the δ-tube at time s. This means $p \notin C_s^i(\delta)$, therefore $f_s(p) \geqslant \delta$. There are two cases: (a) the critical point re-enters the i-th δ-tube at time s' and stays inside the tube until time $t + \epsilon$; and (b) the critical point enters a different δ-tube (i.e. the j-th δ-tube) at time s' and never returns back to the i-th δ-tube, where $s < s' < t + \epsilon$.

In case (a), we consider the particular scenario where $(s' - s)$ approaches zero, by the Lipschitz assumption, $p \in C_t^i(c(s-t))$ and $p \in C_{t+\epsilon}^i(c(t+\epsilon-s))$. Based on Lemma 4, $p \in C_t^i(c(s-t)) \cap C_{t+\epsilon}^i(c(t+\epsilon-s))$. The Lipschitz condition also implies that the function value at p based on straight-line homotopy at time s is, $f_s(p) \leqslant \min\{2c(s-t), 2c(t+\epsilon-s)\}$. The above upper bound achieves its maximum when $s = t + \epsilon/2$, where $f_s(p) \leqslant c\epsilon$. Since $\delta > c\epsilon$, this contradicts the assumption that $f_s(p) \geqslant \delta$. So case (a) is not possible.

In case (b), suppose the critical point leaves $C_s^i(\delta)$ and enters $C_{s'}^j(\delta)$, where $s < s'$. Lemma 4 implies the critical point belongs to $C_s^i(\delta) \cap C_{s'}^j(\delta)$, and in addition, it belongs to any $C_s^i(\delta) \cap C_{s'}^j(\delta)$ as $(s' - s)$ approaches zero. This contradicts the fact that $C_s^i(\delta)$ and $C_{s'}^j(\delta)$ originate from non-intersecting δ-tubes. Therefore case (b) does not hold either. □

Corollary 1 (PL Critical Paths Containment). *For a c-Lipschitz time-varying vector field and any $\delta > c\epsilon$, if a critical path between two δ-robust critical points exists, it will be completely contained within a PL $(\delta + c\epsilon)$-tube between the two time slices f_t and $f_{t+\epsilon}$.*

Proof. This follows directly from Lemma 7, since $C_s^i(\delta) \subseteq \hat{C}_s^i(\delta + c\epsilon)$ for all s and i. Therefore since the critical path is included in the δ-tube, it follows that it is included in the PL $(\delta + c\epsilon)$-tube.

To prove the above inclusion, this property holds at the end points ($s = t$ and $s = t + \epsilon$) of the straight-line homotopy based on the c-Lipschitz assumption and ϵ-sampling. Because the straight-line homotopy is a convex combination of the end points, it holds at any point in between as well. □

Remark 2. The above lemmas prove that regardless of the (possibly unknown) underlying changes of the vector field, the critical paths of the vector fields for robust critical points are contained inside some δ-tubes, which implies that the straight-line homotopy roughly captures the behavior of the critical paths.

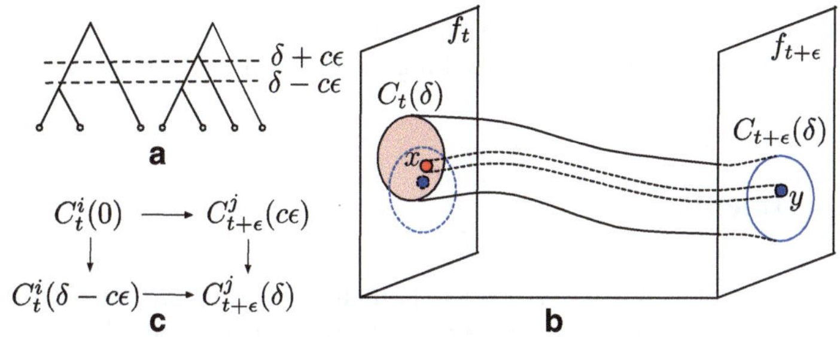

Fig. 8 (**a**) and (**b**): Illustration of Lemma 8. (**c**) Diagram in its proof

Now we have addressed critical points correspondences and critical paths containments, we would like to address the problem of *sublevel set correspondence*, as shown in the following lemma and illustrated in Fig. 8a, b where critical point x in f_t corresponds to y in $f_{t+\epsilon}$.

Lemma 8 (Sublevel Set Unique Correspondence). *For $\delta > c\epsilon$, suppose $C_t^i(\delta)$ and $C_{t+\epsilon}^j(\delta)$ are two components of $C_t(\delta)$ and $C_{t+\epsilon}(\delta)$ respectively such that their intersection contains critical points. If there are no merge events in $[\delta - c\epsilon, \delta + c\epsilon]$ (within the merge trees) between times t and $t + \epsilon$, the map induced between these pairs of components is unique. In other words, the correspondences between connected components in $C_t(\delta)$ and $C_{t+\epsilon}(\delta)$ whose intersections contain critical points are unique.*

Proof. Since there are no merge events in $[\delta - c\epsilon, \delta + c\epsilon]$ between times t and $t + \epsilon$, components can neither merge nor split apart. First we show that each connected component $C_t^i(\delta)$ has intersection with at least one connected component $C_{t+\epsilon}^j(\delta)$. For every i, there exists a j such that we can obtain the diagram in Fig. 8c, where all the maps are inclusions. Three of these inclusions are obvious. We prove the inclusion exists for $C_t^i(\delta - c\epsilon) \to C_{t+\epsilon}^j(\delta)$. Suppose there exists a point $p \in C_t^i(\delta - c\epsilon)$ but $p \notin C_{t+\epsilon}^j(\delta)$. This implies that $f_{t+\epsilon}(p) > \delta$ and $f_t(p) \leqslant \delta - c\epsilon$. This violates the Lipschitz assumption at p. The above diagram implies that for every i, there exists a j such that $C_t^i(\delta) \cap C_{t+\epsilon}^j(\delta) \neq \emptyset$, and their intersection contains critical points (referred to as non-zero intersections). Thus there is a possible correspondence between these two components. To show such a correspondence is unique, we claim that if there were additional intersections, i.e., with a connected component $C_{t+\epsilon}^k(\delta)$, this would imply a merge/splitting event in the required interval (between $C_{t+\epsilon}^j(\delta)$ and $C_{t+\epsilon}^k(\delta)$). Assume that both $C_{t+\epsilon}^j(\delta)$ and $C_{t+\epsilon}^k(\delta)$ have a non-zero intersection with $C_t^i(\delta)$, it follows by Lipschitz assumptions that there exists a path $\gamma \subset C_t^i(\delta)$ connecting $C_{t+\epsilon}^j(\delta)$ and $C_{t+\epsilon}^k(\delta)$, such that $\forall x \in \gamma$, $||f_{t+\epsilon}(x)||_2 \leqslant \delta + c\epsilon$. However this implies that the two components merge in the interval $[\delta - c\epsilon, \delta + c\epsilon]$ which contradicts our assumption.

Therefore, we conclude there cannot be two components $C^j_{t+\epsilon}(\delta)$ and $C^k_{t+\epsilon}(\delta)$ with a non-zero intersection with $C^i_t(\delta)$, making the map unique. □

4 Experiments

We demonstrate robustness-based critical point tracking on three real world datasets, which are extracted from consecutive time slices of 2D time-varying vector fields. The first two datasets, OceanA and OceanB, come from top layers of the 3D simulation of global oceanic eddies [21] for 350 days in the year 2002. We extracted tiles from this simulation data, representing the flow in the central Atlantic Ocean for OceanA (resolution 60×60), south Atlantic Ocean for OceanB (resolution 100×100), and construct standard triangulations on the point samples. We use time slices #21310 and #21311 for OceanA, and #20710 and #20711 for OceanB. Our third dataset CombustionC is taken from the simulation of homogeneous charge compression ignition (HCCI) engine combustion [15]. The domain has periodic boundary and is represented as a 640×640 regular grid. The 2D time-varying vector field consists of 299 time-steps with a time interval of 10^{-5} s. We selected time slices #173 and #174 from this data.

The critical points correspondences based on robustness are shown in Fig. 9. Suppose we use PL interpolation between time slices. Our theoretical results rely on the quantity $c\epsilon$, which depends on our prior knowledge of the datasets, or some form of estimation.

First, suppose $c\epsilon$ is equal to the magnitude of the maximum observed difference in the vector fields of the two adjacent time slices t_1 and t_2, $\delta = \sup_x ||f_{t_1}(x) - f_{t_2}(x)||_2$. That is, let $c\epsilon = \delta$. For OceanA (a) and (b), we illustrate the components of the sublevel set at $c\epsilon$ that contains critical points. Since $c\epsilon$ is relatively large, based on Lemma 5, the components (pointed by white solid arrows) in (a) and (b) have unique intersection and each contains a single critical point, therefore these critical points correspond to one another. On the other hand, the blue points in t_1 and t_2 could all potentially map to one another, creating a many-to-many correspondence scenario. It appears that the single yellow point in t_1 could potentially map to any of the three points in t_2 based on region overlap (or distance proximity, if all three yellow points are moved even closer in distance). However, this is not true as we investigate further in OceanA(c) and (d) by showing the components of the sublevel set at the robustness values of each critical point. It is interesting to point out that the points (pointed by white arrows) end up matching to each other uniquely based on Lemma 5 since their robustness values are higher than $c\epsilon$, while the two unmatched points (in the black components) are considered newly appeared. A similar situation occurs in OceanB(a) and (b). High robustness points have unique correspondences (pointed by solid white arrows) and many low robustness points are matched under many-to-many (usually pair-to-pair) scenarios. Newly appeared critical points (pointed by a hollow white arrow) at t_2 are not matched and are shown

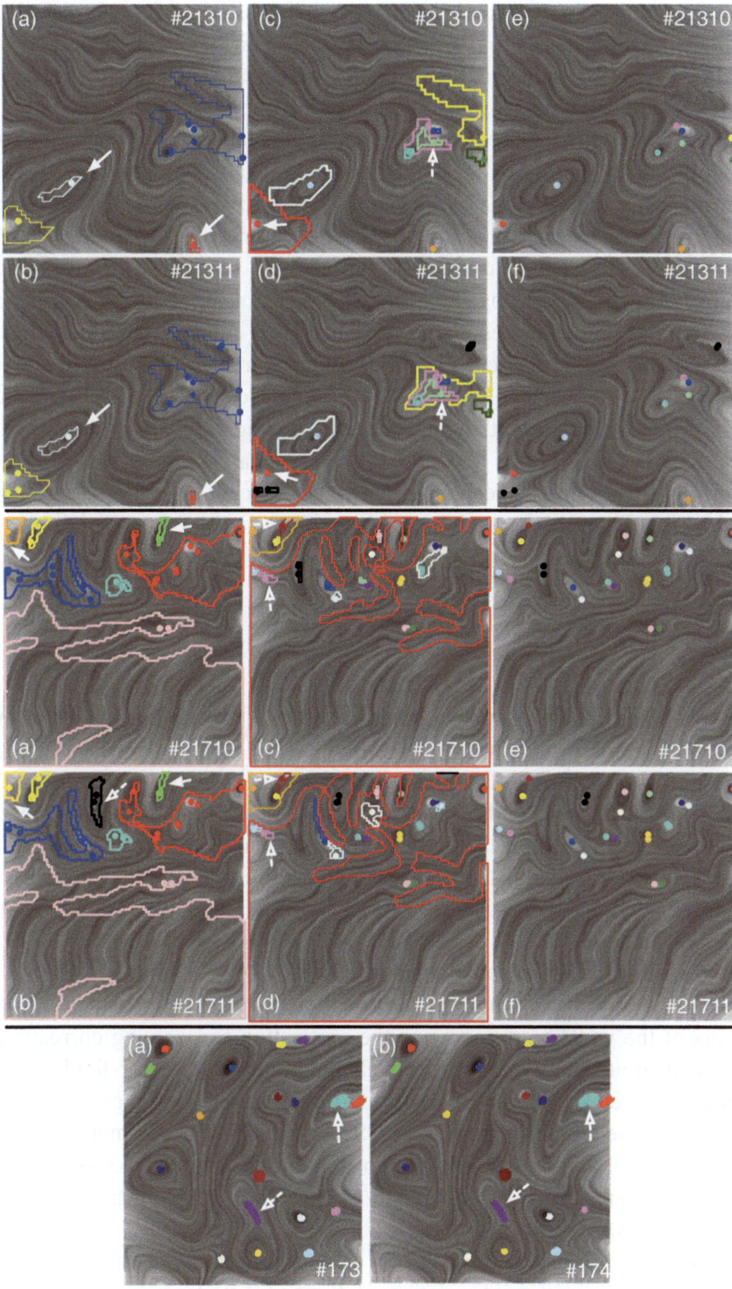

Fig. 9 Tracking critical points correspondences based on robustness for OceanA (*top*), OceanB (*middle*) and CombustionC (*bottom*). In each pair of pictures between time slices (e.g. (**a**) vs (**b**), (**c**) vs (**d**), (**e**) vs (**f**)), the corresponding points are shown by the same color

in black. In CombustionC(a) and (b), $c\epsilon$ is quite small, therefore almost all points have unique matches except for two low robustness pairs which are matched as pairs (pointed by the hollow white arrows). Second, we can also measure the difference in vector fields between t_1 and t_2 in local regions. We compute $\delta^* = \sup_{x\in\Omega} ||f_{t_1}(x) - f_{t_2}(x)||_2$ for some Ω in the local neighborhoods of some critical points. Suppose the critical points in the local regions Ω has robustness values higher than $c\epsilon$, then we could further differentiate some of the many-to-many correspondence scenarios, and create unique correspondences. Such conditions are met by OceanA and OceanB, as shown in Fig. 9 OceanA (c) and (d) and OceanB (c) and (d). For example, in the local regions pointed to by the hollow white arrows, the critical points across t_1 and t_2 obtain unique correspondences since their robustness values are higher than the amount of vector field perturbation in their local neighborhoods. These correspondences are shown without sublevel sets in OceanA (e) and (f) and OceanB (e) and (f) (black marks points which are not matched). Finally, it is interesting to note that when critical points leave the boundary or appear near fold bifurcations, they typically do not find correspondences in the adjacent time slices (e.g. the black points shown in Fig. 9).

5 Discussion

Feature tracking, especially critical point tracking, is crucial for understanding the temporal behavior of time-varying vector fields. The theory of well groups allows us to make rigorous statements under mild assumptions about the correspondences: both when they are unique and when possible ambiguities exist. We infer correspondences between critical points based on their closeness in stability, measured by robustness, instead of just distance proximities within the domain. The correlations among critical points with high robustness values inherently capture some core structures of the time-varying vector field that is otherwise hidden due to the noise associated with region correspondence techniques.

The stability of well diagrams and the bijection between the critical points and the generators of the well groups serve as the motivation for viewing correspondence through well group theory. First, the well diagrams of a vector field and its PL interpolation are close, making it possible to translate the language of well groups from smooth to the PL setting in practice. Second, the bottleneck matching between two well diagrams constructed from two temporally adjacent vector fields gives a bijective mapping between generators of the well groups (which are also the generators of the 0-homology groups).

We show that robust generators are in some sense spatially stable, (namely, the generators must lie in the intersection of the connected components at the two time slices), and therefore there exist correspondences which respect the underlying geometry. The correspondences are not always unique since there may be several possible mappings between the generators of the robust well groups and the robust critical points. Static robustness captures this ambiguity making it possible to give *sufficient* conditions on the uniqueness of correspondences. On the other hand, we constructively show that whenever ambiguity exists, any of the possible correspondences are valid choices. This brings up many interesting questions. For example, does other criteria exist to choose the *best* correspondence from the set of possible correspondences? This may depend on the distance between the critical points, preservation of the topological skeletons, etc.

One future research direction is the constructions of different homotopy. The current construction, while bounded, is not guaranteed to be Lipschitz. It remains an open question how to construct a homotopy with a controlled Lipschitz constant. A second direction is the application of these methods to three dimensional and higher dimensional vector field data. The theorems and proofs are general and generalize to higher dimensions with minimal modification. Finally, the robustness framework gives correspondences a natural sense of scale: if we allow larger perturbations, more correspondences are possible. A natural question which arises is how to best visualize possible correspondences in a clear and intuitive way.

Acknowledgements This work was funded in part by the EU project TOPOSYS (FP7-ICT-318493-STREP), NSF OCI-0906379, NSF OCI- 0904631, DOE/NEUP 120341, DOE/MAPD DESC000192, DOE/LLNL B597476, DOE/Codesign P01180734, and DOE/SciDAC DESC0007 446. The authors would like to thank Guoning Chen and Paul Rosen for developing the tool for robustness-based visualization. We thank Jackie Chen for the combustion dataset. We also thank Mathew Maltude from the Climate, Ocean and Sea Ice Modeling program at Los Alamos National Laboratory (LANL) and the BER Office of Science UV-CDAT team for providing us the ocean datasets. The authors would also like to thank the anonymous reviewers for many useful comments to improve the readability of the paper.

References

1. D. Bauer, R. Peikert, Vortex tracking in scale space, in *Proceedings of the Symposium on Data Visualisation*, Switzerland, 2002, pp. 233–240
2. P. Bendich, H. Edelsbrunner, D. Morozov, A. Patel, Homology and robustness of level and interlevel sets. Homol. Homotopy Appl. **15**(1), 51–72 (2013)
3. G. Carlsson, A. Zomorodian, The theory of multidimensional persistence, in *Proceedings of the 23rd Annual Symposium on Computational Geometry*, Gyeongju, 2007, pp. 184–193
4. H. Carr, J. Snoeyink, U. Axen, Computing contour trees in all dimensions, in *Proceedings of the 11th Annual ACM-SIAM Symposium on Discrete Algorithms*, Philadelphia, 2000, pp. 918–926
5. F. Chazal, A. Patel, P. Skraba, Computing the robustness of roots. Manuscript. http://ailab.ijs.si/primoz_skraba/papers/fp.pdf,2011
6. F. Chazal, P. Skraba, A. Patel, Computing well diagrams for vector fields on $\mathbb{R}^n$. Appl. Math. Lett. **25**, 1725–1728 (2012)
7. H. Edelsbrunner, J. Harer, Jacobi sets of multiple Morse functions, in *Foundations of Computational Mathematics*, ed. by F. Cucker, R. DeVore, P. Olver, E. Sueli (Cambridge University Press, Cambridge, 2004), pp. 37–57
8. H. Edelsbrunner, J. Harer, A. Mascarenhas, V. Pascucci, Time-varying Reeb graphs for continuous space-time data, in *Proceedings of the 20th Annual Symposium on Computational Geometry*, Brooklyn, 2004, pp. 366–372

9. H. Edelsbrunner, D. Letscher, A. Zomorodian, Topological persistence and simplification. Discret. Comput. Geom. **28**, 511–533 (2002)
10. H. Edelsbrunner, D. Morozov, A. Patel, The stability of the apparent contour of an orientable 2-manifold, in *Topological Methods in Data Analysis and Visualization* (Springer, Berlin/Heidelberg, 2010), pp. 27–41
11. H. Edelsbrunner, D. Morozov, A. Patel, Quantifying transversality by measuring the robustness of intersections. Found. Comput. Math. **11**, 345–361 (2011)
12. C. Garth, X. Tricoche, G. Scheuermann, Tracking of vector field singularities in unstructured 3D time-dependent datasets, in *IEEE Visualization*, Washington, DC, 2004, pp. 329–336
13. A. Gerndt, T. Kuhlen, C. Cruz-Neira, Interactive tracking of three-dimensional critical points in unsteady, large-scale CFD datasets, in *Proceedings of the International Conference on Computer Graphics and Virtual Reality*, Las Vegas, 2008, pp. 22–28
14. A. Hatcher, *Algebraic Topology* (Cambridge University Press, Cambridge, 2002)
15. E. Hawkes, R. Sankaran, P. Pébay, J. Chen, Direct numerical simulation of ignition front propagation in a constant volume with temperature inhomogeneities: II. Parametric study. Combust. Flame **145**, 145–159 (2006)
16. M. in den Haak, H.J.W. Spoelder, F.C.A. Groen, Matching of images by using automatically selected regions of interest. *Computer Science in the Netherlands*, pp. 27–40, 1992
17. G. Ji, H.-W. Shen, R. Wenger, Volume tracking using higher dimensional isosurfacing, in *IEEE Visualization*, Seattle, 2003, pp. 209–216
18. H. King, K. Knudson, N. Mramor, Birth and death in discrete Morse theory. Manuscript, 2008
19. T. Klein, T. Ertl, Scale-space tracking of critical points in 3D vector fields. *Topology-Based Methods in Visualization, Mathematics and Visualization* (Springer, Berlin, 2007), pp. 35–49
20. M. Kraus, Visualization of uncertain contour trees. *Proceedings of the International Conference on Information Visualization Theory and Applications*, Angers, 2010, pp. 132–139
21. M. Maltrud, F. Bryan, S. Peacock, Boundary impulse response functions in a century-long eddying global ocean simulation. Environ. Fluid Mech. **10**, 275–295 (2010)
22. F.H. Post, B. Vrolijk, H. Hauser, R.S. Laramee, H. Doleisch, The state of the art in flow visualization: feature extraction and tracking. Comput. Graph. Forum **22**, 775–792 (2003)
23. F. Reinders, F.H. Post, H.J.W. Spoelder, Attribute-based feature tracking, in *Data Visualization*, Vienna, 1999, pp. 63–72
24. F. Reinders, F.H. Post, H.J.W. Spoelder, Visualization of time-dependent data using feature tracking and event detection. Vis. Comput. **17**, 55–71 (2001)
25. F. Reinders, I.A. Sadarjoen, B. Vrolijk, F.H. Post, Vortex tracking and visualisation in a flow past a tapered cylinder. Comput. Graph. Forum **21**, 675–682 (2002)
26. J. Reininghaus, J. Kasten, T. Weinkauf, I. Hotz, Efficient computation of combinatorial feature flow fields. IEEE Trans. Vis. Comput. Graph. **18**(9), 1563–1573 (2011)
27. J. Reininghaus, C. Lowen, I. Hotz, Fast combinatorial vector field topology. IEEE Trans. Vis. Comput. Graph. **17**(10), 1433–1443 (2011)
28. R. Samtaney, D. Silver, N. Zabusky, J. Cao, Visualizing features and tracking their evolution. Computer **27**, 20–27 (1994)
29. A. Shamir, C. Bajaj, B.-S. Sohn, Progressive tracking of isosurfaces in time-varying scalar fields. Technical report, CS & TICAM Technical Report TR-02-4, University of Taxes Austin, 2002
30. D. Silver, X. Wang, Volume tracking, in *IEEE Visualization*, San Francisco, 1996, pp. 157–164
31. D. Silver, X. Wang, Tracking and visualizing turbulent 3D features. IEEE Trans. Vis. Comput. Graph. **3**, 129–141 (1997)
32. D. Silver, X. Wang, Visualizing evolving scalar phenomena. Future Gener. Comput. Syst. **15**, 99–108 (1999)
33. P. Skraba, B. Wang, G. Chen, P. Rosen, 2D vector field simplification based on robustness. SCI Technical report UUSCI-2013-004, 2013
34. B.-S. Sohn, C. Bajaj, Time-varying contour topology. IEEE Trans. Vis. Comput. Graph. **12**, 14–25 (2006)

35. H. Theisel, H. P. Seidel, Feature flow fields, in *Proceedings of the Symposium on Data Visualisation*, Switzerland, 2003, pp. 141–148
36. X. Tricoche, G. Scheuermann, H. Hagen, Topology-based visualization of time-dependent 2D vector fields. *Data Visualization* (Springer, Vienna, 2001), pp. 117–126
37. X. Tricoche, T. Wischgoll, G. Scheuermann, H. Hagen, Topology tracking for the visualization of time-dependent two-dimensional flows. Comput. Graph. **26**, 249–257 (2002)
38. B. Wang, P. Rosen, P. Skraba, H. Bhatia, V. Pascucci, Visualizing robustness of critical points for 2D time-varying vector fields. Comput. Graph. Forum **32**(3), 221–230 (2013)
39. G. Weber, P.-T. Bremer, M. Day, J. Bell, V. Pascucci, Feature tracking using Reeb graphs, in *Topological Methods in Data Analysis and Visualization*, ed. by V. Pascucci, X. Tricoche, H. Hagen, J. Tierny (Springer, Berlin/Heidelberg, 2011)
40. C. Weigle, D.C. Banks, Extracting iso-valued features in 4-dimensional scalar fields, in *IEEE Symposium on Volume Visualization*, New York, 1998, pp. 103–110
41. T. Weinkauf, H. Theisel, A.V. Gelder, A. Pang, Stable feature flow fields. IEEE Trans. Vis. Comput. Graph **17**, 770–780 (2011)
42. A. Yilmaz, O. Javed, M. Shah, Object tracking: A survey. ACM Comput. Surv. **38**, 13 (2006)

Simplification of Morse Decompositions Using Morse Set Mergers

Levente Sipeki and Andrzej Szymczak

Abstract A common problem of vector field topology algorithms is the large number of the resulting topological features. This chapter describes a method to simplify Morse decompositions by iteratively merging pairs of Morse sets that are adjacent in the Morse Connection Graph (MCG). When Morse sets A and B are merged, they are replaced by a single Morse set, that can be thought of as the union of A, B and all trajectories connecting A and B. Pairs of Morse sets to be merged can be picked based on a variety of criteria. For example, one can allow only pairs whose merger results in a topologically simple Morse set to be selected, and give preference to mergers leading to small Morse sets.

1 Introduction

Due to its relevance to computer aided design, meteorology, fluid dynamics and computer vision, there has been increasing interest in vector field topology and its applications in vector field visualization. For time independent vector fields, vector field topology is typically based on trajectories with special properties: stationary points, periodic trajectories and separatrices. While several successful methods to determine these features have been developed, a common challenge that arises in vector field topology is the overwhelmingly large number of topological features detected in the input vector field. This causes the visualization of these features to be too overcrowded to be useful for complex data. The goal of this chapter is to provide a simple and flexible framework for building multiscale representations of vector field topology. We build upon earlier work on Morse decompositions for piecewise constant (PC) vector fields. Morse decompositions provide a natural

L. Sipeki • A. Szymczak (✉)
Colorado School of Mines, Golden, CO 80401, USA
e-mail: levente.sipeki@gmail.com; aszymcza@mines.edu

P.-T. Bremer et al. (eds.), *Topological Methods in Data Analysis and Visualization III*, Mathematics and Visualization, DOI 10.1007/978-3-319-04099-8_3,

way to construct multiscale representations of vector field topology, since they support *merger operations*. that generalize cancellation operations known in scalar field topology [1] to the case of general, non-gradient, vector fields. Conceptually, merging two Morse sets is equivalent to replacing them by a single Morse set, containing both of the initial ones and all trajectories that connect them. Our algorithm works by iteratively applying such merger operations in a certain order that can be controlled by the user. This leads to a sequence of Morse decompositions of decreasing complexity.

The rest of the chapter will be organized as follows. Section 2 contains a discussion of related work. Section 3 reviews the work on Morse decompositions and Morse Connection Graphs that we build upon. Morse set mergers and their properties are discussed in Sect. 4. Section 5 discusses our approach to building the hierarchy of Morse decompositions. Finally, experimental results are presented in Sect. 6.

2 Prior Work

Vector field visualization has been an active research topic during the past two decades [2–4]. In most cases, the focus of work on vector field topology has been on computing basic features such as stationary points, periodic orbits and separatrices. Stationary points can be found using the technique of [5]. Periodic orbits can be computed by following trajectories until they converge to a limit cycle [6] or by intersecting stream surfaces [7]. In the 2D case, separatrices can be obtained by following trajectories along unstable and stable directions of saddles [8, 9]. In [10], stationary points, periodic orbits and connecting trajectories are found using a Morse decomposition. An approach to 3D vector field visualization based on the concept of saddle connectors, i.e. isolated trajectories connecting repelling and attracting saddles, is introduced in [11]. In parameter-dependent vector fields, features such as stationary points or vortices can be tracked using feature flow fields [12, 13].

Numerical instability (sensitivity to a numerical method used to approximate trajectories) intrinsically associated with vector field topology defined in terms of individual trajectories is discussed in [14]. The same paper proposes to use Morse decomposition and the Morse Connection Graph (MCG), that represents trajectories connecting the Morse sets, as a more robust representation of vector field topology. An adaptive refinement scheme for Morse decompositions that can lead to more efficient and more precise analysis was recently introduced in [15]. The PC approximation based approach to computing Morse decomposition, which is faster and yields finer results, has been introduced in [16]. The CVPC framework designed to support computation of Morse decompositions of user-prescribed stability with respect to perturbation of the vector field was introduced in [17]. An approach to vector field topology based on discrete vector fields motivated by Forman's discrete Morse theory [18] is proposed in [19]. While this is an attractive approach because it is purely combinatorial, it has poor approximation properties. Trajectories of a discrete vector field can only move along edges of the mesh or its dual graph and, in general, do not converge to the trajectories of the original vector field as triangle sizes go to zero. The edge map approach recently developed in [20–22] provides another elegant solution to the consistency issues related to numerical integration. The basic idea is to represent the way trajectory segments contained in a single mesh triangle Δ connect its boundary points as a mapping of the boundary of Δ into itself. That mapping can be approximated by a piecewise linear (possibly quantized), mapping with arbitrary accuracy. As a result, trajectories can be followed by applying a series of linear interpolations rather than numerical integration. Most algorithmic tools developed for PC vector fields (in particular, the concept of transition graph, MCG, and the simplification scheme introduced in this chapter) can be combined with edge maps. However, edge maps are more difficult to implement and their extension to higher dimensions appears to be hard. In contrast, some aspects of the PC framework have recently been extended to the 3D case in [23]. On the other hand, the advantage of edge maps is their consistency with the standard piecewise linear interpolation scheme for vector field data.

Prior work on multi-scale topology for non-gradient vector fields includes topology simplification schemes based on merging nearby stationary points to reduce their number [24] and cancellations of stationary points connected by a separatrix [25]. A method of merging critical points based on Conley index theory has been described in [26] and extended to handle periodic trajectories in [10]. An adaptive refinement scheme for Morse decompositions based built upon the τ-map graph representation is described in [15]. In contrast to the approach described here, their method is top-down, i.e. selectively *refines* a coarse Morse decompsition rather than coarsens a fine one. A hierarchy of stable Morse decompositions, which is controlled by stability of the decomposition with respect to perturbation of the vector field is described in [27]. In this case, the construction is bottom up and driven by a simple and natural criterion (stability). However, this construction is computationally expensive. Multi-scale analysis has also been incorporated into combinatorial vector field based algorithms [19]. Note that most vector field topology simplification schemes attempt to *alter the vector field* in an explicit manner to reduce its topology. A notable exception is [28], where topological features are modeled as convex subsets of the domain and analyzed by examining inflow and outflow regions on the boundary. Simplification is achieved by merging these convex sets into larger ones. In our framework, topology simplification is achieved by simplifying the Morse decomposition itself, through Morse set mergers. Morse set mergers can be viewed as generalizations of critical point cancellations in scalar field topology [1, 29].

3 Morse Decompositions and Morse Connection Graphs

Morse decompositions and Morse connection graphs are generally computed from a finite directed graph representation of the flow induced by the input vector field. At least three suitable graph representations have been used for vector field visualization purposes: the geometry based graph [10, 14], the τ-map graph [14] and the transition graph for piecewise constant vector fields [16].

In this chapter, we build upon the method of [16, 30], designed for piecewise constant (PC) vector fields. In addition to being robust and efficient, this approach also supports a simple and reliable way to classify Morse sets based on fixed point index and stability (i.e. categorization as repelling, attracting or neither repelling nor attracting). This classification scheme provides a natural way to specify the Morse set merger constraints (Sect. 5).

3.1 PC Vector Fields: Basics

Formally, a PC vector field defined on a triangulated manifold surface is a function that assigns a vector $f(\Delta)$ parallel to the Δ's plane to every mesh triangle Δ. Inside Δ, trajectories of the PC vector field move with constant velocity equal to $f(\Delta)$. Additionally, trajectories are allowed to slide along any *exploding* or *imploding* edge e, for which vectors in the two incident triangles both point away from or toward it, with constant velocity $f(e)$ determined as follows. First, rotate one of the incident triangles around e to make it co-planar with the other. Then, define $f(e)$ as the convex combination of vectors assigned to the incident triangles, with weights chosen to make it parallel to e (Fig. 1, top box). Trajectories of a PC vector field can also stay at *stationary vertices*, for any amount of time. Stationary vertices are mesh vertices that are determined by sector analysis: a vertex is stationary if and only if it has a parabolic sector or its number of hyperbolic sectors is other than 2. Two examples of sector structure of a vertex are shown in Fig. 1, bottom box.

To sum up, any trajectory of a PC vector field can be obtained by a concatenation of linear *simple segments*, either moving through a triangle Δ with velocity $f(\Delta)$, sliding along an exploding or imploding edge with velocity $f(e)$ or staying at a stationary vertex. In particular, this means that trajectories can be constructed without numerical integration. In contrast to the standard continuous vector fields, there may be more than one trajectory passing through a point on the surface. For example, infinitely many trajectories originating from the lowermost vertex on the exploding edge in Fig. 1, top, can be constructed by following that edge for some time and then turning into one of the incident triangles at any point along the edge. Despite this, the multivalued flow defined using trajectories described above can be analyzed using topological tools [16].

3.2 Transition Graph

For PC vector fields, a *transition graph* is used to represent its trajectories. Nodes of a transition graph are in one to one correspondence with *n-sets*. N-sets are either mesh vertices or edge pieces, that result from subdividing mesh edges into

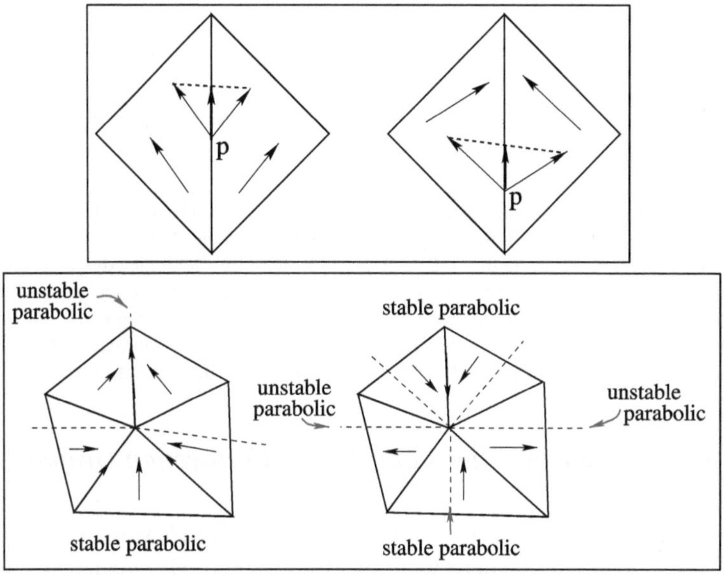

Fig. 1 *Top box*: an exploding and an imploding edge. In both cases, trajectories can move along the edge. The process of determining the velocity of trajectories moving along the edge can be visualized by moving vectors assigned to both incident triangles so that they are anchored at the same point p on the edge. The velocity vector is the vector from p to the intersection point of the edge and the line connecting the endpoints of the vectors (*dashed line*). *Bottom box*: examples of sector analysis of two vertices. The one on the left is non-stationary: it has one large stable parabolic sectors (on the bottom, between the two nearly *horizontal dashed half-lines*), one unstable parabolic sector running along the edge leading to the uppermost neighbor and two hyperbolic sectors between them. The vertex on the right is stationary: it has two stable and two unstable parabolic sectors and four hyperbolic sectors between them

shorter line segments. Two n-sets are required to be connected by an arc of the transition graph if they are directly connected by a simple segment (Fig. 2). Since any trajectory can be obtained by joining simple segments, an encoding of a trajectory of the PC vector field as a path in the transition graph can be obtained by recording the consecutive n-sets visited by that trajectory. Such a sequence of n-sets has to be a path in any valid transition graph. Note that graphs obtained by inserting new arcs into a transition graph are also transition graphs and therefore can be used to compute Morse decompositions.

An attractive feature of the transition graph is that it can be refined to provide more precise information on the flow, by simply splitting an edge piece e into shorter ones and connecting them with the neighbors of e in the graph according to the local flow structure. Details of this process as well as refinement strategies to increase the precision of Morse decomposition and Morse Connection Graph are described in [16] and [30].

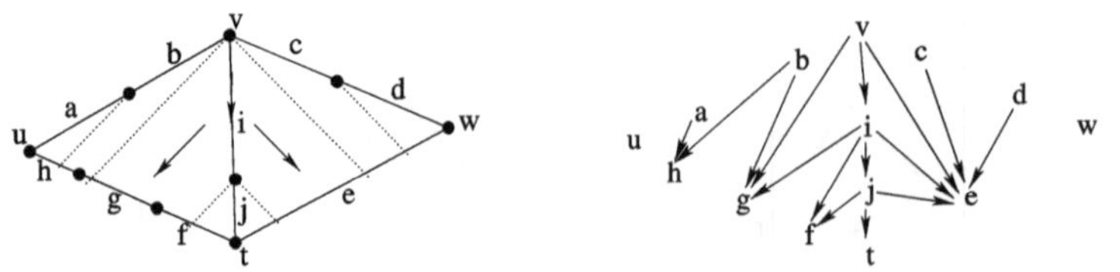

Fig. 2 *Left*: two adjacent triangles in the mesh, with vectors assigned to them shown as the *arrows*. The edge $\bar{vt}$ is exploding, with flow along it moving down. The contribution of the two triangles to the transition graph is shown on the *right*. For example, the flow along the edge $\bar{vt}$ is represented by the path $v \rightarrow i \rightarrow j \rightarrow t$. Since some simple segments traversing the left triangle connect b to h and g, the arcs $b \rightarrow h$ and $b \rightarrow g$ are in the graph. However, the arc $b \rightarrow f$ is not since there is no simple segment starting in b and ending in f. Note that for any stationary vertex, the graph also contains the loop arc connecting the n-set corresponding to that vertex to itself

3.3 Morse Decompositions and Morse Connection Graphs

A Morse decomposition is derived from a transition graph $\mathcal{G}$. Morse sets are defined by strongly connected components of $\mathcal{G}$. They capture all circulation present in the flow, in particular all stationary points and periodic trajectories (since every cycle in the graph is contained in a strongly connected component). Morse sets are classified by means of fixed point index and stability. Morse sets of index i are referred to as Morse sets of type $(i, +)$ if they are repelling, $(i, -)$ if attracting and $(i, 0)$ if neither attracting nor repelling. This classification scheme provides a simple way to distinguish Morse sets similar to classical topological features: sinks (Morse sets of type $(1, -)$), sources $(1, +)$, saddles $(-1, 0)$, attracting periodic trajectories $(0, -)$ and repelling periodic trajectories $(0, +)$. Morse sets of type $(0, 0)$ are *trivial*. In practice, they represent regions where circulation cannot be excluded because of insufficient resolution of the transition graph or clusters of features that can be removed by applying a local perturbation to the vector field. Morse sets of types discussed above are called *simple*. All other Morse sets are *complex*. In practice, simple Morse sets are easier to understand since the flow near them resembles flow near classical features (or laminar flow in the case of trivial Morse sets).

A Morse connection graph (MCG) is also determined from a transition graph $\mathcal{G}$ and represents potential connections between different Morse sets. Its nodes are the strongly connected components of $\mathcal{G}$. Here, we define the MCG in two steps, following [31]. First, define $\trianglelefteq$ as the 'is connected to' relationship on the strongly connected components of $\mathcal{G}$. More precisely, for two components A and B, $A \trianglelefteq B$ if and only if there is a path in $\mathcal{G}$ connecting a node in A to a node in B. By definition of strongly connected components, existence of such a path is equivalent to existence of a path from any node $p \in A$ to any node $q \in B$ in $\mathcal{G}$. Note that $\trianglelefteq$ is a partial order [32]. The MCG obtained from $\mathcal{G}$, denoted by $MCG(\mathcal{G})$, can then be defined as the Hasse diagram of $\trianglelefteq$. More precisely, an arc $A \rightarrow B$ is in the MCG if and only if B is an immediate successor of A (or, equivalently, A is an immediate predecessor of B) in the partial order $\trianglelefteq$, i.e. $A \trianglelefteq B$ and there is no C such that $A \trianglelefteq C \trianglelefteq B$.

An alternative definition of arcs of $MCG(\mathcal{G})$, which will be convenient in Sect. 4.3, can be stated directly in terms of properties of $\mathcal{G}$. Strongly connected components A and B are connected by the arc $A \to B$ in $MCG(\mathcal{G})$ if and only if the set $Conn_{\mathcal{G}}(A, B)$, defined as the union of all paths in $\mathcal{G}$ connecting a node in A to a node in B, is nonempty and intersects no strongly connected components of $\mathcal{G}$ other than A and B. It is easy to see that this definition and the one based on the Hasse diagram are equivalent.

4 Morse Set Mergers

In this section, we describe a basic Morse set merger operation that our hierarchy construction is built upon. Recall that Morse sets are represented by strongly connected components of a transition graph $\mathcal{G}$ (Sect. 3). We describe how Morse set (or, on the transition graph level, strongly connected component) mergers can be modeled by inserting new arcs into $\mathcal{G}$ in Sect. 4.1. Later on, we show how these merger operations affect the MCG in Sect. 4.2 and provide a proof of consistency between the two in Sect. 4.3.

4.1 Mergers on the Transition Graph Level

A particularly simple way to execute a merger of strongly connected components A and B, such that the arc $A \to B$ is in $MCG(\mathcal{G})$, is to pick arbitrary nodes $a \in A$ and $b \in B$ and insert the arc $b \to a$ into $\mathcal{G}$ (Fig. 3). In the resulting graph $\mathcal{G}'$, A, B and all paths that connect them, are in the same strongly connected component C. Since A and B are adjacent in the MCG, no other strongly connected component is affected by this operation (since paths connecting A and B do not intersect any other component).

4.2 Impact of Mergers on the MCG

We now describe how the merger of strongly connected components A and B described in the previous section changes the MCG. This is useful since the MCG is much smaller than a transition graph and therefore certain properties of the Morse sets involved in the merger are cheaper to compute from the MCG than from the transition graph (examples are discussed in Sect. 5 below).

To update the MCG, we proceed in two steps illustrated in Fig. 4. First, we contract the arc $A \to B$. Nodes A and B in the MCG are replaced by a single node C, representing the new strongly connected component in $\mathcal{G}'$. For each node D such that $A \to D$ or $B \to D$ is in the MCG prior to the merger, the arc $C \to D$

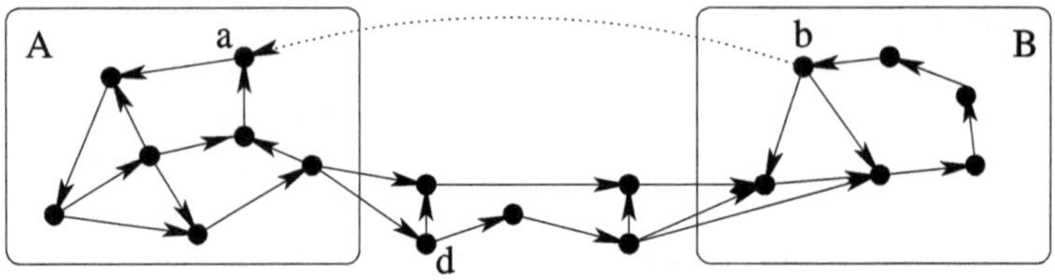

Fig. 3 Merger of two strongly connected components A and B in the transition graph, connected by the arc $A \rightarrow B$ in the MCG. Since the two nodes are adjacent in the MCG, paths connecting them exist in the graph (they pass through nodes outside the two boxes). None of these paths intersects a strongly connected component other than A and B. After the merger, i.e. after inserting the *long dotted arc* $b \rightarrow a$, the graph contains a loop through every node shown in the figure. For example, to obtain a loop containing d, follow a path from a to d, then to b and back to a using the new arc

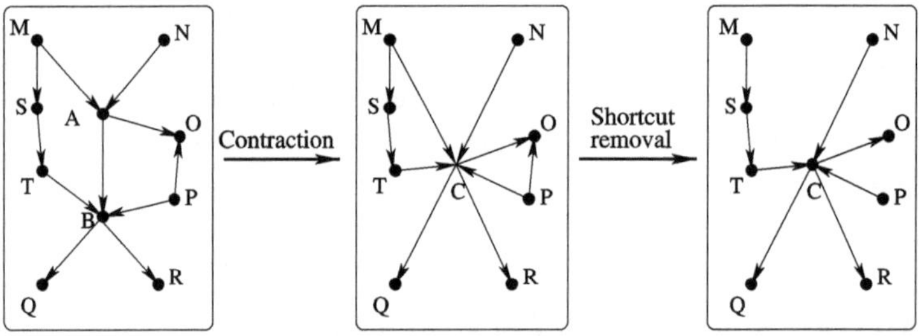

Fig. 4 Morse set merger on the MCG level. Arc contraction step: nodes corresponding to the strongly connected components A and B are merged to a single node representing the new component C. The starting vertices of arcs into the new node are the same as starting vertices of arcs into A or B in the original graph (here, M, N, P and T). Similarly, endpoints of arcs out of C are the same as the endpoints of arcs out of A or B in the graph before the merger (O, Q and R). Shortcut removal step deletes arcs that are shortcuts of paths through C, introduced by the contraction step. In this case, two arcs are removed, $P \rightarrow O$ and $M \rightarrow C$

is added to the resulting graph. Analogously, for each node D such that $D \rightarrow A$ or $D \rightarrow B$ is in the MCG prior to the merger, the arc $D \rightarrow C$ is added to the updated graph.

The second step is the shortcut removal. Since an MCG is a Hasse diagram of a partial ordering of strongly connected components, it is not allowed to possess shortcut arcs. Such shortcuts (but only for paths passing through C) may result from the edge contraction step described above (Fig. 4). Thus, any arc $D \rightarrow E$ representing a shortcut for a path passing through C (possibly, starting or ending at C) is removed from the graph.

4.3 Consistency

This section focuses on a formal proof that the MCG update (Sect. 4.2) is consistent with the transition graph update (Sect. 4.1). In what follows, by $\mathcal{H}$ we denote the graph $MCG(\mathcal{G})$ with the arc $A \to B$ contracted as described in the previous section. We start with the proposition stating that no arcs of $MCG(\mathcal{G}')$ are missed by our algorithm.

Proposition 1. *$MCG(\mathcal{G}')$ is a subgraph of $\mathcal{H}$.*

Proof. Since the two graphs have the same vertices, we need to show that every arc in $MCG(\mathcal{G}')$ is also an arc of $\mathcal{H}$. Let $E \to F$ be any arc of $MCG(\mathcal{G}')$. As discussed at the end of Sect. 3, this means that $Conn_{\mathcal{G}'}(E, F)$ is nonempty and does not intersect any strongly connected component of $\mathcal{G}'$ except for E and F. Consider the following three cases.

Case 1. Neither E nor F is the same as C, the node resulting from the contraction. In this case, E and F are also strongly connected components of $\mathcal{G}$ and $Conn_{\mathcal{G}}(E, F) = Conn_{\mathcal{G}'}(E, F)$ (since $\mathcal{G}'$ has only one arc more than $\mathcal{G}$, and that arc connects a node in C, which is disjoint with $Conn_{\mathcal{G}'}(E, F)$). Since every strongly connected component of $\mathcal{G}$ is contained in the strongly connected component of $\mathcal{G}'$, $Conn_{\mathcal{G}}(E, F)$ is disjoint with any strongly connected component in $\mathcal{G}$ except for E and F, and therefore $E \to F$ belongs to $MCG(\mathcal{G})$ and therefore also to $\mathcal{H}$.

Case 2. $E = C$. Sine there is a path σ in $\mathcal{G}'$ connecting C to F, there is a path in $\mathcal{G}$ connecting A to F. Such a path can be obtained by following a path from a node in A to the last node of σ in C, and then following σ to a node in F. This means that $Conn_{\mathcal{G}}(A, F) \neq \emptyset$. Since $\mathcal{G}$ is a subgraph of $\mathcal{G}'$, any path connecting A and F in $\mathcal{G}$ is also a path connecting C to F in $\mathcal{G}'$. Since such paths cannot intersect any strongly connected components in $\mathcal{G}'$ other than C and F, we conclude that $Conn_{\mathcal{G}}(A, F)$ cannot intersect components of $\mathcal{G}$ other than A, B or F. If it does not intersect B, then $A \to F$ is in $MCG(\mathcal{G})$. If it does, then $Conn_{\mathcal{G}}(B, F)$ is nonempty and the only components it can intersect are B and F and therefore $B \to F$ is in $MCG(\mathcal{G})$. We conclude that in both cases, $C \to F$ belongs to $\mathcal{H}$.

Case 3. $F = C$. The argument in this case is analogous to the one for case 2, with the directions of any arc involved reversed. □

It remains to show that all shortcuts to be removed from $\mathcal{H}$ to obtain $MCG(\mathcal{G}')$ are for paths passing through C. This is a simple consequence of the following proposition.

Proposition 2. *If the arc $E \to F$ belongs to $\mathcal{H}$ and there is a path from E to F in $\mathcal{H}$ consisting of more than 1 arc, then there is a path from E to C and from C to F in $\mathcal{H}$.*

Proof. The proposition is trivially satisfied if either E or F is equal to C. Thus, from now on, we assume neither E nor F is the same as C. This means that the arc $E \to F$ is in $MCG(\mathscr{G})$. Since it is a Hasse diagram, $MCG(\mathscr{G})$ cannot contain a path of length greater than 1 connecting E to F. Such a path can arise in $\mathscr{H}$ only if there are paths from E to B and from A to F in $MCG(\mathscr{G})$. But then, paths form E to C and from C to F exist in $\mathscr{H}$. □

5 Construction of the Hierarchy

Our algorithm works in a bottom up fashion. We start off by computing a fine Morse decomposition and Morse connection graph from the transition graph constructed using the method similar to [30]. First, the coarse transition graph is constructed. Then, the following process is iterated a user-specified number of times:

- Compute the strongly connected components of the transition graph; let S be the union of these components.
- Compute the union U of all paths in the graph that pass through a node in a strongly connected component corresponding to a Morse set that is neither attracting nor repelling (note that these nodes belong to *generalized separatrices* [30]; by subdividing these nodes we make them more precise).
- Refine all edge pieces in $U \cup S$, by subdividing each of them into two of equal length.

Then, we initialize a priority queue that controls the Morse set mergers by inserting all *admissible* MCG arcs into it. In general, our procedure merges only pairs of Morse sets connected by an admissible arc. The precise definition of an admissible arc can be set arbitrarily by the user of the system. The *cost* of an admissible arc is also computed using an arbitrary user-provided formula.

To build the hierarchy of Morse decompositions, we execute a series of Morse mergers. We extract the arc $A \to B$ of lowest cost from the queue. We merge A and B by adding one arc to the transition graph as described in Sect. 4.1. We also update the MCG to keep it consistent with the transition graph resulting from the merger (Sect. 4.2), and update the priority queue by removing all arcs that are no longer in the MCG and inserting all new MCG arcs (all of which are incident to the new node, C in Sect. 4.2) that are admissible. This process is continued until the queue becomes empty, yielding a complete hierarchy.

A compact representation of the hierarchy consists of the initial transition graph and a sequence of arcs that were inserted into it for each Morse set merger. To compute the Morse decomposition after n mergers, we add the first n arcs to the graph and then run the generic algorithm of [16]. Note that the arc additions can be viewed as counterparts of gradient path reversals in discrete Morse theory [33].

5.1 Admissibility and Cost: An Example

Our procedure is general enough to use an arbitrary user-provided definition of admissibility and cost. However, its running time can be highly dependent on these definitions. For complex data, it is beneficial to be able to determine the cost and admissibility directly from the Morse Connection Graph, perhaps with a small amount of additional information stored at the nodes. Computing them from the transition graph is typically significantly more expensive since it is much larger than the MCG.

A natural goal that one can set for the hierarchy construction is to keep the Morse sets topologically simple throughout the process. In our algorithm, with each MCG node we keep the corresponding Morse set type. This information is easy to update as Morse sets are merged. If A and B are merged into C, the index of C is equal to the sum of indices of A and B. If the domain of the vector field is a manifold surface with no boundary, then C is attracting (repelling) if and only if there are no arcs in the updated MCG out of (respectively, into) C. Let us stress that this may not be true for domains with boundary: for example, if all paths starting at a strongly connected component leave the domain through the boundary, there is no arc out of that component in the MCG which may cause the component to appear attracting even when it is not. A way to deal with this issue will be described in future work.

Admissibility of an MCG edge $A \to B$ can be defined in terms of types of Morse sets defined by A and B and the type of Morse set that would result from the merger of A and B (as usual, let us call it C). In this chapter, we define $A \to B$ as admissible if the following two conditions are satisfied:

(i) A is repelling or B is attracting; the motivation is to suppress the growth of non-attracting or non-repelling sets that are generally harder to understand if they are large and have complex geometry
(ii) C is of a simple type, i.e. $(0, 0)$, $(1, +)$, $(1, -)$, $(-1, 0)$, $(0, +)$ or $(0, -)$: trivial or similar to one of the classical vector field features (sink, source, saddle or attracting or repelling periodic orbit). Such Morse sets are easier to understand since they behave in a way similar to the classical features.

While there are many ways to define a sensible cost function to control the Morse set merger order, also in this case efficiency should be taken into account. A natural goal that one may set for the process is to keep the Morse sets not only topologically simple, but also small. Size measure can be based on bounding box size for efficiency. In our implementation, each MCG node of the initial MCG is equipped with the bounding box information for the corresponding Morse set. This bounding box estimate is propagated as the Morse sets are merged. Specifically, if A and B are merged into C then we use the smallest axis oriented box containing bounding boxes of A and B as an estimate of the size of bounding box of C. Clearly, this estimate is inaccurate since it ignores the contribution of paths connecting A and B (which are also included in C). Nevertheless, it can be used as a useful estimate of the size of C. In our implementation, the diameter of the bounding box of C is used as the cost of the edge $A \to B$.

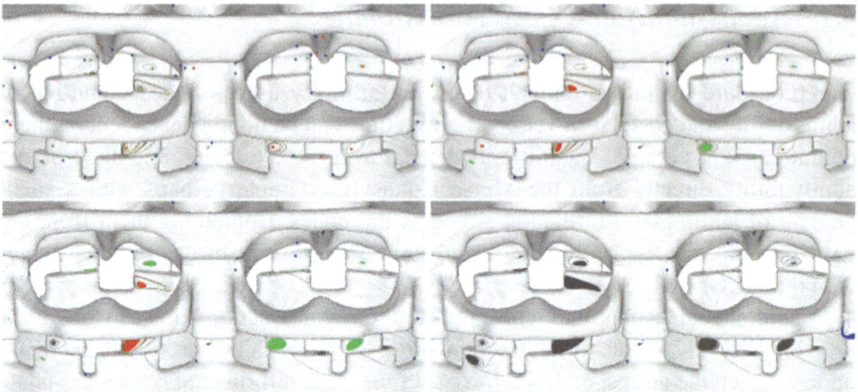

Fig. 5 Closeup of results for the cooling jacket dataset. The original Morse decomposition obtained using 10 refinement operations is shown in the *top left* (note it contains 723 Morse sets, 139 of which are trivial). Consecutive images show the simplified Morse decomposition after 300, 400 and 500 Morse set mergers. Notice that features near small vortex regions are successfully cancelled (i.e. combined into trivial Morse sets) after 500 mergers

6 Experimental Results

Our algorithm provides a simple yet effective way to simplify Morse decompositions. A series of snapshots of the simplification process for the cooling jacket dataset [34] is shown in Fig. 5. We use the standard color-coding of Morse sets according to their type (trivial – grey, $(1,+)$ and $(0,+)$ – red, $(1,-)$ and $(1,+)$ – green and $(-1,0)$ – blue; in figures shown here, there are no nontrivial Morse sets since mergers they would result from are not admissible). Note that color figures are available only in the electronic version of this chapter. It is interesting to note that in the Morse decomposition for 500 mergers (Fig. 6) there are only 2 repelling Morse sets, one on the inlet and one on top of the jacket on the side of the inlet, and 3 attracting sets, one on the outlet and two close to it. All the other Morse sets are either trivial or of type $(-1,0)$ (saddle-like). This can be seen as an indication of a good design, since large and complex attracting or repelling sets near the cylinder heads would mean potential hot spots (cf. [27, 34]). The algorithm is also effective in simplifying Morse decomposition for the gas engine dataset (Fig. 7).

In general, the running time of our algorithm is dominated by the time needed to compute the initial MCG. For example, for the cooling jacket discussed above, the initial Morse decomposition took 2 min to construct, while the order of all mergers was determined in 1.5 s. For the gas engine dataset, the construction of the initial decomposition took 10 s and the merger order was computed in 0.08 s. Since the transition graph is large, the geometric model of Morse sets for a decomposition in the hierarchy is relatively expensive to compute: it can be determined within 1.5 s for the gas engine dataset and within 14 s for the cooling jacket dataset.

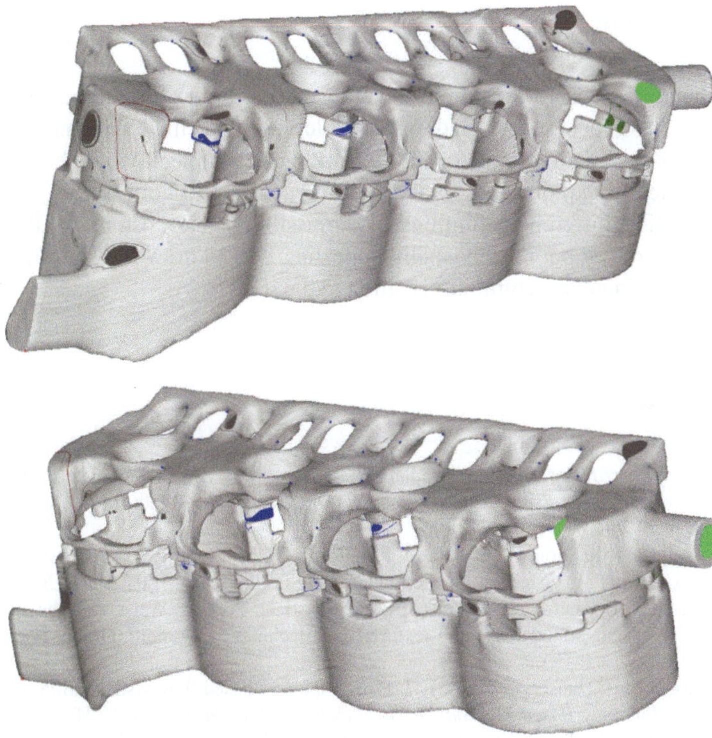

Fig. 6 Two views of the Morse decomposition after 500 mergers

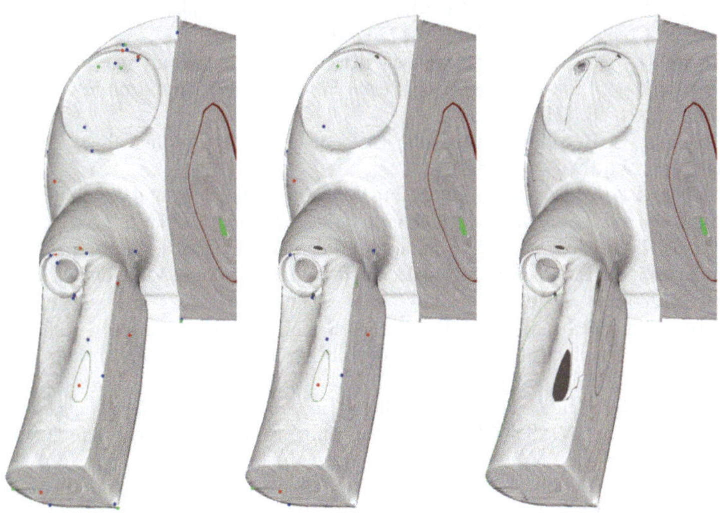

Fig. 7 Simplification stages for the gas engine dataset. The first image shows the starting Morse decomposition, consisting of 62 Morse sets. The following images show the Morse decomposition after 20 and 40 mergers

7 Conclusion

In this chapter we describe a simple and efficient framework for building hierarchies of Morse decompositions based on flexible, user-specified criteria. There are number of potential directions for future work. For example, it would be interesting to systematically explore and compare different ways to control the mergers, in particular using different definitions of admissibility and cost. It would also be interesting to develop a more visually appealing way to draw the simplified MCG to convey important structural information to the user.

References

1. H. Edelsbrunner, J. Harer, A. Zomorodian, Hierarhical Morse-Smale complexes for piecewise linear 2-manifolds, in *Symposium on Computational Geometry*, Medford (ACM, New York, 2001), pp. 70–79
2. T. McLouglin, R.S. Laramee, R. Peikert, F.H. Post, M. Chen, Over two decades of integration-based geometric flow visualization. Comput. Graph. Forum **29**(6), 1807–1829 (2010)
3. R.S. Laramee, H. Hauser, H. Doleisch, B. Vrolijk, F.H. Post, D. Weiskopf, The state of the art in flow visualization: dense and texture-based techniques. Comput. Graph. Forum **23**(2), 203–221 (2004)
4. R.S. Laramee, H. Hauser, L. Zhao, F.H. Post, Topology-based flow visualization, the state of the art, in *Topology-Based Methods in Visualization*, ed. by H. Hauser, H. Hagen, H. Theisel (Proceedings of the TopoInVis 2005) (Springer, Berlin/Heidelberg, 2007), pp. 1–19
5. J.L. Helman, L. Hesselink, Representation and display of vector field topology in fluid flow data sets. IEEE Comput. **22**(8), 27–36 (1989)
6. T. Wischgoll, G. Scheuermann, Detection and visualization of planar closed streamline. IEEE Trans. Vis. Comput. Graph. **7**(2), 165–172 (2001)
7. H. Theisel, T. Weinkauf, Grid-independent detection of closed stream lines in 2D vector fields, in *Proceedings of the Conference on Vision, Modeling and Visualization 2004 (VMV 04)*, Stanford, 2004, pp. 421–428
8. J.L. Helman, L. Hesselink, Visualizing vector field topology in fluid flows. IEEE Comput. Graph. Appl. **11**(3), 36–46 (1991)
9. A. Globus, C. Levit, T. Lasinski, A tool for visualizing the topology of three-dimensional vector fields, in *Proceedings of the 2nd Conference on Visualization '91*, San Diego, 1991, pp. 33–40
10. G. Chen, K. Mischaikow, R.S. Laramee, P. Pilarczyk, E. Zhang, Vector field editing and periodic orbit extraction using Morse decomposition. IEEE Trans. Vis. Comput. Graph. **13**(4), 769–785 (2007)
11. H. Theisel, T. Weinkauf, H.C. Hege, H.P. Seidel, Saddle connectors – an approach to visualizing the topological skeleton of complex 3D vector fields, in *IEEE Visualization*, Seattle, 2003, pp. 225–232
12. H. Theisel, H.P. Seidel, Feature flow fields, in *Proceedings of the Symposium on Data Visualisation 2003*. VISSYM'03, Grenoble. (Eurographics Association, Aire-la-Ville, 2003), pp. 141–148
13. T. Weinkauf, H. Theisel, A.V. Gelder, A. Pang, Stable feature flow fields. IEEE Trans. Vis. Comput. Graph. **17**, 770–780 (2011)
14. G. Chen, K. Mischaikow, R.S. Laramee, E. Zhang, Efficient Morse decompositions of vector fields. IEEE Trans. Vis. Comput. Graph. **14**(4), 848–862 (2008)

15. G. Chen, Q. Deng, A. Szymczak, R.S. Laramee, E. Zhang, Morse set classification and hierarchical refinement using Conley index. IEEE Trans. Vis. Comput. Graph. **18**(5), 767–782 (2012)
16. A. Szymczak, E. Zhang, Robust Morse decompositions of piecewise constant vector fields. IEEE Trans. Vis. Comput. Graph. **18**(6), 938–951 (2012)
17. A. Szymczak, Stable Morse decompositions for piecewise constant vector fields on surfaces. Comput. Graph. Forum **30**(3), 851–860 (2011)
18. R. Forman, Combinatorial vector fields and dynamical systems. Mathematische Zeitschrift **228**, 629–681 (1998)
19. J. Reininghaus, C. Lowen, I. Hotz, Fast combinatorial vector field topology. IEEE Trans. Vis. Comput. Graph. **17**(10), 1433–1443 (2010)
20. H. Bhatia, S. Jadhav, P.T. Bremer, G. Chen, J.A. Levine, L.G. Nonato, V. Pascucci, Edge maps: representing flow with bounded error, in *Pacific Visualization Symposium (PacificVis) 2011*, Hong Kong, 2011, pp. 75–82
21. H. Bhatia, S. Jadhav, P. Bremer, G. Chen, J. Levine, L. Nonato, V. Pascucci, Flow visualization with quantified spatial and temporal errors using edge maps. IEEE Trans. Vis. Comput. Graph. **18**(9), 1383–1396 (2012)
22. J.A. Levine, S. Jadhav, H. Bhatia, V. Pascucci, P.T. Bremer, A quantized boundary representation of 2D flows. Comput. Graph. Forum **31**(3pt1), 945–954 (2012)
23. A. Szymczak, N. Brunhart-Lupo, Nearly-recurrent components in 3D piecewise constant vector fields. Comput. Graph. Forum **31**(3pt3), 1115–1124 (2012)
24. X. Tricoche, G. Scheuermann, H. Hagen, A topology simplification method for 2D vector fields, in *Proceedings IEEE Visualization 2000*, Salt Lake City, 2000, pp. 359–366
25. X. Tricoche, G. Scheuermann, Continuous topology simplification of planar vector fields, in *Proceedings IEEE Visualization 2001*, San Diego, 2001, pp. 159–166
26. E. Zhang, K. Mischaikow, G. Turk, Vector field design on surfaces. ACM Trans. Graph. **25**(4), 1294–1326 (2006)
27. A. Szymczak, Hierarchy of stable Morse decompositions. IEEE Trans. Vis. Comput. Graph. **19**(5), 799–810 (2013)
28. T. Weinkauf, H. Theisel, K. Shi, H.C. Hege, H.P. Seidel, Extracting higher order critical points and topological simplification of 3D vector fields, in *Proceedings of the IEEE Visualization 2005*, Minneapolis, 2005, pp. 559–566
29. A. Gyulassy, V. Natarajan, V. Pascucci, B. Hamann, Efficient computation of Morse-Smale complexes for three-dimensional scalar functions. IEEE Trans. Vis. Comput. Graph. **13**(6), 1440–1447 (2007)
30. A. Szymczak, Morse connection graphs for piecewise constant vector fields on surfaces. Comput. Aided Geom. Des. **30**(6), 529–541 (2013)
31. W.D. Kalies, K. Mischaikow, R.C.A.M. VanderVorst, An algorithmic approach to chain recurrence. Found. Comput. Math. **5**(4), 409–449 (2005)
32. D. Kozen, *The design and analysis of algorithms* (Springer, New York, 1991)
33. R. Forman, Morse theory for cell complexes. Adv. Math. **134**(1), 90–145 (1998)
34. R.S. Laramee, C. Garth, H. Doleisch, J. Schneider, H. Hauser, H. Hagen, Visual analysis and exploration of fluid flow in a cooling jacket, in *Proceedings of the IEEE Visualization 2005*, Minneapolis, 2005, pp. 623–630

Toward the Extraction of Saddle Periodic Orbits

Jens Kasten, Jan Reininghaus, Wieland Reich, and Gerik Scheuermann

Abstract Saddle periodic orbits are an essential and stable part of the topological skeleton of a 3D vector field. Nevertheless, there is currently no efficient algorithm to robustly extract these features. In this chapter, we present a novel technique to extract saddle periodic orbits. Exploiting the analytic properties of such an orbit, we propose a scalar measure based on the finite-time Lyapunov exponent (FTLE) that indicates its presence. Using persistent homology, we can then extract the robust cycles of this field. These cycles thereby represent the saddle periodic orbits of the given vector field. We discuss the different existing FTLE approximation schemes regarding their applicability to this specific problem and propose an adapted version of FTLE called *Normalized Velocity Separation*. Finally, we evaluate our method using simple analytic vector field data.

1 Introduction

In the field of scientific visualization, the analysis of three dimensional vector fields is often connected to its topology. In many cases the structures that are extracted consist only of sources, sinks, saddles, saddle connectors, and the separation surfaces. However, there are more structures contained in these fields such as attracting, repelling, and saddle-like periodic orbits. In divergence-free flows, there are also center-like periodic orbits. All orbits are global structures that cannot be extracted using local analysis.

J. Kasten (✉) • W. Reich • G. Scheuermann
Leipzig University, Leipzig, Germany
e-mail: kasten@informatik.uni-leipzig.de; reich@informatik.uni-leipzig.de; scheuermann@informatik.uni-leipzig.de

J. Reininghaus
Institute of Science and Technology Austria, Klosterneuburg, Austria
e-mail: Jan.Reininghaus@ist.ac.at

P.-T. Bremer et al. (eds.), *Topological Methods in Data Analysis and Visualization III*, Mathematics and Visualization, DOI 10.1007/978-3-319-04099-8_4,

The extraction of attracting and repelling periodic orbits in three-dimensional vector fields is comparatively easy, since their connected manifolds are volumetric. This has been previously explored by Wischgoll in [23]. Conversely, the manifolds connected to a saddle periodic orbit are surfaces. Nearly all stream lines that come close to a saddle periodic orbit will never reach it, but instead diverge. It is therefore nearly impossible to detect saddle periodic orbits directly by stream line integration. They have to be found indirectly.

Saddle periodic orbits can be extracted by intersecting the attracting and repelling separation surface, i.e., distinguished stream surfaces that are similar to separatrices in 2D. This was shown by Peikert et al. [14]. Unfortunately, there are two major problems with this approach: First, the integration of separation surfaces is started at local structures, i.e., the saddle points. However, not all global structures have to be connected to this type of critical points. In particular, there are vector fields with saddle periodic orbits but no saddle points, see for example Fig. 4. In addition, the integration of separation surfaces is a numerically complex task. Many publications have been written on this topic. Especially at saddle-like structures, the correct integration of the surfaces becomes problematic.

In this chapter, we propose a different approach to extract the saddle periodic orbits. According to their nature, a significant amount of stream line separation in both time directions can be detected in their vicinity. The finite-time Lyapunov exponent (FTLE) is the square root of the largest eigenvalue of the Cauchy-Green deformation tensor and measures that separation by computing the strain of a very small volume placed inside the flow. It is typically used to analyze time-dependent flow fields, since vector field topology fails here to extract the interesting structures. Nevertheless, in many cases, FTLE reveals the separatrices for two-dimensional stationary vector fields. We propose a new measure called *Normalized Velocity Separation (NVS)* based on the FTLE. In three dimensions, the fields of forwards and backwards integrated NVS indicate the repelling and attracting surfaces. We propose to use the minimum of both fields to indicate a high amount of separation and convergence using only one scalar field. The closed curves of this field with a high value therefore indicate the presence of a saddle periodic orbit. Since the NVS fields can be very complex we make use of persistent homology [6] to robustly extract the persistent cycles.

2 Related Work

While the notion of flow topology has been introduced by Helman and Hesselink [10] to the visualization community over two decades ago, the particular interest in detecting closed stream lines was initiated by Wischgoll and Scheuermann [23]. Their approach is to find cycles of cells in a discrete vector field in two dimensions that get crossed repeatedly by flow particles under numerical integration. This method requires the closed stream line to be either attracting or repelling. For the same purpose, Theisel et al. [21] transform planar, two-dimensional vector fields into an appropriate 3D vector field and detect closed stream lines as a intersection curve of stream surfaces.

Chen et al. [4] provided a combinatorial approach to extract the topological skeleton of a planar vector field. Their method is based on the construction of a graph from the maps that are induced by particle transport from cell to cell. Closed stream lines are represented by cycles in that graph. There are approaches in generalizing that technique to three dimensions [16], which are computationally inefficient at the present.

A tool that is able to determine the stability properties of closed stream lines is the Poincaré-map [11]. In some cases, this map can be used to extract periodic orbits. For instance, Sanderson et al. [18] extracted orbits in domain for a magnetic field by exploiting a natural Poincaré section. In general, such a natural Poincaré section may not exist, see Fig. 6. An alternative method is also the generation of transition matrices by a similar combinatorial method from Dellnitz et al. [5]. Periodic orbits are a special case of a stationary state of such a matrix, i.e. an eigenvector to the eigenvalue of 1.

In this chapter, we use the fact that saddle-like closed stream lines can be described by the intersection of two separating manifolds. Therefore, we can use the finite-time Lyapunov exponent, or brief FTLE, as introduced by Haller [9] to compute manifolds of separation and convergence. The value of FTLE for flow visualization was especially shown by Sadlo and Peikert[17], Garth et al. [8] and Kasten et al. [12]. Interested readers are also referred to surveys concerning flow visualization by Weiskopf et al. [22] and Post et al. [15].

3 Foundation

3.1 Periodic Orbits

In an arbitrary vector field, there can be distinguished closed stream lines, i.e., periodic orbits. They are characterized by attracting, repelling, center-like or saddle-like behavior. A periodic orbit is called isolated periodic orbit if it has an open neighborhood that does not contain any other periodic orbit. Note that periodic orbits cannot occur in fields that are the gradient of an appropriate scalar field. For further details, we refer to the book of Asimov [1].

3.2 Saddle Periodic Orbits

A saddle-like periodic orbit is characterized by two distinguished manifolds: an attracting and a repelling one. Therefore, saddle-like periodic orbits are similar to saddle connectors [20], but, in contrast, there is no critical point at the intersection

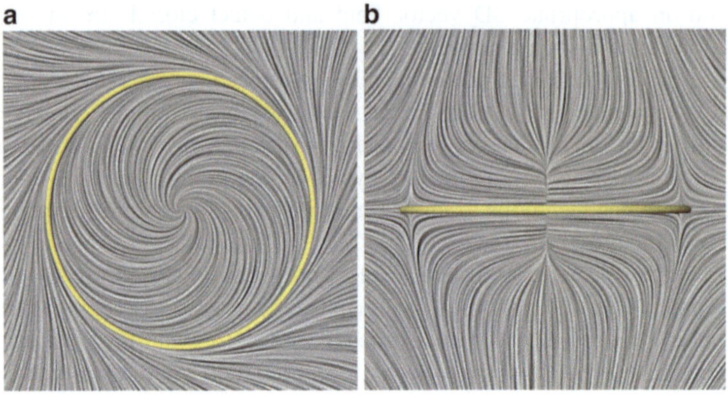

Fig. 1 2D views of a simple saddle periodic orbit (*yellow*). (**a**) xy-View. (**b**) yz-View

of these manifolds. For saddle periodic orbits, one of the characterizing manifolds has to have a tube-like structure and the other one has to have a planar structure in a sufficiently small neighborhood. In Fig. 1, line integral convolution images are shown for two planes intersecting a periodic orbit. In Fig. 1a, the attracting manifold of this periodic orbit can be seen. In this plane, the three-dimensional periodic orbit looks like a two-dimensional attracting periodic orbit with a source at the origin of the domain. In contrast, Fig. 1b shows that the periodic orbit has a distinguished manifold where the particles diverge. Note that we chose a simple saddle periodic orbit for the explanation. In a general setting, separation or convergence does not have to occur at every point of the periodic saddle. It is sufficient that the saddle periodic orbit is characterized by separation surfaces, which are global structures. Thus, a local analysis can never indicate a saddle periodic orbit.

3.3 Finite-Time Lyapunov Exponent (FTLE)

To detect separation within a vector field, the first idea is to use the FTLE. The finite-time Lyapunov exponent (FTLE) measures the separation of infinitesimally close particles over a finite time period T. If the FTLE is calculated backwards, it measures the convergence of particles. We therefore call the separation measure FTLE^{+} and the convergence measure FTLE^{-}.

There are different approaches to calculate the FTLE. In this section, we will discuss two of them: the original method based on the flow map (F-FTLE) and a method based on accumulating the Jacobian, i.e., the vector gradient, that is called localized FTLE (L-FTLE). First, we will describe both methods in the general context of time-dependent flow fields.

F-FTLE – Let $\mathbf{v} : \mathbb{R}^d \times I \to \mathbb{R}^d$ be a d-dimensional, time-dependent flow field. The advection of a particle with the flow for a time T can be described using the flow map

$$\phi : \mathbb{R}^d \times I \times I' \to \mathbb{R}^d. \tag{1}$$

It maps a particle at position $\mathbf{x}_0$ and time t_0 onto its advected position $\phi(\mathbf{x}_0, t_0, T) = \phi_{t_0}^T(\mathbf{x}_0)$ at time T. The gradient of the flow map

$$\nabla\phi_{t_0}^T : \mathbb{R}^d \to \mathbb{R}^{d\times d} \tag{2}$$

characterizes the local flow deformation of a particle neighborhood. Maximum stretching of nearby particles is given by the spectral norm $||.||_\lambda$ of $\nabla\phi_{t_0}^T$. Flow map FTLE (F-FTLE) is defined as the normalized maximal separation

$$\text{F-FTLE}^+(\mathbf{x}_0, t_0, T) = \frac{1}{T}\ln(||\nabla\phi(\mathbf{x}_0, t_0, T)||_\lambda). \tag{3}$$

In practice, the flow map is mostly computed by sampling particles on regular grids.

L-FTLE – Consider a path line $\mathbf{p}(t) = \mathbf{p}(\mathbf{x}_0, t_0, t)$ for a particle started at space-time location $(\mathbf{x}_0, t_0)$. The deviation of trajectories of infinitesimally close particles started at $(\mathbf{x}_0 + \delta_0, t_0)$, with $\delta_0 \to 0$, is given by the differential equation

$$\dot{\delta}(t) = \mathbf{J}_\mathbf{v}(\mathbf{p}(t), t_0 + t)\delta(t). \tag{4}$$

Solving the differential equation yields

$$\delta(t) = exp\left(\int_0^t \mathbf{J}_\mathbf{v}(\mathbf{p}(\tau), t_0 + \tau)d\tau\right)\delta_0. \tag{5}$$

Given a finite time span T, the matrix

$$\Psi_T(\mathbf{p}) = exp\left(\int_0^T \mathbf{J}_\mathbf{v}(\mathbf{p}(t), t_0 + t)dt\right) \tag{6}$$

expresses the mapping of a neighborhood at the starting point $\mathbf{p}(0)$ onto its deviations at the end point $\mathbf{p}(T)$. Compared to the flow map approach, this matrix corresponds to the gradient of the flow map. Defining a temporal discretization of the path line $T = \Delta t \cdot N$, where Δt is the length of one time step and N the number of steps, Eq. (6) can be approximated as

$$\Psi_T(\mathbf{p}) \simeq \prod_{i=0}^{N-1} \exp\left(\mathbf{J}_\mathbf{v}(\mathbf{p}(i\Delta t), t_0 + i\Delta t)\cdot\Delta t\right). \tag{7}$$

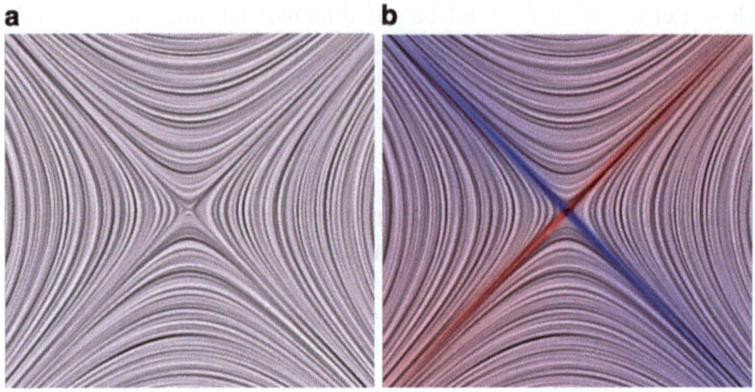

Fig. 2 FTLE (*left*) and *Normalized Velocity Separation (NVS)* (*right*) computed for a linear saddle. *Red* (*blue*) coloring depicts forward (backward) integration. The forward and backward FTLE field is constant and therefore the separatrices cannot be detected by FTLE. By normalizing the field for the flow map computation, NVS can detect the separatrices correctly

L-FTLE is now defined as the largest separation of this mapping. It is computed as

$$\text{L-FTLE}^{+}(\mathbf{x}_0, t_0, T) = \frac{1}{T}\ln(||\Psi_T(\mathbf{p})||_\lambda), \tag{8}$$

where $||.||_\lambda$ represents the spectral norm of the resulting matrix.

Comparison – Kuhn et al. [13] analyzed different FTLE methods with the conclusion that L-FTLE yields better results for long integration times. Indeed, L-FTLE measures the flow behavior using a single path line and therefore resembles the separation for longer separation times better – in contrast to the flow map or renormalization approaches. L-FTLE measures the separation for a single particle point-wise. In contrast, F-FTLE is calculated using a discrete flow map. Thus, the separation is measured for a small volume element of the size of the discretization. For the detection of separation, this has a severe impact: Finding separation surfaces is nearly impossible using a point-wise evaluation. It is only possible to detect actual particle separation, if the flow behavior is analyzed using small volume elements. Therefore, the flow map approach seems to be the method of choice for our purpose.

4 Normalized Velocity Separation

In the last section, we have seen that the F-FTLE approach is able to detect particle separation for small volume elements. However, even F-FTLE is not perfectly suited to detect all separation surfaces or separatrices. For instance, for a linear saddle, the FTLE will be constant in both cases, see Fig. 2. Thus, no separatrices would

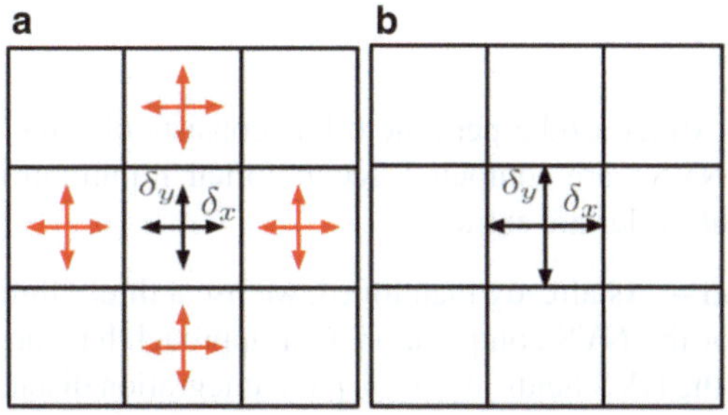

Fig. 3 A common idea when using FTLE (or NVS) is that the sampling distance of the particles to compute the flow map has to be decreased to gain better results. Unfortunately, this worsens the detection of separation in the field, since particle separation may be missed locally (*left*). The resolution of the flow map should to be at least as equally coarse as the resolution of the FTLE or NVS field

be highlighted by this measure. Instead of particle separation, we are interested in stream line separation, i.e., we need to detect if two stream lines spatially diverge. Therefore, we introduce a slightly different measure that lifts the aforementioned limitation.

Similar to the F-FTLE approach, we compute the flow map. Instead of using a standard integration of the stream lines that are parametrized by the particle advection time, we normalize the vectors of the underlying vector field since we are only interested in the limiting behavior of the particles and not in their velocity. After computing this flow map of the normalized vector field, we need to compute its gradient. Note that the newly defined flow map is not smooth. In fact, it is not even continuous. Nevertheless, is has bounded variation, since $T < \infty$. We can therefore evaluate the distributional derivative on each volume element of a given grid. In practice, this means that standard finite difference approaches are still mathematically well founded in this setting. We call the spectral norm of this measure *Normalized Velocity Separation (NVS)*. Similar to FTLE, NVS can be computed in forward and backward direction.

Note that if we compute the flow map with a smaller resolution than the final NVS field, not all separation might be detected. If we choose the same discretization for the flow map as for the final NVS field, then we can assure that we measure every separation contained in the field, see Fig. 3. In this chapter, we use NVS with the same resolution for the flow map as for the final NVS field.

Note that for instationary vector fields, the FTLE is used to extract Lagrangian coherent structures (LCS). In our context, we use NVS to detect separation and convergence in a stationary field. Thus, the resulting structures do not necessarily have to be connected to LCS.

5 Method

Our approach to extract saddle periodic orbits consists of three main steps: First, both NVS^+ and NVS^- are computed, second, their minimum is computed, and, third, the persistent cycles are extracted.

NVS computation – As already mentioned, we use a three-dimensional flow map based approach for the NVS computation. Our approach has three parameters: the grid resolution of the NVS fields $[\delta_x, \delta_y, \delta_z]$, the integration distance T, and the step size for each integration step Δt. We first compute the flow map at each point of the grid. The resolution of the flow map is the same as for the final NVS field. The integration is done using a simple Runge-Kutta integrator of fourth order without step size control. We support periodic data sets with our integrator, which enables unbounded evaluation.

The difference for the integrator between the FTLE computation and the NVS computation is that we normalize the used vectors. The normalization assures that shear does not impact our result.

After the flow map is computed, at each grid point, the gradient of the flow map is computed. We use a central differences approach here. Using the flow map gradient, the eigenvalues are computed using the method of Smith [19].

Feature field – As we already noted, saddle periodic orbits can be extracted as an intersection of attracting and repelling manifolds. In the context of NVS, these manifolds are characterized by high values in the NVS^+ and NVS^- fields. The intersection can be mimicked by computing the minimum of both fields. The minimum of both NVS fields is therefore called

$$\mathrm{NVS}^{min} = min\left(\mathrm{NVS}^+, \mathrm{NVS}^-\right),$$

and its closed curves with a high value indicate the presence of saddle periodic orbits.

Note that the integration time plays an important role in the detection of saddle periodic orbits. Since the separation might only occur at one single point along the orbit, the integration has to cover at least one period of the orbit. Therefore, with a integration time T, only saddle periodic orbits of arc length $L \leqslant T$ can be detected.

Cycle extraction – Since the NVS^{min} field can be very complex in practice, we propose to employ persistent homology [6] to robustly extract its dominant cycles. Especially, derivative-based methods would be difficult to use, since the derivatives of the minimum field might not be well-defined. Besides, we will see in the results section that persistence filters out most cycles in the NVS^{min} field that are not connected to saddle periodic orbits.

The basic idea in persistent homology is to measure the lifetime of homological features, i.e. components, tunnels, or voids, with respect to the sub-level threshold. As the sub-level threshold is increased, these features get born or may die. Using fast algorithms [3], one can efficiently track these events and thereby compute the long living topological features.

In our case we not only need the information of the presence of persistent cycles, but also a geometric representative of the cycle itself. We therefore make use of the extension to the persistence algorithm proposed in [7]. In this chapter, we make use of the open source library PHAT [2] to quickly compute the persistent cycles.

6 Results

In this section, we evaluate our method using three data sets. The first two data sets contain only a single saddle periodic orbit and are therefore simple analytic data sets. The third data set is constructed from a truncated Fourier series with random coefficients. It is much more complex and contains a lot of different topological structures.

6.1 *Single Saddle Periodic Orbit No. 1*

The first data set is given by the equation

$$u = x - y - 0.2 \cdot x \cdot \left(x^2 + y^2 + z^2\right) \tag{9}$$

$$v = x + y - 0.2 \cdot y \cdot \left(x^2 + y^2 + z^2\right) \tag{10}$$

$$w = 0.2 \cdot z \cdot \left(x^2 + y^2 + z^2\right). \tag{11}$$

For this data set, the position of the saddle periodic orbit is known.

In Fig. 4a, we visualized the stream lines of this data set using illuminated lines. We restricted the field to a bounding box $[-4, 4]^3$. However, the stream line integration is possible for an infinite length. In the figure, it is hard to see the saddle periodic orbit. Anyhow, one can observe some stream lines that indicate a saddle-like behavior.

Figures 4c+d show the NVS fields of this data set as a volume rendering. NVS^+ is shown in red and NVS^- is shown in blue. The amount of opacity is determined by the data value. The NVS fields are computed with parameters $T = 10$ and $\Delta t = 0.1$ in a volume of $[-4, 4]^3$ with a discretization of 256 steps in each direction. In the NVS^+ field, a *wall* of separation can be seen. In the NVS^- field, the expected *tube* of convergence can be seen. Intersecting these objects by computing the minimum of the NVS fields leads to the rendering shown in Fig. 4e. The field is colored in grey and the amount of opacity is determined by the data value. Here, we can already see the saddle periodic orbit as a cycle in the NVS^{min} field. This cycle can be easily extracted using persistent homology.

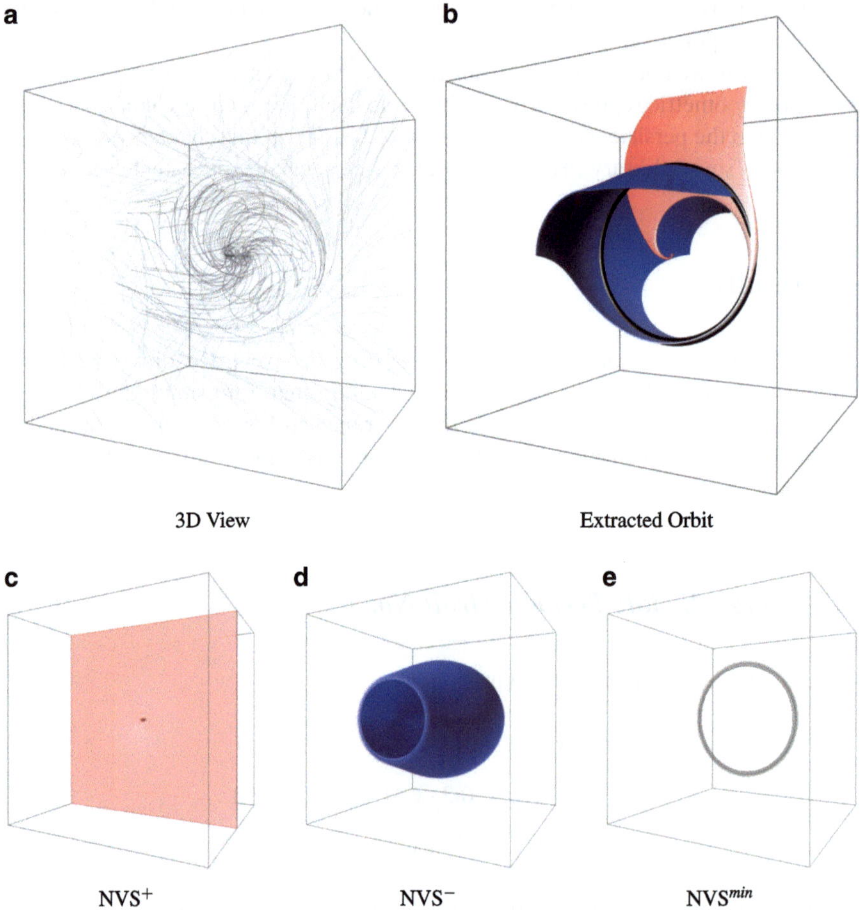

Fig. 4 Data set No 1: (**a**) Stream line visualization of the vector field; (**c+d**) Volume rendering of the NVS fields indicating the attracting and repelling separation surfaces; (**e**) The minimum of both NVS fields; (**b**) Using persistent homology, the dominant cycle in the minimum field is extracted and is depicted by a *black line*. We additionally seeded stream lines in the vicinity of the periodic orbit and colored them according to an arc length parametrization

6.2 Single Saddle Periodic Orbit No. 2

The second data set is given by the equation

$$u = -y + x \cdot (1 - r) \tag{12}$$

$$v = x + y \cdot (1 - r) \tag{13}$$

$$w = z \cdot (2 - r), \tag{14}$$

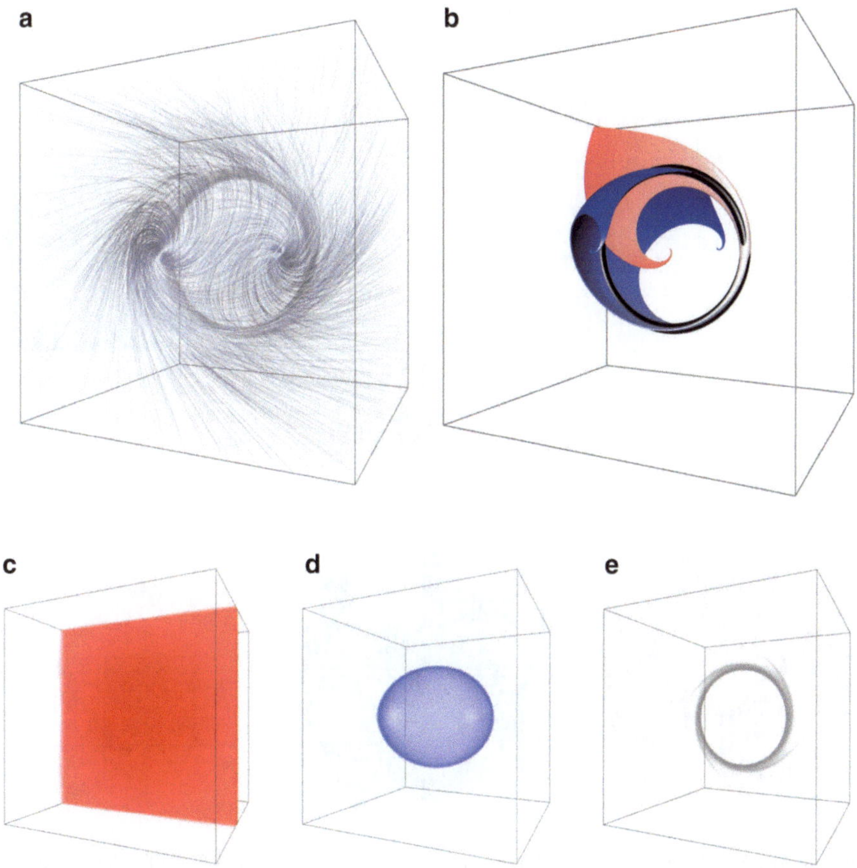

Fig. 5 Data set No 2: see caption of Fig. 4 for details. (**a**) 3D View. (**b**) Extracted Orbit. (**c**) NVS^+. (**d**) NVS^-. (**e**) NVS^{min}

where $r = \sqrt{x^2 + y^2 + z^2}$ is the distance to the origin. As for the first data set, the position of the saddle periodic orbit is known.

In Fig. 5a, the stream lines of this data set are shown as illuminated lines. Again, we restricted the visualization to a bounding box of $[-2, 2]^3$, while the stream line integration can be done for an unbounded length. In contrast to the first data set, we can see that there is some kind of bubble enclosing the inner flow. We cannot see any indication for a saddle periodic orbit using the stream line visualization.

Figures 5c+d show the NVS fields of the second data set as a volume rendering. Again, NVS^+ is shown in red and NVS^- is shown in blue and the amount of opacity is determined by the data value. The NVS fields are computed with parameters $T = 5$ and $\Delta t = 0.1$ in a volume of $[-2, 2]^3$ with a discretization of 256 steps in each direction. The NVS fields look quite similar to the NVS fields of data set no. 1. In particular, the NVS^+ field again shows a plane as separation manifold.

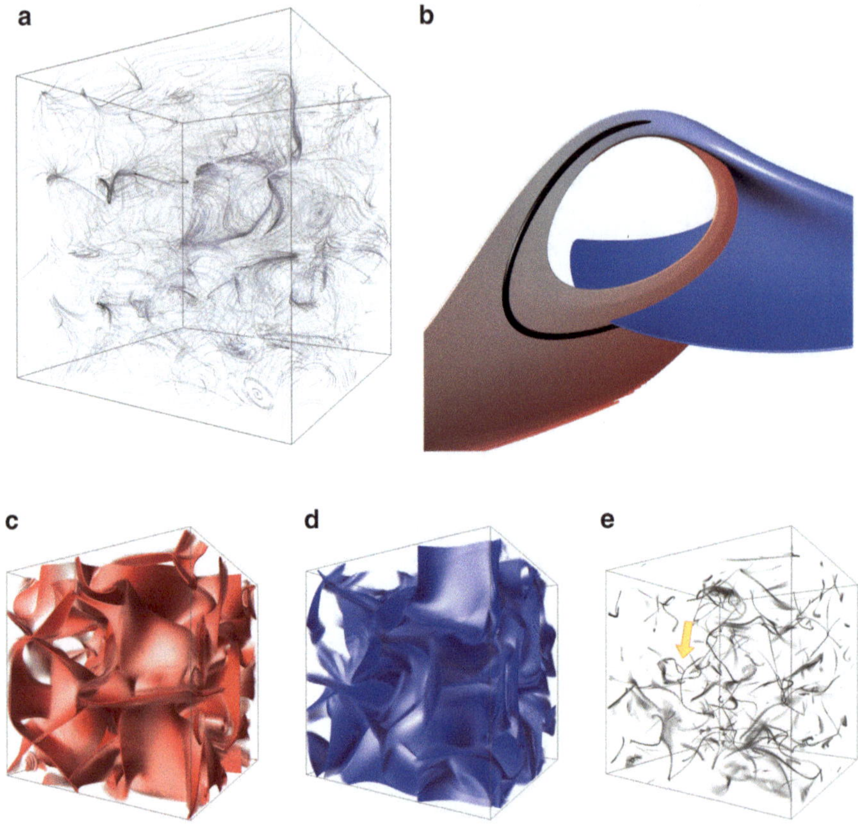

Fig. 6 Truncated Fourier series with random coefficients: see caption of Fig. 4 for details. (**a**) 3D View. (**b**) Extracted Orbit. (**c**) NVS^+. (**d**) NVS^-. (**e**) NVS^{min}

In contrast, the NVS^- field does not show a tube but an ellipsoid. Note that the ellipsoid is not solid. Similar to the first data set, the minimum field in Fig. 5e shows a ring structure. This is the expected saddle periodic orbit.

6.3 *Truncated Fourier Series with Random Coefficients*

In Fig. 6a, the stream lines of the third data set are shown. This data set resembles the complexity of real world data. Each component of the vector field consists of a sum of a few sin and cos functions with random coefficients. Note that due to the periodicity of the data set, the stream lines can be traced for an unbounded time. We sampled this smooth function using 256 steps in each direction. In the stream line visualization of this vector field, we can see a lot of attracting and repelling structures, but the contained saddle periodic orbit cannot be found.

Figures 6c+d show the NVS fields for this data set. The coloring is chosen as for the data sets aforementioned. We computed the NVS with parameter $T = 0.2$ and $\Delta t = 0.01$ for the whole domain with a discretization of 256 steps in each direction.

The separation surfaces are clearly visible in the NVS$^+$ and NVS$^-$ fields. The complexity of both fields is much higher than for the simple analytic data sets. In Fig. 6e, the NVS^{min} field is shown. This image shows that the NVS^{min} not only indicates the saddle periodic orbits, but also the saddle connectors. Also, the finite approximation of the field introduces noise. Both effects generate a few false positives, when we extract the persistent cycles from the NVS^{min} field, since saddle connectors can form loops. The amount of extracted saddle connector loops can be reduced by filtering with respect to minimal vector magnitude, while the effect of noise can be controlled by focussing on highly persistent cycles. Using these filtering techniques, we were able to extract a saddle periodic orbit indicated by an arrow, since the resulting candidate set was small enough to allow for a manual selection. A zoom-in of this structure is shown in Fig. 6b.

7 Conclusion

In this chapter, we presented a novel method to extract saddle periodic orbits in a three-dimensional vector field. We introduced a new measure called *Normalized Velocity Separation (NVS)* which is based on the FTLE measure. The minimum of both fields indicates the presence of saddle periodic orbits. We have seen that NVS based on F-FTLE fits best for this problem, since we can measure the separation in small volumes instead of a point-wise computation. We extracted the dominant cycles in this feature field using persistent homology. This makes our method applicable to relatively complex data sets as we have shown in the results section.

It should be noted that we are not able to extract the periodic orbit shown in Fig. 6b in a fully automatic manner. Our method depends crucially on the choice of suitable parameters and generates false positives in the case of complex data sets. Also, due to the numerical approximation we cannot rule out the presence of false negatives.

While most topological structures of three-dimensional vector fields are efficiently extractable, saddle periodic orbits remain a challenging problem and a robust and parameter-free method has yet to be proposed.

Acknowledgements First, we thank the reviewers of this paper for their ideas and critical comments. In addition, we thank Ronny Peikert and Filip Sadlo for a fruitful discussions. This research is supported by the European Commission under the TOPOSYS project FP7-ICT-318493-STREP, the European Social Fund (ESF App. No. 100098251), and the European Science Foundation under the ACAT Research Network Program.

References

1. D. Asimov, Notes on the topology of vector fields and flows. Technical Report RNR-93-003, NASA Ames Research Center, 1993
2. U. Bauer, M. Kerber, J. Reininghaus, PHAT: persistent homology algorithm toolbox. http://phat.googlecode.com/
3. C. Chen, M. Kerber, Persistent homology computation with a twist, in *27th European Workshop on Computational Geometry (EuroCG 2011)*, 2011. Extended abstract
4. G. Chen, K. Mischakow, R.S. Laramee, E. Zhang, Efficient morse decompositions of vector fields. IEEE Trans. Vis. Comput. Graph. **14**, 848–862 (2008)
5. M. Dellnitz, O. Junge, On the approximation of complicated dynamical behavior. SIAM J. Numer. Anal. **36**, 491–515 (1999)
6. H. Edelsbrunner, D. Letscher, A. Zomorodian, Topological persistence and simplification. Discret. Comput. Geom. **28**, 511–533 (2002)
7. H. Edelsbrunner, A. Zomorodian, Computing linking numbers of a filtration, in *Algorithms in Bioinformatics*. LNCS, vol. 2149 (Springer, Berlin/Heidelberg, 2001), pp. 112–127
8. C. Garth, G.-S. Li, X. Tricoche, C.D. Hansen, H. Hagen, Visualization of coherent structures in transient 2D flows. In *Topology-Based Methods in Visualization II* (Springer, Berlin/Heidelberg, 2007), pp. 1–13
9. G. Haller, Distinguished material surfaces and coherent structures in three-dimensional fluid flows. Physica D **149**, 248–277 (2001)
10. J. L. Helman, L. Hesselink, Visualizing vector field topology in fluid flows. IEEE Comput. Graph. Appl. **11**(3), 36–46 (1991)
11. M. Hirsch, S. Smale, R. Devaney, *Differential Equations, Dynamical Systems and an Introduction to Chaos*, 2nd edn. (Academic, 2004)
12. J. Kasten, C. Petz, I. Hotz, B. Noack, H.-C. Hege, Localized finite-time lyapunov exponent for unsteady flow analysis, in *Proceedings Vision, Modeling and Visualization 2008*, Braunschweig, 2009, pp. 265–274
13. A. Kuhn, C. Rössl, T. Weinkauf, H. Theisel, A benchmark for evaluating FTLE computations, in *Proceedings IEEE Pacific Visualization 2012*, Songdo, 2012, pp. 121–128
14. R. Peikert, F. Sadlo, Topologically relevant stream surfaces for flow visualization, in *Proceedings of Spring Conference on Computer Graphics*, Budmerice, 2009, pp. 171–178
15. F.H. Post, The state of the art in flow visualization: feature extraction and tracking. Comput. Graph. Forum **22**(4), 775–792 (2003)
16. W. Reich, D. Schneider, C. Heine, A. Wiebel, G. Chen, G. Scheuermann, Combinatorial vector field topology in three dimensions, in *Topological Methods in Data Analysis and Visualization II* (Springer, Berlin/Heidelberg, 2012)
17. F. Sadlo, R. Peikert, Visualizing lagrangian coherent structures and comparison to vector field topology, in *Topology-Based Methods in Visualization II* (Springer, Berlin/Heidelberg, 2009)
18. A. Sanderson, G. Chen, X. Tricoche, E. Cohen, Understanding quasi-periodic fieldlines and their topology in toroidal magnetic fields, in *Topological Methods in Data Analysis and Visualization II, Mathematics and Visualization*, ed. by R. Peikert, H. Hauser, H. Carr, R. Fuchs, (Springer, Berlin/Heidelberg, 2012) pp. 125–140.
19. O.K. Smith, Eigenvalues of a symmetric 3×3 matrix. Commun. ACM **4**(4), 168 (1961)
20. H. Theisel, T. Weinkauf, H.-C. Hege, H.-P. Seidel, Saddle connectors – an approach to visualizing the topological skeleton of complex 3D vector fields, in *Proceedings of IEEE Visualization 2003*, ed. by G. Turk, J.J. van Wijk, R. Moorhead, Seattle, Oct 2003, pp. 225–232
21. H. Theisel, T. Weinkauf, H.-C. Hege, H.-P. Seidel, Grid-independent detection of closed stream lines in 2D vector fields, in *Proceedings Vision, Modeling and Visualization 2008*, Konstanz, 2004, pp. 421–428

22. D. Weiskopf, B. Erlebacher, Overview of flow visualization, in *The Visualization Handbook*, (Academic, 2005), pp. 261–278
23. T. Wischgoll, G. Scheuermann, Detection and visualization of closed streamlines in planar flows. IEEE Trans. Vis. Comput. Graph. **7**(2), 165–172 (2001)

Part II
Efficient Computation of Topology

Computational Topology via Functional Programming: A Baseline Analysis

David Duke and Hamish Carr

Abstract Computational topology is of interest in visualization because it summarizes useful global properties of a dataset. The greatest need for such abstractions is in massive data, and to date most implementations have opted for low-level languages to obtain space and time-efficient implementations. Such code is complex, and is becoming even more so with the need to operate efficiently on a range of parallel hardware. Motivated by rapid advances in functional programming and compiler technology, this chapter investigates whether a shift in programming paradigm could reduce the complexity of the task. Focusing on contour tree generation as a case study, the chapter makes three contributions. First, it sets out the development of a concise functional implementation of the algorithm. Second, it shows that the sequential functional code can be tuned to match the performance of an imperative implementation, albeit at some cost in code clarity. Third, it outlines new possiblilities for parallelisation using functional tools, and notes similarities between functional abstractions and emerging ideas in extreme-scale visualization.

1 Introduction

The contour tree, Reeb graph and Morse-Smale complex are of interest to computational scientists because they provide useful, well-founded, abstractions that facilitate analysis of massive datasets. Efficient algorithms to generate these abstractions are already difficult to implement, and the need for variants that can exploit a growing range of parallel hardware is an open challenge.

Most algorithms for contour tree generation start with the requirement to traverse samples in order, with the result that a global property of the data (the ordering)

D. Duke (✉) • H. Carr
School of Computing, University of Leeds, Leeds, UK
e-mail: D.J.Duke@leeds.ac.uk; H.Carr@leeds.ac.uk

P.-T. Bremer et al. (eds.), *Topological Methods in Data Analysis and Visualization III*, Mathematics and Visualization, DOI 10.1007/978-3-319-04099-8_5,

limits opportunity to apply simple parallel operations on spatially localised regions. Approaches that require explicit management and coordination of parallel tasks then add further complexity to implementations. As extreme-scale computation looks toward exa-scale architectures and new trade-offs between computation and communication costs, there is an irony that one of the main tools needed for massive data analysis may be one of the more awkward to parallelize.

One escape route is to find an alternative derivation for the structure, one that is independent of global assumptions. The Joint Contour Net [8] may provide such a trap-door, by taking a constructive, *computational* view of the contour tree as a set of relationships between discrete geometric regions. Rather than move to a different algorithm, this chapter investigates whether a different programming paradigm, (pure) functional programming, could simplify the implementation of topological computations, and potentially provide a simpler path to parallel computational topology. The *prima facie* case for investigating the functional paradigm is:

1. The ability to describe algorithms at a very high level of abstraction, effectively providing a concise executable specification.
2. Evidence that FP can yield runtime performance commensurate (or sometimes even better than) that obtained from imperative languages, for example [9, 11].
3. Rapid advances in compiler and run-time technologies, in particular the emergence of lightweight abstractions for opportunistic parallel evaluation [13, 14].

In the longer term, these points are equally important: we seek programs that combine effective exploitation of parallelism with clear expression of the underlying algorithm. This chapter addresses the first point, and investigates whether FP can match imperative performance on a topological computation. Our functional code is written in Haskell [21], a pure functional language with support multiple forms of parallelism. Space precludes substantial presentation of parallel approaches, so on this point the chapter reports only initial findings and opportunities.

Section 2 briefly reviews the relevant literature on approaches to parallelizing computational topology. Section 3 reviews pure functional programming and introduces the expression of these concepts within Haskell. Three stages of development are then set out, addressing each of the three claims above. Section 4 describes an initial contour tree implementation that favours simplicity over performance. Section 5 then derives an efficient sequential implementation, before Sect. 6 identifies routes to functional parallelization. Section 7 then stands back to consider the findings against the objectives of this study.

2 Computational Topology and Functional Programming

Computational topology in visualization has focussed on three data structures – the contour tree, the Reeb graph, and the Morse-Smale complex. Each of these has posed its own challenges in terms of algorithmic development, particularly with respect to performance, memory consumption, and parallelization. For the contour tree, Pascucci and Cole-McLaughlin [18] described a block-wise parallel implementation in which contour trees (or merge trees) were computed for individual blocks, then merged hierarchically in a standard fan-in approach. In a related approach, Arge and Revsbaek [1] showed how to implement an out-of-core version of Carr et al.'s approach [3]. More recently, Maadasamy et al. [12] have described a hybrid approach relying on GPGPU programming. Similarly, Pascucci et al. [19] described a practically efficient streaming algorithm for Reeb graph computation, which tackles large data sets by minimising the memory overhead on a single CPU. Shivashankar et al. [20] also use GPGPU, to implement a paralell algorithm for computing 2D Morse-Smale complexes, while Guenther et al. [10] propose an efficient parallel algorithm for computing MS complexes from 3D grayscale images.

Common to all of these approaches however, has been the difficulty of moving from a relatively simple serial algorithm to a parallel algorithm, and this is where functional languages become of interest. Functional languages have been a useful incubator for programming concepts that have only later appeared in more widespread languages; examples here are parametric polymorphism, and lambda expressions. Deliberately or coincidentally, a similar pattern can be observed in emerging software architectures for extreme-scale visualization. The Dax toolkit [16] replaces per-filter iteration with 'worklets' whose execution is coordinated within a single traversal. Haskell and its libraries exploit 'fusion', both for simple types such as lists and more recently bulk array operations; pushing fusion deeper into the datatype level, rather than components, may create opportunities for tighter optimization. At a different level, the Functor/Iterator model of EAVL [15] bears similarities to the use of type-directed optimization of array operations [11].

3 Functional Programming

In a von Neumann architecture computation proceeds through the application of instructions that result in the modification of data stored in a (global) memory. Imperative languages provide layers of abstraction over the raw architecture, capturing common patterns of computation as control structures, and providing data representations that mediate between flat machine memory and the conceptual structures of problem domains.

Computational models typically trade-off control over the raw machine against level of abstraction. In place of a specific machine architecture, functional programming starts from Church's mathematical lambda calculus, with computation viewed as the reduction of an expression to a normal form. In place of assignment, the fundamental operation in a functional program is the application of a function to an argument (also called β-reduction). As an introduction to the Haskell language [21] used in this chapter the following code defines two functions, *take* and *filter*, and uses these to define a program that returns the first three even numbers:

$$
\begin{aligned}
&\mathit{take} :: \mathit{Int} \to [a] \to [a]\\
&\mathit{take}\ 0\ \_ \quad\quad = [\,]\\
&\mathit{take}\ \_\ [\,] \quad\quad = [\,]\\
&\mathit{take}\ n\ (v : vs) = v : \mathit{take}\ (n-1)\ vs\\
&\mathit{filter} :: (a \to \mathit{Bool}) \to [a] \to [a]\\
&\mathit{filter}\ \_\ [\,] = \ [\,]\\
&\mathit{filter}\ p\ (v : vs)\ |\ p\ v \qquad\quad = v : \mathit{filter}\ p\ vs\\
&\qquad\qquad\qquad\ |\ \mathit{otherwise} = \quad \mathit{filter}\ p\ vs\\
&\mathit{take}\ 3\ (\mathit{filter}\ \mathit{even}\ [1\,..])
\end{aligned}
$$

The first line of each function gives its *type*; all Haskell definitions are strongly typed, but declarations are optional as types can be inferred. Haskell's rich type system is beyond the scope of this chapter; for the present all we need are basic types (*Int*, *Float*, *Bool*, ..), type *constructors* for lists ([...]) and functions (→), and parametric polymorphism. The first line states that *take* is a function that given an integer and a list of values (of some arbitrary type *a*), returns a list of the same type *a*. Haskell functions are *curried*: rather than taking a tuple of parameters, they take one argument at a time, allowing a new function to be constructed by partially applying a function to some of its arguments. Lines 2–4 implement *take*. The equations use *pattern matching*: if the first argument is the constant 0, the result is an empty list; if the first argument is non-zero and the second argument is the empty list, the result is the empty list; otherwise the function returns the result of cons-ing the head of the list to the result of taking one fewer item from the tail.

The second function, *filter*, is *higher order*; its first argument is itself a function (from some type *a* to boolean), and it applies this function to each value in a list, retaining just those in the output for which the first argument returns *True*. Higher-order functions take the place of control structures in imperative languages, capturing common patterns of computation, with the advantage that new patterns can be introduced as desired.

An essential component of this expressive power is use of *lazy* evaluation, where an argument to a function is only evaluated to the extent that it is needed by the computation. Most imperative languages evaluate arguments, then pass the result to the function via a stack. This *eager* evaluation presumes that the arguments are needed, limiting the extent to which functions can be 'glued together' into new abstractions.[1] Fixed control abstractions are necessary in imperative code in part because they impose particular requirements on evaluation order (e.g. we need to evaluate the predicate of a conditional statement before deciding which of the branches to perform). In the last line, the sub-expression *filter even* [1 ..] applies the filter function to two arguments, a function *even* from numbers to boolean, and

[1]This is a simplification. Technically, lazy evaluation is the combination of normal order reduction with overwriting the root of reducible expressions to ensure that arguments are evaluated at most once.

the infinite list of numbers $1, 2, 3, \ldots$. Under eager evaluation, the call to *filter* would never return, but under lazy semantics, evaluation proceeds as follows:

$$\begin{aligned}
&\mathit{take}\ 3\ (\mathit{filter\ even}\ [1\,..])\\
&= \mathit{take}\ 3\ (\mathit{filter\ even}\ [2\,..])\\
&= \mathit{take}\ 3\ (2 : \mathit{filter\ even}\ [3\,..])\\
&= 2 : \mathit{take}\ 2\ (\mathit{filter\ even}\ [3\,..])\\
&= 2 : \mathit{take}\ 2\ (\mathit{filter\ even}\ [4\,..])\\
&= 2 : \mathit{take}\ 2\ (4 : \mathit{filter\ even}\ [5\,..])\\
&\ldots
\end{aligned}$$

Once three values have been generated, the first equation of *take* will apply, and the expression will reduce to $2 : 4 : 6 : [\,]$, or more simply $[2, 4, 6]$. Haskell also provides n-ary tuples (e.g. $(1, \mathit{True}, \text{“}\mathit{foo}\text{''})$ is a 3-tuple of an integer, a boolean and a string), type synonyms, and user-defined data types. For example, a type to represent 2D points in either polar or rectangular coordinates can be defined by

$$\begin{aligned}
&\textbf{type}\ \mathit{Angle} &&= \mathit{Float}\\
&\textbf{data}\ \mathit{Coord} &&= \mathit{Rectangular\ Float\ Float} \mid \mathit{Polar\ Float\ Angle}
\end{aligned}$$

The expression *Rectangular* 1.5 2.5 constructs a value of type *Coord*. *Rectangular* and *Polar* are called *constructors*, and can be used on the left-hand side of function definitions to pattern-match against arguments. Indeed the list type used in *take* and *filter* is just a data type with two constructors, cons (written :) and nil ([]). Components of a constructor can be accessed by pattern matching or by naming the fields, with the field names then available as *accessor* functions,

$$\begin{aligned}
\textbf{data}\ \mathit{Coord} &= \mathit{Rectangular}\ \{x :: \mathit{Float}, y :: \mathit{Float}\}\\
&\mid\ \mathit{Polar}\ \{\mathit{rho} :: \mathit{Float}, \mathit{theta} :: \mathit{Angle}\}
\end{aligned}$$

Here $\mathit{rho} :: \mathit{Coord} \to \mathit{Angle}$, and if $x = \mathit{Polar}\ 30.0\ 45.0$, then $\mathit{rho}\ x = 30.0$.

The functions described so far have been *pure*, that is, evaluation of the function produces a single result. In practice most computation involves side-effects, such as input-output, concurrency, exceptions, or writing to memory. Haskell has adopted an elegant solution that not only accommodates side effects, but allows programmers to define new computational models within the language. A *Monad* is a data type that captures the operational behaviour of an effect using a small number of fundamental operations: the ability to create an effect from a value, and to bind the output of one effect into the input of another. Haskell's *IO* monad, for example, captures operations such as reading a file or writing output; other standard monads deal with mutable state or non-determinism. As a monad is just a data type supporting a set of standard operators, monadic expressions can be written just as other functions; for example the following counts the number of words in a file:

$$\mathit{readFile}\ \texttt{"test"} \gg\!= \lambda c \to \mathit{return} \circ \mathit{length} \circ \mathit{words}\ \$\ c$$

The function '$\gg=$' takes the IO operation of reading a file, and binds it to an IO action that breaks the content into a list of words, and returns the length of the list. Right-associative function application '\$' is used in place of parentheses: $f \; \$ \; g \; v$ is equivalent to $f \; (g \; v)$. Haskell also provides 'syntactic sugar' to hide the 'plumbing' needed to pass the out output from one action to the next; the above can be written

```
do c ← readFile "test"
   let len = length ∘ words $ c
   return len
```

This may look like a block of imperative code, but the compiler de-sugars this expression into the form above, chaining monadic actions explicitly. For further information on Haskell, see [17, 21].

4 Step I: The *Functional* Contour Tree

The first goal was to implement a contour tree algorithm, aiming for code simplicity and where possible exploiting the functional paradigm, e.g. using higher-order functions in place of explicit recursion. The implementation is based on the contour tree algorithm described in [3], and in particular the relatively unoptimized code used by Carr in his original implementation. All implementations require basic infrastructure for representing the sampled data, the tree, and other working structures. The intention was to explore performance using 8-bit volumetric data from the 'volvis' repository,[2] but it was straightforward to define a flexible representation that could also support other simple synthetic test cases. A *dataset* holds three items, the dimensionality, a function *value* that reports the scalar value at a given vertex in the mesh, and a second function, *mesh*, that maps each vertex to a list of its immediate neighbours.

```
type Dimension = (Int, Int, Int)  -- dimensionality of a 3D mesh
type Vertex    = Int              -- unique index of each mesh vertex
type Coord     = (Int, Int, Int)  -- position in a 3D mesh
data Dataset a
   = DS {dim   :: Dimension          -- spatial dimension
        ,mesh  :: Vertex → [Vertex]  -- local structure
        ,value :: Vertex → a         -- scalar value at vertex
        }
type ScalarField = Dataset Word8
```

[2]www.volvis.org

There is no commitment, at this point, to where or how the data is stored. Nor does the representation explicitly record subdivision into cells; it simply provides a function to compute neighbourhood information needed by the algorithm.

A concrete dataset is constructed by providing a list of values and dimensionality; the list is converted into an array for direct access to values, and a Dataset value is constructed where the *value* function is simply the function that indexes the array. As '!', the array indexing operator, has type *Array i e* → *i* → *e*, partial application of this operator to an array of samples (i.e., *array*! below), results in a function of type *i* → *e* that can be used to lookup a value in that specific dataset. The second parameter to *dataset*, *nbhood*, provides adjacency information of vertex coordinates in a 3D mesh; this is transformed into a function that operates with vertex ids.

```
dataset :: Dimension → (Coord → [Coord]) → [a] → Dataset a
dataset dimension nbhood samples
   = let array = listArray (0, size dimension − 1) samples
     in DS dimension (neighbours dimension nbhood) (array!)
```

Most tree structures are straightforward in Haskell. However the contour tree algorithm constructs an *unrooted* tree, and for this we needed a general graph representation. Unfortunately graphs are not (easily) defined as inductive functional datatypes. Our implementation makes use of both a simple edge list, and an adjacency list structure stored in an *IntMap*, an efficient updatable mapping implemented via Patricia trees [6]. *IntMap* is also used for the disjoint set structure.

```
type Node      = Int               -- graph nodes
type EdgeList  = [(Node, Node)]    -- edge list representation
data Adjacency = Adj {incoming :: [Node]
                     , outgoing :: [Node]
                     }
type Graph = I.IntMap Adjacency
```

Contour tree generation involves two sweeps through the data, to produce the join and split trees. Algorithmically these are identical, and take two arguments: a list of mesh *vertices* ordered with respect to samples, and the neighbourhood mapping.

```
join_tree :: [Vertex] → (Vertex → [Vertex]) → EdgeList
join_tree vertices adjacent
   = tree ∘ for_each vertices (init_JoinS vertices) $ λjs v →
        for_each (adjacent v) (mark v js) $ λjs n →
          let components = disj js
              lowest     = base js
              comp_v     = find v components
              comp_n     = find n components
              u          = lowest ? comp_n
```

$$\begin{array}{l}
\textbf{in if } n \text{ `}S.member\text{` } (seen\ js) \wedge comp_v \not\equiv comp_n \\
\quad \textbf{then } js\ \{ disj\ = union\ comp_n\ comp_v\ components \\
\qquad\qquad , tree\ = (u,v) : (tree\ js) \\
\qquad\qquad , base = set\ (set\ lowest\ comp_v\ v)\ comp_n\ v \\
\qquad\qquad , seen = S.insert\ n\ (seen\ js) \\
\qquad\qquad \} \\
\quad \textbf{else } js
\end{array}$$

Some brief points of explanation:

- *init_JoinS* wraps the information used by the join algorithm into a data constructor with four fields, a disjoint set (*disj*), a list of edges (*tree*), a map giving the lowest vertex in each component (*base*), and a set (*seen*) to track (i.e. 'mark') the vertices seen during traversal.
- *for_each* is an example of building a new control structure. The definition is

$$\begin{array}{l}
for_each :: [a] \to b \to (b \to a \to b) \to b \\
for_each\ xs\ init\ op = foldl'\ op\ init\ xs
\end{array}$$

  Starting from a seed (*init*), we take each value in turn from a list (*xs*), and apply a function to the seed and the value to obtain a new seed.

The initial sweeps produce join and split trees stored as edge lists, which are converted into adjacency lists for the merge phase of the algorithm:

$$\begin{array}{l}
contour_tree :: Graph \to Graph \to EdgeList \\
contour_tree\ jtree\ stree \\
\quad = unfoldr\ step\ (leaves, jtree, stree) \\
\qquad \textbf{where} \\
\qquad\quad step\ (q, jt, st) \\
\qquad\qquad \mid sizeQ\ q > 1 = Just\ (edge, (q', jt', st')) \\
\qquad\qquad \mid otherwise\ \ = Nothing \\
\qquad\qquad\qquad \textbf{where} \\
\qquad\qquad\qquad\quad (xi, q1) = leaveQ\ q \\
\qquad\qquad\qquad\quad xj \qquad = choose\ (jt\ ?\ xi)\ (st\ ?\ xi) \\
\qquad\qquad\qquad\quad jt' \qquad = jt /\!-\!/ xi \\
\qquad\qquad\qquad\quad st' \qquad = st /\!-\!/ xi \\
\qquad\qquad\qquad\quad q' \qquad = \textbf{if } leaf\ (jt'\ ?\ xj) \not\equiv leaf\ (st'\ ?\ xj) \\
\qquad\qquad\qquad\qquad\qquad \textbf{then } enterQ\ xj\ q1\ \textbf{else } q1 \\
\qquad\qquad\qquad\quad edge \quad = \textbf{if } incoming\ (jt\ ?\ xi) \equiv [\,] \\
\qquad\qquad\qquad\qquad\qquad \textbf{then } (xi, xj)\ \textbf{else } (xj, xi) \\
\qquad\quad \text{-- Initial leaves:} \\
\qquad\quad leaves = fromListQ\ [v \mid v \leftarrow I.keys\ jtree, starter\ (jtree\ ?\ v)\ (stree\ ?\ v)] \\
\qquad\quad \text{-- Test whether up-degree J(v) + down-degree S(v) == 1 by} \\
\qquad\quad \text{-- pattern-matching on adjacency structures.}
\end{array}$$

$$
\begin{aligned}
&\mathit{starter}\ (\mathit{Adj}\ []\ \_)\quad (\mathit{Adj}\ [\_]\ \_) = \mathit{True}\\
&\mathit{starter}\ (\mathit{Adj}\ [\_]\ \_)\ (\mathit{Adj}\ []\ \_)\quad = \mathit{True}\\
&\mathit{starter}\ \_\qquad\qquad\ \_\qquad\qquad\quad = \mathit{False}
\end{aligned}
$$

-- Choose arc (xi,xj) from J if xi is an upper-leaf,
-- else return the arc (xi,xj) from S.

$$
\begin{aligned}
&\mathit{choose}\ \mathit{aj}\ \mathit{as}\ |\ \mathit{leaf}\ \mathit{aj} \quad = \mathit{head} \circ \mathit{outgoing}\ \$\ \mathit{aj}\\
&\qquad\qquad\qquad |\ \mathit{otherwise} = \mathit{head} \circ \mathit{outgoing}\ \$\ \mathit{as}\\
&\mathit{leaf} = \mathit{null} \circ \mathit{incoming}
\end{aligned}
$$

As with the split/join tree, we exploit a higher-order pattern, in this case *unfold*; starting from an initial seed, unfold generates a sequence of values by applying a 'step' function to the seed. This either produces '*Just*' the next item for the output and a new seed, or '*Nothing*', terminating the computation. Here the 'seed' contains the join and split trees, and the queue of nodes to be considered for transfer into the contour tree. This implementation outputs the augmented contour tree; a further function converts this to a reduced tree, eliminating internal nodes.

Runtime: When executed on the 'nucleon' dataset ($41 \times 41 \times 41$), the program took 48 minutes of user time[3] to generate the contour reduced tree. By comparison, Carr's C++ implementation with debug flags enabled performed the same task in 0.5 s.

5 Step II: The *Fast* Functional Contour Tree

The obvious response to 48 min runtime is "...but where does all that time go?". Lazy evaluation complicates the task of profiling, as evaluation of one function may be spread over a substantial period of time, as more of its output is forced by the calling context. The problem is further complicated by the extensive series of code transformations and optimization steps involved in mapping Haskell into executable code. However the Glasgow Haskell Compiler (ghc) suite provides profiling tools, and there are known causes, and fixes, for runtime performance problems [17].

The GHC approach to profiling involves identifying a set of *cost centres* within the code; these are (by default) major expressions such as top level functions. During execution, statistics are gathered on the number of entries to, and time spent within, these blocks, along with the amount of heap space allocated against cost centres and data types. Heap usage is particularly important: by default, Haskell data and function closures are allocated on the heap. Heap 'churn' was likely to be part of the slow observed runtime, caused by updates to persistent structures like the integer

[3]Timings were performed on an Apple MacBook, measured using the unix 'time' command. Programs were compiled with `-O2 -prof -rtsopts -auto-all -rtsopts`, enabling optimization and run-time profiling.

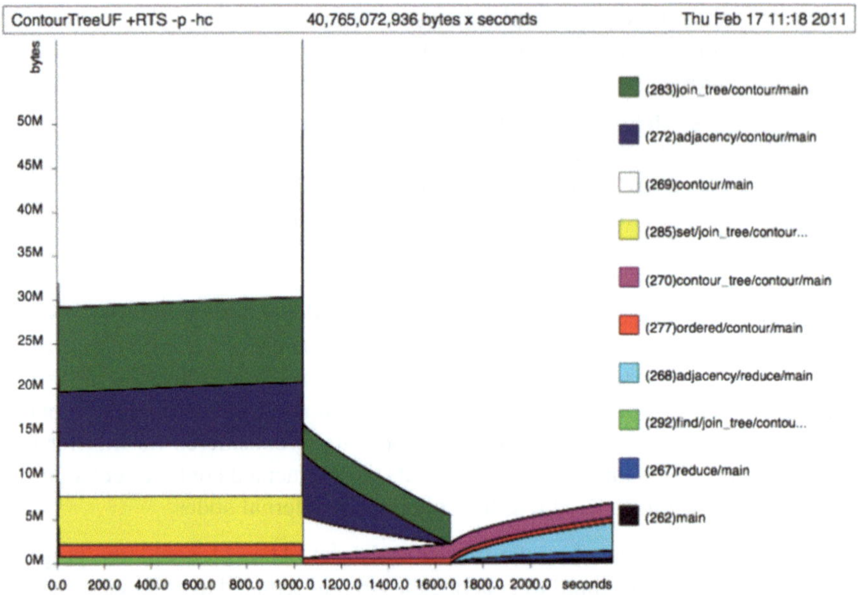

Fig. 1 Heap usage, per cost centre, of pure program

maps used for graph adjacency lists, and for the disjoint set structure. From the runtime profile and heap statistics, the following stood out:

- Twenty percent of runtime was spent in the *find* operation of the disjoint set structure. With path compression, this operation involves multiple lookups and updates to the *IntMap* storing the vertex set partition.
- Ten percent of time was spent in a function, *adjacency*, used to convert from the graph edge-list to adjacency list representation.
- As shown in Fig. 1, the program was churning over 40 GB of heap space, with peak occupancy of 30 MB; this usage could be tracked to the lists and maps used in the principal functions (*join_tree*, *contour*), and in *adjacency*.

We start by reducing heap churn. Most Haskell functions operate on *boxed* representations of data, where values are allocated in heap cells. There are two reasons for this. First, structures such as lists are produced lazily, so rather than a concrete value, the head and/or tail of a list may be a thunk, i.e. the code to generate the value when it becomes needed. Second, polymorphic functions cannot know the shape/storage of arguments. Passing parameters as references to heap cells rather than values ensures uniformity of size and access.

We reduce overheads by making data constructors fields *strict*; such a fields will always be evaluated, never stored as a thunk. A directive then encourages the compiler to unbox strict fields. Informed by runtime statistics, we replaced the lazy adjacency list representation of graphs with a strict *AdjNode* structure:

```
data AdjNode   = Link ! Node ! AdjNode | Nil
data Adjacency = Adj { indeg    :: !Int
                     , outdeg   :: !Int
                     , incoming :: !AdjNode
                     , outgoing :: !AdjNode
                     }
```

After adding explicit fields for in- and out-degree, these changes improved runtime to 10 min 52 s, still a factor of 20× slower than C. Unboxing other structures, e.g. coordinate lists defining neighbourhoods, yielded only modest gains. Runtime profiling indicated that update time and heap churn from the graph and disjoint set structures still contributed substantial overhead.

These costs could be reduced by in-place update of data. The solution is to use monads, specifically the state (*ST*) and unboxed array (*STUArray*), which allow us to embed effects in an otherwise pure computation, somewhat like a 'sandpit' for unsafe code. The price paid is code clarity – here for example is the monadic equivalent to the *join_tree* function:

```
join_tree :: Int → UArray Int Vertex → (Vertex → [Vertex])
              → Array Node Adjacency
join_tree nrV verts adj
   = runSTArray $ do
       js ← initJ nrV
       forM_ [nrV − 1, nrV − 2 .. 0] $ λi → do
         let v = verts `atA` i
         putA (seen js) v True
         comp_v ← findU v $ disj js
         forM_ (adj v) $ λn → do
           comp_n ← findU n $ disj js
           u      ← (base js) `getA` comp_n
           m      ← (seen js) `getA` n
           when (m ∧ comp_v ≢ comp_n) $
             do unionU comp_n comp_v $ disj js
                add_edge u v $ tree js
                putA (base js) comp_n v
       return $ tree js
```

The function parameters are the number of vertices (for counting iterations), an array of vertices (sorted on the underlying field), and the neighbourhood. In place of the 'pure' *for_each*, the function uses a generic monadic iterator, *forM_*, which applies a monadic operation which folds the initial seed over the array of vertex indices. We replaced the default array indexing operators, which perform bounds checking, to 'unsafe' versions which do not, and reimplented the sorting operation as a one-buffer merge sort. The heap statistics in Fig. 2 highlight the dramatic improvement, with peak memory use now <600 K.

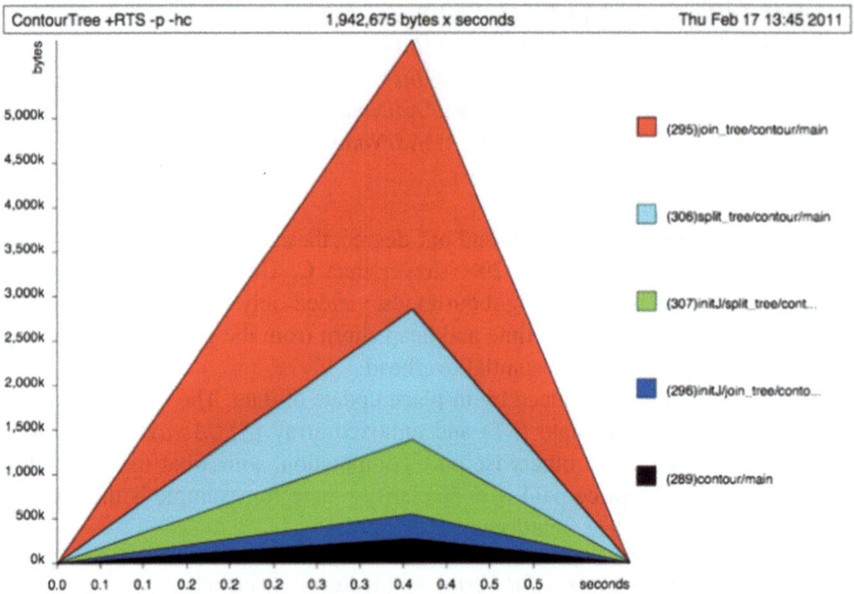

Fig. 2 Heap usage, per cost centre, of monadic program. Program runtime is $\leq$0.5 s; the horizontal axis shows a longer runtime as profiling code contributes non-neglible overhead

Table 1 Haskell and C++ performance (runtime, in seconds)

Dataset	Dimensions	Haskell	C++	Speedup (H/C)
Nucleon	$41 \times 41 \times 41$	0.36	0.10	3.6
Silicium	$98 \times 34 \times 34$	0.74	0.15	4.9
Neghip	$64 \times 64 \times 64$	1.42	0.34	4.2
Hydrogen	$128 \times 128 \times 128$	9.87	2.71	3.6
Lobster	$301 \times 324 \times 56$	40.80	9.60	4.3
Aneurism	$256 \times 256 \times 256$	74.90	19.92	3.8

Runtime: Applying the above changes reduces runtime on the 'nucleon' dataset to 0.36 s, slightly faster than the unoptimised C++ code. With optimisation turned on (-O3 instead of -g), the C++ code executed in 0.10 s. Table 1 shows test results using a MacBook 2.4 GHz Core 2 Duo with 4 GB of RAM, OS X 10.6.8, ghc version 7.4.1 and g++ version 4.2.1. With optimisation turned on, the C++ code was consistently three to four times faster than the Haskell code. The C++ code included modest low-level optimisations, such as block allocation of memory. While the Haskell is slower, it indicates that optimised Haskell can be brought within sight of C++ efficiency.

6 Step III: Towards *Parallel* Functional Contour Trees

Haskell provides two low-level primitives for parallelism. For expressions *a* and *b*,

- *a* '*par*' *b* marks *a* and *b* as candidates for sparking as separate threads;
- *a* '*pseq*' *b* forces evaluation of *a* to a normal form before evaluation of *b*.

As interaction between these operations and lazy evaluation can be subtle, higher-level tools for parallel computation, built on top of these operators, have been developed; see [13, 14]. As a preliminary experiment, we identified three opportunities to parallelize our code: (i) running the split and join tree computations in parallel; (ii) exploiting parallelism in early stages, sorting and initialisation; and (iii) performing parallel updates on the three merge-phase data structures. Implementing this using evaluation strategies [13], parallelization of our most efficient join/split computation is expressed as follows (*rpar* and *rseq* are monadic versions of the *par* and *pseq*) :

```
contour :: Ord a => Dataset a -> Graph
contour ds = runEval $ do ascending <- rpar $ ordered ds
                          descending <- rpar $ reverse ascending
                          jt <- rpar $ join_tree descending (mesh ds)
                          st <- rpar $ join_tree ascending (mesh ds)
                          ct <- rseq $ contour_tree jt st
                          return ct
```

Programs were compiled with the multi-threaded runtime system, and executed on a four-core linux machine. On the datasets Table 1 we experienced at best modest speedup (around 1.5) on two cores, with negligible improvement on further cores. Run-time profiles show that (a) the join and split computations are running effectively in parallel, but (b) the data structure updates are too fine-grained: iteration over leaf nodes during the merger phase forces evaluation of the 'seed' state and threads sparked to update the structures 'fizzle' as there is insufficient time to execute the thread before the runtime system forces evaluation of the entire structure. Further work on larger datasets and alternative parallel strategies are needed before we can draw conclusions on the utility of *parallel* functional topology.

7 Conclusions and Prospects

We have described development of a functional implementations of a contour tree algorithm, and observed the following:

- A pure description of the CT is pleasingly concise, and arguably close(r) to a specification of the algorithm than imperative code;
- However, its run time performance is prohibitive for anything but trivial problems;
- The monadic version, reached through incremental transformation of the code guided by profiling data reaches performance within sight of C++ code, and
- Optimising Haskell code is nearly as opaque as optimising C++ code.

Modest gains from tentative steps toward parallelization are entirely expected given Amdahl's law. Looking forward, the following comment by Backus may be germane: "The assignment statement is the von Neumann bottleneck of programming languages *and keeps us thinking in word-at-a-time terms in much the same way the computer's bottleneck does*." (our emphasis) [2]. Although our functional code, particularly the non-monadic versions, exploits higher order functions, the implementation shadows the published algorithm, and our own thinking has been consciously to follow the structure of an existing (sequential) algorithm. Further work is needed to understand how to exploit parallelism in a functional setting, and we see three roles that functional programming might usefully play in advancing computational topology:

- The pure version of the code is close to an executable specification, and Haskell facilitates early testing of components. In other work, we have found it very useful to have test cases generated by an implementation that is relatively slow but where we have some confidence in its correctness; see e.g. [7]. Increasing concern with verification and validation of visualization may make Haskell and its tool chain attractive.
- Functional languages are particularly well suited to hosting embedded DSLs (where the DSL is constructed as an expression in the host language) and deploying generative programming. The Diderot project [5], for example, is exploiting this to map high-level programs for tensor visualization onto multiple parallel architectures.
- We have examined a single algorithm for contour tree computation. Other algorithms may be more amenable to parallelism, and could exploit recent Haskell libraries for array processing (REPA [11]) and GPGPU (Accelerate [4]).

Acknowledgements Support for this work was provided by the UK Engineering and Physical Sciences Research Council under Grant EP/J013072/1.

References

1. L. Arge, M. Revsbaek, Robust on-line computation of reeb graphs: simplicity and speed. Algorithms Comput. **5878**, 1155–1165 (2009)
2. J. Backus, Can programming be liberated from the von Neumann style? A functional style and its algebra of programs. Commun. ACM **21**(8), 75–94 (1978)

3. H. Carr, J. Snoeyink, U. Axen, Computing contour trees in all dimensions. Comput. Geom. **24**(2), 75–94 (2003)
4. M. Chakravarty, G. Keller, S. Lee, T. McDonell, V. Grover, Accelerating haskell array codes with multicore GPUs, in *Proceedings of the Declarative Aspects of Multicore Programming*, Austin (ACM, 2011), pp. 3–14
5. C. Chiw, G. Kindlmann, J. Reppy, L. Samuels, N. Seltzer, Diderot: a parallel DSL for image analysis and visualization, in *Proceedings of the Conference on Programming Language Design and Implementation*, Beijing (ACM, 2012), pp. 111–120
6. Data.IntMap, www.haskell.org/ghc/docs/7.4-latest/html/libraries/containers-0.4.2.1/Data-IntMap.html. Last accessed Nov 2012
7. D. Duke, R. Borgo, C. Runciman, M. Wallace, Huge data but small programs: visualization design via multiple embedded DSLs, in *Proceedings of the Practical Applications of Declarative Languages*, Savannah, vol. 5418 (Springer, 2009), pp. 31–45
8. D. Duke, H. Carr, A. Knoll, N. Schunck, H. Nam, A. Staszczak, Visualizing nuclear scission through a multifield extension of topological analysis. Trans. Vis. Comput. Graph. **18**(12), 2033–2040 (2012)
9. D. Duke, M. Wallace, R. Borgo, C. Runciman, Fine-grained visualization pipelines and lazy functional languages. Trans. Vis. Comput. Graph. **12**(5), 973–980 (2006)
10. D. Günther, J. Reininghaus, H. Wagner, I. Hotz, Efficient computation of 3D morse-smale complexes and persistent homology using discrete morse theory. Vis. Comput. **28**(10), 959–969 (2012)
11. B. Lippmeier, M. Chakravarty, G. Keller, S. Peyton Jones, Guiding parallel array fusion with indexed types, in *Proceedings of the Haskell Symposium*, Copenhagen (ACM, 2012), pp. 25–36
12. S. Maadasamy, H. Doraiswamy, V. Natarajan, A hybrid parallel algorithm for computing and tracking level set topology, in *International Conference on High Performance Computing*, Pune (IEEE, 2012)
13. S. Marlow, P. Maier, H.W. Loidl, M. Aswad, P. Trinder, Seq no more: better strategies for parallel haskell, in *Proceedings of the Haskell Symposium*, Baltimore (ACM, 2010), pp. 91–102
14. S. Marlow, R. Newton, S. Peyton Jones, A monad for deterministic parallelism, in *Proceedings of the Haskell Symposium*, Tokyo (ACM, 2011), pp. 71–82
15. J. Meredith, S. Ahern, D. Pugmire, R. Sisneros, EAVL: the extreme-scale analysis and visualization library, in *Eurographics Symposium on Parallel Graphics and Visualization*, Cagliari, 2012
16. K. Moreland, U. Ayachit, B. Geveci, K.L. Ma, Dax toolkit: a proposed framework for data analysis and visualization at extreme scale, in *Symposium on Large Data Analysis and Visualization*, Providence (IEEE Computer Society, 2011), pp. 97–104
17. B. O'Sullivan, J. Goerzen, D. Stewart, *Real World Haskell* (O'Reilly Media, Sebastopol, 2008)
18. V. Pascucci, K. Cole-McLaughlin, Parallel computation of the topology of level sets. Algorithmica **38**(1), 249–268 (2004)
19. V. Pascucci, G. Scorzelli, P.T. Bremer, A. Mascarenhas, Robust on-line computation of reeb graphs: simplicity and speed. Trans. Graph. **26**(3), 58 (2007)
20. N. Shivashankar, M. Senthilnathan, V. Natarajan, Parallel computation of 2D morse-smale complexes. Trans. Vis. Comput. Graph. **18**(10), 1757–1770 (2012)
21. The Haskell Programming Language, www.haskell.org. Last accessed Nov 2012

Distributed Contour Trees

Dmitriy Morozov and Gunther H. Weber

Abstract Topological techniques provide robust tools for data analysis. They are used, for example, for feature extraction, for data de-noising, and for comparison of data sets. This chapter concerns contour trees, a topological descriptor that records the connectivity of the isosurfaces of scalar functions. These trees are fundamental to analysis and visualization of physical phenomena modeled by real-valued measurements.

We study the parallel analysis of contour trees. After describing a particular representation of a contour tree, called local–global representation, we illustrate how different problems that rely on contour trees can be solved in parallel with minimal communication.

1 Introduction

To make sense of the world around us, it is common in natural sciences to encode physical phenomena as scalar functions. Sometimes this is done directly, as when pressure measurements are recorded during physical experiments. Other times such functions are derived from the data, e.g., when the geometry of a shape is encoded in its distance function. It is rarely feasible to understand such functions directly: the data sets have become too large. Data analysis and visualization techniques, therefore, focus on extracting salient features that elucidate interesting behavior in the data. In this context, topological techniques are particularly attractive because they provide robust descriptors and help quantify the significance of detected patterns.

D. Morozov (✉) • G.H. Weber
Computational Research Division, Lawrence Berkeley National Laboratory,
One Cyclotron Road, Berkeley, CA 94720, USA
e-mail: dmitriy@mrzv.org; GHWeber@lbl.gov

P.-T. Bremer et al. (eds.), *Topological Methods in Data Analysis and Visualization III*, Mathematics and Visualization, DOI 10.1007/978-3-319-04099-8_6,

Broadly speaking, topological data analysis can be viewed as a two-step process. First, we compute a topological descriptor that summarizes the given data set. Then we use this descriptor to extract the relevant information. In this work we focus on a way to describe the connectivity of isosurfaces of a scalar function. Familiar in two dimensions as contours on a topographic map, isosurfaces consist of all points in the domain of a function that have the same value. Isosurface extraction [10, 13, 15] is a versatile technique in visualization and data analysis. One reason is that isosurfaces often have an immediate physical interpretation. For example, isosurfaces for certain charge densities in molecular simulations indicate boundaries of atoms or molecules; in combustion simulations, an isotherm (isosurface of temperature) can represent the location of the flame.

By varying the function value and computing the corresponding isosurfaces, it is possible to explore the behavior of the data. In this process, it is useful to know the values where interesting changes occur. A contour tree [2] is a standard tool to describe such changes. It is a graph whose nodes represent extrema and saddles where the number of connected components of the isosurface changes. Applications of the contour trees include speeding up isosurface extraction by identifying and extracting each one of their connected components by region growing [22]; manipulating individual connected components of "flexible isosurfaces": for example, hiding a connected component that encloses relevant portions of the isosurface [3]; and segmenting data for volume rendering [24]. Arge and Revsbaek [1] consider the problem of I/O-efficient contour tree simplification. Contour trees are a special case of Reeb graphs [18], which have been used, among many other applications, for shape matching [9], tracking burning regions in combustion simulations [23], and identifying pore structures in porous media [21].

In the above two-step view of topological data analysis, the first step is performed once—there is only one contour tree, Morse–Smale complex, persistence diagram, etc. associated with a scalar function—but the second step, such as the extraction of a specific isosurface, depends on additional parameters. Therefore, a user usually repeats it over and over again (often in an interactive setting) as she explores the structure hidden in the input.

As the size of the available data grows, it is natural for researchers to turn to larger, parallel computers to meet their analysis needs. Most of the work on parallelization of topological computation focuses on the first step of the above process, on using multiple processors to compute a particular topological descriptor. Cole-McLaughlin and Pascucci [16] study parallelization of merge tree computation. Gyulassy et al. [7] study parallel computation of Morse–Smale complexes. Shivashankar et al. [19,20] consider the same problem in shared memory (and on GPUs). Without focusing on the details of those procedures, we note that their outcome is always the same: a single, monolithic topological descriptor.[1]

[1]The algorithm of [7] is an exception. It computes many descriptors, Morse–Smale complexes of smaller portions of the domain. However, this information is not sufficient to resolve the Morse–Smale complex of the entire function. In particular, [7] ignores how one would use

Such a representation makes it difficult to take advantage of the multiple available processors during the analysis step.

We suggest taking a holistic view and develop techniques that focus not only on how long it takes to compute a descriptor, but also on how efficiently we can analyze it in parallel. In an earlier work [14], we introduced a *local–global representation* of a merge tree. Explained in detail in Sect. 3, it combines local information (the immediate responsibility of a given processor) with extra global information that specifies where the local data fits globally. In this chapter, we show that having such representations of two merge trees serves as a local–global representation of a contour tree. Specifically, it provides enough information to answer queries about level sets of a function on a simply connected domain.

2 Background

Scalar functions. The central object of our study is a continuous scalar function, $f : \mathbb{X} \to \mathbb{R}$. We assume its domain $\mathbb{X}$ is simply connected, meaning any loop inside it can be contracted to a point. For computational purposes, we restrict our view further. We assume that $\mathbb{X}$ is a simplicial complex, and function f is defined on its vertices and is linearly interpolated on the interiors of its simplices.

Given a scalar function $f : \mathbb{X} \to \mathbb{R}$, we say that two points $x, y \in \mathbb{X}$ are related, $x \sim y$, if they belong to the same component of the level set $f^{-1}(f(x)) = f^{-1}(f(y))$. A *Reeb graph* is a quotient space of $\mathbb{X}$ with respect to this relation, $\mathbb{X}/{\sim}$; in other words, we construct a Reeb graph by continuously contracting the level set components of f to points. Intuitively, a Reeb graph tracks connectivity of level sets of the function—how they merge and split—as we vary the level set threshold. When the domain of our function is simply connected, the Reeb graph is a tree, called a *contour tree* [2].

If instead of the level sets of a function, we examine its sublevel sets, i.e. the sets of the form $f^{-1}(-\infty, a]$, we get a merge tree. Specifically, we say that two points x and y are related, if they have the same function value, and they belong to the same component of the sublevel set $f^{-1}(-\infty, f(x)] = f^{-1}(-\infty, f(y)]$. The quotient space of $\mathbb{X}$ with respect to this relation is called a *merge tree* of the function. Intuitively, it tracks the evolution of sublevel sets of f as we vary the defining threshold parameter a. (Symmetrically,[2] we can consider the merge tree of $-f$, which tracks the evolution of the superlevel sets of f.) Below we use $\mathbb{X}_a = f^{-1}(-\infty, a]$ and $\mathbb{X}^a = f^{-1}[a, \infty)$ to denote the sublevel and superlevel sets of f, respectively.

---

such a representation for the actual analysis. In our terminology, these descriptors are the local representations.

[2]Sometimes authors make distinction between merge trees of super- and sub-level sets, calling the former join and the latter split trees. We prefer a unified terminology in this chapter.

The full merge tree of the function f consists of the following nodes. Its leaves represent the minima of the function; its root is the global maximum. Its internal nodes correspond to those saddles of the function where multiple sublevel set components merge. It is often convenient to implicitly add each of the remaining vertices to the branch of the merge tree that represents the component of the sublevel set where the vertex first appears. Such additional vertices always have degree two in the merge tree.

Carr et al. [2] give a linear-time algorithm that combines merge trees of functions f and $-f$ into a contour tree of function f.

Parallel setup. We assume that we have a collection $\mathcal{U} = \{U_i\}$ of sets that cover the domain of our function, $\mathbb{X} = \cup \mathcal{U}$. We assume that $|\mathcal{U}| = p$, and we are given p processors, each responsible for a single set in $\mathcal{U}$. Specifically, we use the same indexing for processors as for the cover sets and say that processor P_i is responsible for the set U_i. We note that this formulation fits especially well with the so-called in situ analysis, where an external code decides how the input data is split among the different processors. In this case, the initial set U_i is the union of the in situ blocks assigned to processor P_i; we make no assumptions about the topology of U_i. The processors communicate by message passing.

3 Local–Global Representation

Let $T_{\mathbb{X}}$ denote the full merge tree of the function f. In an earlier work [14], we introduced a local–global representation of a merge tree. Its key idea is the following. In a distributed setting, where the domain of the function is split among many processors, in addition to the information about the connectivity of the sublevel sets restricted to the local portion of the domain, one may record how these sublevel sets fit into the domain globally. Such global information creates only minor overhead; it can be computed efficiently in parallel; and, most importantly, it minimizes the need for communication during analysis. In this section, we review this representation.

Local–global merge tree. A local–global representation of the full tree $T_{\mathbb{X}}$ is defined with respect to a subset $U \subseteq \mathbb{X}$; we denote it by $T_{\mathbb{X}}(U)$. For each vertex $x \in U$, we record all the sublevel set components that contain it. We represent each component by the minimum of the function inside it. At first, x belongs to the component of m_0 in $\mathbb{X}_{f(x)}$. As we increase the threshold of the function, this component grows until eventually, at s_1, it merges into the component of m_1, which, in turn, merges into the component of m_2 at s_2, and so on. We get a sequence of minima and saddles, $m_0, s_1, m_1, s_2, m_2, \ldots, m_n$, with

$$m_n < m_{n-1} < \ldots < m_0 \leqslant x \leqslant s_1 < \ldots < s_n.$$

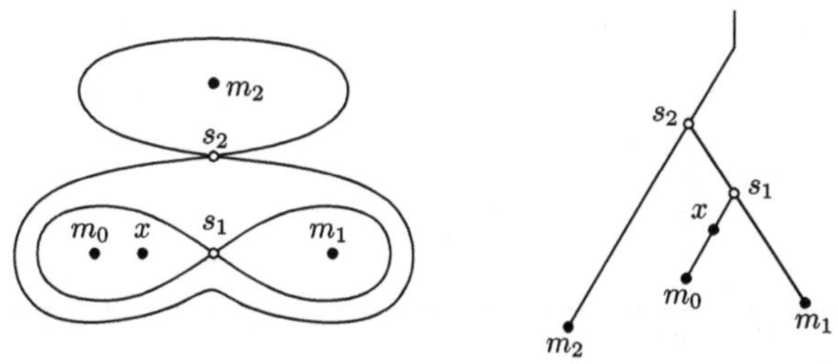

Fig. 1 Local–global representation of a vertex x. On the *left*, the lines represent the contours of the relevant saddles. On the *right*, the local–global subtree induced by x

Setting $s_0 = x$ for convenience, in the sublevel sets $\mathbb{X}_a$, for $a \in [s_i, s_{i+1})$, x belongs to the component with the lowest minimum m_i; see Fig. 1. It is important to note that the minima m_i and the saddles s_i do not necessarily belong to U—hence, the "global" part of the representation. A reader familiar with the theory of persistent homology [6] will notice that the pairs (m_i, s_{i+1}) belong to the 0–dimensional persistence diagram of function f.

Naturally, all the relevant information appears in the merge tree $T_{\mathbb{X}}$, which, after all, records all the components of all the sublevel sets of the function. We can think of the local–global representation of a vertex as a subtree of $T_{\mathbb{X}}$ (to be precise, a subdivision of this representation is a subtree of $T_{\mathbb{X}}$). Accordingly, we can represent the local–global representations of multiple vertices compactly as a subtree of $T_{\mathbb{X}}$, thus avoiding duplication of the minima whose components contain many of the local vertices. By definition, the local–global representation of $T_{\mathbb{X}}$ with respect to U is the tree formed as the union of (subdivided) representations of all the vertices in U.

Construction. It is easy to extract a local–global representation from the full merge tree $T_{\mathbb{X}}$. To do so, we perform a post-order traversal that identifies the branches of the tree that contain the vertices in U as well as the branches (with deeper minima) that they merge into.

An important contribution of [14] is an algorithm that computes the local–global representation in parallel without assembling the entire tree $T_{\mathbb{X}}$ first. In $\log p$ iterations, the processors can interleave merging and sparsification steps to compute $T_{\mathbb{X}}(U_i)$ with respect to their local subsets of the domain. Another significant argument in favor of this representation is its size. In all our experiments, the local–global tree $T_{\mathbb{X}}(U)$ is only slightly larger than the merge tree T_U of $f_{|U}$, the function restricted to U. In other words, most of the global merging data is the same for the vertices in the local domain and, thus, creates only minor overhead; see Table 1.

Analysis routine. To understand why the local–global representation is useful, consider the following problem. Given a threshold t, we would like to find the volume of the component of the sublevel set $\mathbb{X}_t$ that contains a given point x. This component may be distributed across many processors, all but one of which

Table 1 Maximum local and local–global tree sizes on any processor. In all cases, we count only the critical nodes in the trees. The time is listed as the time to compute the local tree plus the extra time to compute the local-global tree using our parallel algorithm [14]. Data from http://volvis.org: prone16 is a CT scan of an abdomen; vertebra16 is a rotational angiography scan of a head with an aneurysm; backpack16 is a CT scan of a backpack

		Processors			
Dataset	Measure	32	64	128	256
Prone16	Local (nodes)	272,241	143,659	75,335	41,418
$512 \times 512 \times 463$	Local–global (nodes)	289,735	154,018	82,598	45,778
	Time (s)	33.4 + 17.2	15 + 13	6.6 + 9.6	2.6 + 7.8
Vertebra16	Local (nodes)	69,693	38,236	19,708	10,277
$512 \times 512 \times 512$	Local–global (nodes)	77,939	43,565	23,684	12,901
	Time (s)	40 + 6.3	15.6 + 5.1	7.4 + 3.8	2.9 + 3.4
Backpack16	Local (nodes)	203,792	102,019	67,698	39,951
$512 \times 512 \times 373$	Local–global (nodes)	211,324	121,723	72,391	43,147
	Time (s)	23.6 + 10.5	10.8 + 8.3	5.2 + 6.1	2.1 + 6.3

know nothing about x. However, very little communication is actually required. The processor P_i responsible for x, i.e., $x \in U_i$, identifies the minimum m in the component of $\mathbb{X}_t$ that contains x. It does this by first marching up from x towards the root of the tree $T_{\mathbb{X}}(U_i)$ until it finds a saddle s with $f(s) \leqslant t$, but $f(s') > t$ for its parent s'. P_i then finds the lowest minimum m in the subtree of s. (We will re-use this operation multiple times in the following section, so from now on we call it COMPONENT$(x, t, T_{\mathbb{X}}(U_i))$.)

Processor P_i broadcasts m (and t) to the rest of the processors. Each processor P_j finds all the vertices in U_j that fall into the component of $\mathbb{X}_t$ that contains m. To do so, it traverses up from m until it identifies the last saddle s before the traversal crosses the threshold t. (Naturally, if m is not in $T_{\mathbb{X}}(U_j)$, P_j has nothing to do.) The processor then counts the number of the local vertices in the subtree rooted at s. The processors combine their counts via a standard reduction.

The entire process has only two communication steps: the initial broadcast and the final reduction; the rest of the work is performed by the processors independently.

4 Contour Tree

Carr et al. [2] give an algorithm to combine merge trees of f and $-f$ into a contour tree of f. However, since we are not trying to compute the full contour tree anyway, we do not need to combine the trees. Instead, we use the algorithm of [14] on each processor P_i to compute the local–global merge trees $T^+_{\mathbb{X}}(U_i)$ and $T^-_{\mathbb{X}}(U_i)$ of the functions f and $-f$ with respect to the cover set U_i.

Chiang and Lu [4] use an implicit representation of a contour tree as two merge trees. It is, however, unclear how to use their scheme in our (distributed) setting,

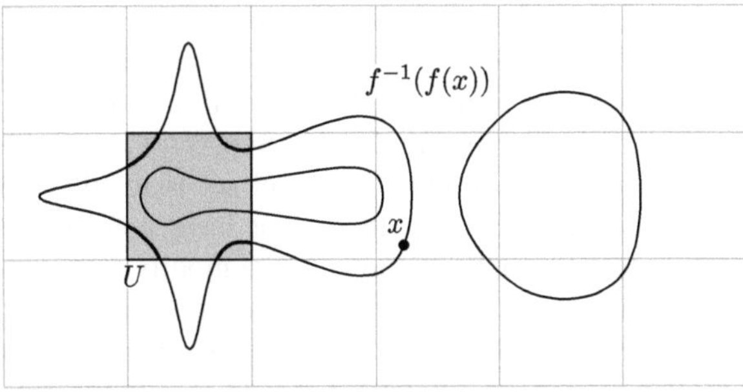

Fig. 2 The level set $f^{-1}(f(x))$ consists of three components. *Bold line* highlights the intersection of the component that contains point x with the cover set U. Of the five components of the level set inside U only four belong to this intersection

since they label edges of the contour tree by the edges of the merge trees. To do so in a distributed setting, one would have to assemble the entire merge tree on a single processor, an expensive operation we carefully avoid. Below, instead, we use extrema as level set labels.

In this section, we describe three problems efficiently solved using contour trees; one can interpret them as operations on "flexible isosurfaces" [3]. We explain how to solve these problems in the distributed setting using the local–global representation. In all cases, our emphasis is on minimizing the communication between processors.

4.1 Levelset Component

Imagine a user interactively exploring a data set. She picks a point x and wants to see the component of the level set $f^{-1}(f(x))$ that contains x. Imagine that the data set is so large that it does not fit in the memory of a single processor; therefore, it is distributed across multiple compute nodes.

Each processor must find the intersection of its local domain with the level set component that contains the given point x. As Fig. 2 illustrates, this intersection need not be connected. We solve this problem in three steps:

1. First, we identify the component of the level set that contains point x. Suppose that $x \in U_i$. The processor P_i identifies the component of the sublevel set $\mathbb{X}_{f(x)}$ and the component of the superlevel set $\mathbb{X}^{f(x)}$ that contain x. As before, it identifies the component by its minimum (or, symmetrically, maximum), which it finds by the traversals of the subtrees rooted at x, $\text{COMPONENT}(x, f(x), T_{\mathbb{X}}^{+}(U_i))$ and $\text{COMPONENT}(x, f(x), T_{\mathbb{X}}^{-}(U_i))$. We denote these extrema by $\min_x$ and $\max_x$. Processor P_i broadcasts this pair to the rest of the processors.
2. Each processor P_j records all the local points in the subtree of $T_{\mathbb{X}}^{+}(U_j)$ at level $f(x)$ that contains $\min_x$. We denote these points by $\text{Sub}_x(U_j)$; they are exactly

the vertices of the intersection of the set U_j with the sublevel set component that contains x. Similarly, P_j records all the points in the subtree of $T^-_{\mathbb{X}}(U_j)$ at level $f(x)$ that contains $\max_x$ as $\mathrm{Sup}_x(U_j)$. Naturally, if $\min_x$ or $\max_x$ do not belong to their respective trees, the sets $\mathrm{Sub}_x(U_j)$ or $\mathrm{Sup}_x(U_j)$ are empty.

3. A simple (brute-force) way to extract a level set is to filter all the maximal simplices of the domain detecting those that have both a vertex below the prescribed threshold $f(x)$ and one above it. We modify this procedure and find all those maximal simplices in U_j that contain a vertex in $\mathrm{Sub}_x(U_j)$ and a vertex in $\mathrm{Sup}_x(U_j)$. Every maximal simplex that has such a pair of vertices is not only intersected by the level set $f^{-1}(f(x))$, but it intersects the component of this level set that contains x.

We note that the only required communication is the broadcast of the identification of the queried component, namely, the pair $(\min_x, \max_x)$. The rest of the procedure identifying the local contribution to a level set is carried out completely independently. Naturally, an extra fourth step is necessary to somehow collect the data, but its details depend on our specific goal. Computing total volume of the level set component requires a single reduction; compositing such a distributed level set for rendering is a well-studied topic in visualization [11, 12, 17].

Correctness. Why does the above procedure find on each processor P_j the intersection of U_j with the component of the level set $f^{-1}(f(x))$ that contains x? The following theorem implies the answer.

Theorem 1. *If $\mathbb{X}$ is simply connected, then a component of the sublevel set $\mathbb{X}_b$ can intersect a component of the superlevel set $\mathbb{X}^a$ in at most one component of the interlevel set $f^{-1}[a,b]$.*

Proof. Informally, the statement and the proof of the theorem are simple: if a component of $\mathbb{X}_b$ intersected a component of $\mathbb{X}^a$ in two components of $f^{-1}[a,b]$, then we could take two paths from the first component to the second. The first path would lie entirely in $\mathbb{X}_b$, while the second path would lie entirely in $\mathbb{X}^a$. Composed together these paths would form a non-contractible loop in $\mathbb{X}$, contradicting the assumption that $\mathbb{X}$ is simply connected.

To make this proof formal, we turn to the Mayer–Vietoris long exact sequence, a standard tool in algebraic topology. To proceed, we need the notion of homology, which we have not defined. Fortunately, we only need its very basic form. Using coefficients in a field, the 0–th homology group of a space Y, denoted by $\mathsf{H}_0(Y)$, is a vector space spanned by the components of Y. Similarly, the first homology group, $\mathsf{H}_1(Y)$, keeps track of 1–dimensional cycles. If $\mathbb{X}$ is simply connected, its first homology group is 0, $\mathsf{H}_1(\mathbb{X}) = 0$.

The necessary portion of the Mayer–Vietoris sequence has the following form:

$$\ldots \to \mathsf{H}_1(\mathbb{X}) \xrightarrow{\partial^*} \mathsf{H}_0(f^{-1}[a,b]) \xrightarrow{(i^*,j^*)} \mathsf{H}_0(\mathbb{X}_b) \oplus \mathsf{H}_0(\mathbb{X}^a) \to \ldots$$

The linear map ∂^* is induced by the intersection of 1–dimensional cycles in $\mathbb{X}$ with the interlevel set $f^{-1}[a,b]$; the maps i^* and j^* are induced by the inclusions of the interlevel set into the respective sub- and super-level sets.

The Mayer–Vietoris sequence is exact, which, by definition, means that the image of the map ∂^* is equal to the kernel of the map (i^*, j^*). Since $\mathsf{H}_1(\mathbb{X}) = 0$, the image of ∂^* is also 0. Therefore, the kernel of (i^*, j^*) is 0, meaning that the inclusion of components of $f^{-1}[a,b]$ into the components of $\mathbb{X}_b$ and $\mathbb{X}^a$ is an injection. Were a component of $\mathbb{X}_b$ to intersect a component of $\mathbb{X}^a$ in two components, the inclusion of these components back into $\mathbb{X}_b$ and $\mathbb{X}^a$ would not be injective, i.e., we would get a contradiction. □

Let Lvl_x denote the component of the level set $f^{-1}(f(x))$ that contains x. We denote its intersection with U_j by $\mathrm{Lvl}_x(U_j)$. The following corollary implies the correctness of our three-step algorithm. Recall that we assume that f is linearly interpolated on the interiors of the simplices.

Corollary 1. *A simplex $\sigma \in U_j$ has a vertex u in $\mathrm{Sub}_x(U_j)$ and a vertex v in $\mathrm{Sup}_x(U_j)$ if and only if the point y on the edge (u, v) with $f(y) = f(x)$ belongs to $\mathrm{Lvl}_x(U_j)$.*

Proof. Let $a = b = f(x)$. Suppose $u \in \sigma$ belongs to the component of x in $\mathbb{X}_b$, and $v \in \sigma$ belongs to the component of x in $\mathbb{X}^a$. Let y be the point on the edge (u, v) with $f(y) = f(x)$ (such a point exists because the function is continuous; it is unique because the function is linearly interpolated). Point y also belongs to the components of x in $\mathbb{X}_b$ and in $\mathbb{X}^a$. Since these two components can intersect in at most one component of $f^{-1}(f(x))$, y belongs to the same component of the level set as x.

Conversely, if point y on the edge (u, v), with $f(y) = f(x)$ and $f(u) < f(v)$, belongs to $\mathrm{Lvl}_x(U_j)$, then y belongs to the boundaries of the components of $\mathbb{X}^a$ and $\mathbb{X}_b$ that contain x. Since the function is linear on the edge (u, v), vertex u belongs to the component of $\mathbb{X}_b$ that contains x and, therefore, to $\mathrm{Sub}_x(U_j)$. Similarly, vertex v belongs to the component of $\mathbb{X}^a$ that contains x and, therefore, to $\mathrm{Sup}_x(U_j)$. □

Remark 1. We note that one cannot uniquely identify a branch of the contour tree of f by a minimum–maximum pair in the respective merge trees $T^+_{\mathbb{X}}$ and $T^-_{\mathbb{X}}$. However, what Theorem 1 and Corollary 1 imply is that, at a fixed level a, such a pair uniquely identifies a point on the contour tree.

Component labeling. A variation on the problem of finding a single component is the consistent assignment of labels to all components of a level set. For example, the user may want to decorate each component of a level set $f^{-1}(a)$ with a unique color.

To solve this problem, each processor P_i extracts all the different components of the level set $f^{-1}(a)$ that intersect its local domain U_i. It identifies each component as the intersection of components in the sub- and the super-level sets by finding the minimum in $\mathbb{X}_a$ and the maximum in $\mathbb{X}^a$ that identify each component. As noted before (and as Fig. 2 illustrates), locally disconnected components may get the same identification.

There are multiple ways to assign consistent labels to the components. Perhaps the simplest approach—the one requiring no communication—is for each processor to simply hash its minimum–maximum pairs. The components get consistent values across all the processors (since the extrema pairs are global); with high probability, all the assigned values are unique.

4.2 *Interlevel Set*

Imagine that we are interested in a branch of the contour tree, or, more generally, a monotone path in the contour tree between two points x and y (we assume that such a path exists). Often such paths correspond to interesting features in the data [24].

Theorem 1 suggests that each point on this path is uniquely identified by a minimum–maximum pair as well as the function value. Assume $f(x) < f(y)$. Processor P_i responsible for point x, i.e., $x \in U_i$, finds the local–global representation of x in the tree $T^+_{\mathbb{X}}(U_i)$. In other words, it finds all the minima m^x_i and saddles s^x_i that describe the components of the sublevel sets of the function that contain point x. Similarly, processor P_j responsible for point y finds its local–global representation in the tree $T^-_{\mathbb{X}}(U_j)$, the maxima m^y_i and the saddles s^y_i. The two processors broadcast these sequences of critical points.

The two representations together uniquely identify each point on the path from x to y in the contour tree. Specifically, let z be a point on this path. Let $a = f(z)$ and suppose that $a \in [s^x_i, s^y_j]$. The component of z in the level set $f^{-1}(a)$ is the unique intersection of the component of $\mathbb{X}_a$ that contains the minimum m^x_i and of the component of $\mathbb{X}^a$ that contains the maximum m^y_j.

Having received the broadcasts from P_i and P_j, each processor identifies its local contribution to the path x–y. As before, all the processors compute independently, except for the initial broadcast and the final collective operation (for example, to compute the total volume of the feature).

4.3 *Contour Tracking*

Consider another problem. The user has extracted a level set $f^{-1}(s)$ for some threshold s. The software has identified its different components and highlighted them with distinct colors. Now the user would like to vary the threshold a little and extract a nearby level set $f^{-1}(t)$, with $t > s$. When visualizing it, we would like to preserve the colors of the different components as much as possible, to maintain consistency with the level set $f^{-1}(s)$.

To solve this problem, we want to match the components of the two level sets. Specifically, we want to find which components of the lower and of the higher level

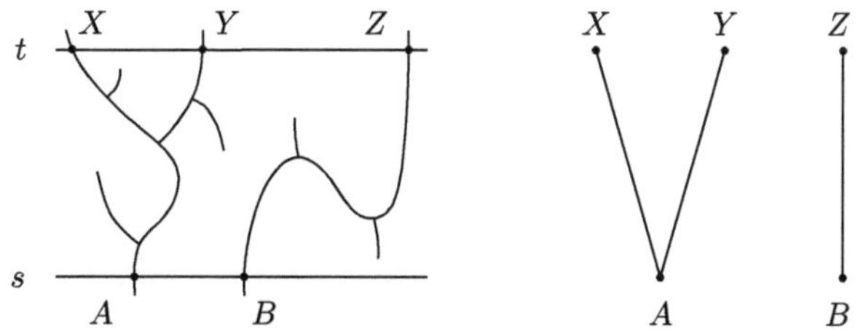

Fig. 3 An interlevel set of a contour tree (*left*). The graph of paired components (*right*)

sets map into the same components of the interlevel set $f^{-1}[s,t]$. Put another way, we want to find the components of $f^{-1}(s)$ and $f^{-1}(t)$ connected in the contour tree restricted to the interlevel set $f^{-1}[s,t]$; see Fig. 3.

Each component x of the level set $f^{-1}(s)$ is identified by a minimum–maximum pair $(\min_x, \max_x)$. Similarly, each component y of the level set $f^{-1}(t)$ is identified by the pair $(\min_y, \max_y)$. Recall that Theorem 1 tells us that if a component of $\mathbb{X}_t$ intersects a component of $\mathbb{X}^s$, then it does so in at most a single component of $f^{-1}[s,t]$. Therefore, to check if the component x and the component y belong to the same component of $f^{-1}[s,t]$, it suffices to check if x belongs to the component of $\mathbb{X}_t$ identified by $\min_y$ and if y belongs to the component of $\mathbb{X}^s$ identified by $\max_x$. To be precise, we test the following equalities:

$$\begin{aligned} \min_y &= \text{COMPONENT}(x, t, T^+_{\mathbb{X}}(U_*)); \\ \max_x &= \text{COMPONENT}(y, s, T^-_{\mathbb{X}}(U_*)). \end{aligned}$$

If both equalities are true, we know that the component of x in the superlevel set $\mathbb{X}^s$ intersects the component of y in the sublevel set $\mathbb{X}_t$. Each processor can find all such intersections locally. The result is a bipartite graph on the components of the two level sets, which is constructed without any communication. As in Fig. 3, this graph may not be a matching, so an auxiliary rule is necessary to break the ties.

5 Experiments

To experiment with the local–global representation, we implemented the algorithm of Sect. 4.1. Given a local–global representation of merge trees $T^-_{\mathbb{X}}$ and $T^+_{\mathbb{X}}$, the processor that contains a given point x locates the minimum–maximum pair that identifies the connected component of x in its level set and broadcasts the pair to the rest of the processors. Each processor finds the intersections of the components of the sublevel and superlevel sets that contain point x with its local domain. If neither one of these intersections is empty, the processor iterates over every tetrahedron of the Freudenthal triangulation of its local domain U_i. For each tetrahedron, we check whether it contains a point in the sublevel set component and another point in the superlevel set component. If it does, we find and record its intersection with the level set $f^{-1}(f(x))$.

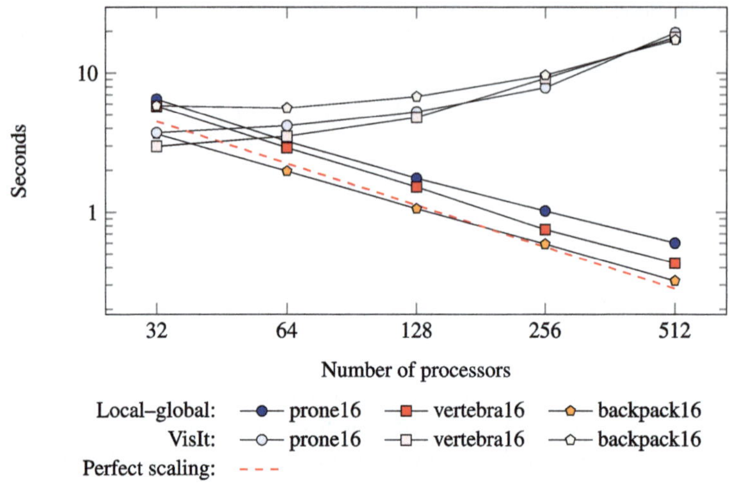

Fig. 4 Times to extract a level set component that contains a prescribed point using the local–global representation, and the times to label all the components of a level set in VisIt (Local–global representation times are taken as the average of ten runs each; VisIt times show the best of ten runs)

As a reference for the running times, we extract the same level sets and label their connected components using VisIt [5], a state-of-the-art visualization tool. VisIt uses the algorithm described by Harrison et al. [8] for the parallel connected component labeling. Although the two procedures are seemingly different—extracting a component of the level set that contains a given point versus labeling all the components of the level set—the comparison is not absurd. In VisIt, to extract a prescribed component, one must first label all the components and then filter out all but one of them. In other words, we measure a lower bound for the running time of this operation. At the same time, using local–global representation, labeling the components requires no communication, as Sect. 4.1 explains. Accordingly, specific component extraction is a more involved (and, therefore, interesting) procedure.

Figure 4 shows the times it takes to perform these procedures for the same data sets as Table 1. The clear trend is the steady decline of the running times, for the local–global representation, as we increase the number of processors.

6 Conclusion

We have presented the idea of local–global representations of contour trees and explained how it can be used for fast parallel analysis of the level sets of a function on a simply connected domain. These representations are small, only slightly larger than the merge trees of the local domain. Our earlier work [14] presents an efficient parallel algorithm to compute them.

Local–global representations scale down well as we increase the number of processors and, thus, really stand out when it comes to analysis. As our Sect. 4 explains, because they incorporate all the necessary global information, these compact representations let us perform variety of tasks with minimal communication.

The most logical directions for future work are extending our construction to the more general case of Reeb graphs and devising a seed point scheme, similar to the work of van Kreveld et al. [22], compatible with the local–global representation. For the latter, there is a natural map from the local to the local–global merge tree. By storing both trees and this map, we believe it is possible to improve the parallel algorithms for level set component extraction.

Acknowledgements This work was supported by the Director, Office of Advanced Scientific Computing Research, Office of Science, of the U.S. DOE under Contract No. DE-AC02-05CH11231 (Lawrence Berkeley National Laboratory) through the grant "Topology-based Visualization and Analysis of High-dimensional Data and Time-varying Data at the Extreme Scale," program manager Lucy Nowell.

References

1. L. Arge, M. Revsbaek, I/O-efficient contour tree simplification, in *Proceedings of the International Symposium on Algorithms and Computation, Honolulu*. LNCS 5878 (Springer, Berlin/Heidelberg, 2009), pp. 1155–1165
2. H. Carr, J. Snoeyink, U. Axen, Computing contour trees in all dimensions. Comput. Geom. Theory Appl. **24**(2), 75–94 (2003)
3. H. Carr, J. Snoeyink, M. van de Panne, Flexible isosurfaces: simplifying and displaying scalar topology using the contour tree. Comput. Geom. Theory Appl. **43**(1), 42–58 (2010)
4. Y.-J. Chiang, X. Lu, Progressive simplification of tetrahedral meshes preserving all isosurface topologies. Comput. Graph. Forum **22**(3), 493–504 (2003)
5. H. Childs, E. Brugger, B. Whitlock, J. Meredith, S. Ahern, D. Pugmire, K. Biagas, M. Miller, C. Harrison, G.H. Weber, H. Krishnan, T. Fogal, A. Sanderson, C. Garth, E.W. Bethel, D. Camp, O. Rübel, M. Durant, J.M. Favre, P. Navrátil, VisIt: an end-user tool for visualizing and analyzing very large data, in *High Performance Visualization—Enabling Extreme-Scale Scientific Insight* (CRC, Hoboken, 2012), pp. 357–372
6. H. Edelsbrunner, J. Harer, *Persistent Homology—A survey*. Volume 453 of Contemporary Mathematics (AMS, Providence, 2008), pp. 257–282
7. A. Gyulassy, V. Pascucci, T. Peterka, R. Ross, The parallel computation of Morse–Smale complexes, in *IEEE IPDPS*, Shanghai, 2012, pp. 484–495
8. C. Harrison, H. Childs, K.P. Gaither, Data-parallel mesh connected components labeling and analysis, in *Proceedings of the 11th EG PGV*, Switzerland, 2011, pp. 131–140
9. M. Hilaga, Y. Shinagawa, T. Kohmura, T.L. Kunii, Topology matching for fully automatic similarity estimation of 3D shapes, in *Proceedings of the 28th Annual Conference on Computer Graphics and Interactive Techniques, SIGGRAPH '01*, Los Angeles, 2001, pp. 203–212
10. W. E. Lorensen, H.E. Cline, Marching cubes: a high resolution 3D surface construction algorithm. Comput. Graph. **21**(4), 163–169 (1987)
11. K.-L. Ma, J.S. Painter, C.D. Hansen, M.F. Krogh, Parallel volume rendering using binary-swap compositing. IEEE Comput. Graph. Appl. **14**(4), 59–68 (1994)
12. S. Molnar, M. Cox, D. Ellsworth, H. Fuchs, A sorting classification of parallel rendering. IEEE Comput. Graph. Appl. **14**(4), 23–32 (1994)

13. C. Montani, R. Scateni, R. Scopigno, A modified look-up table for implicit disambiguation of marching cubes. Vis. Comput. **10**(6), 353–355 (1994)
14. D. Morozov, G.H. Weber, Distributed merge trees, in *Proceedings of the ACM Symposium Principles and Practice of Parallel Programming*, Shenzhen, 2013, pp. 93–102
15. G. Nielson, On marching cubes. IEEE Trans. Vis. Comput. Graph. **9**(3), 341–351 (2003)
16. V. Pascucci, K. Cole-McLaughlin, Parallel computation of the topology of level sets. Algorithmica **38**(1), 249–268 (2003)
17. T. Peterka, D. Goodell, R. Ross, H.-W. Shen, R. Thakur, A configurable algorithm for parallel image-compositing applications, in *Proceedings of the SC*, Portland, 2009, pp. 4:1–4:10
18. G. Reeb, Sur les points singuliers d'une forme de pfaff complètement intégrable ou d'une fonction numérique. CR Acad. Sci. **222**, 847–849 (1946)
19. N. Shivashankar, V. Natarajan, Parallel computation of 3D Morse–Smale complexes. Comput. Graph. Forum **31**, 965–974 (2012)
20. N. Shivashankar, M. Senthilnathan, V. Natarajan, Parallel computation of 2D Morse–Smale complexes. IEEE Trans. Vis. Comput. Graph. **18**(10), 1757–1770 (2012)
21. D.M. Ushizima, D. Morozov, G.H. Weber, A.G. Bianchi, J.A. Sethian, E.W. Bethel, Augmented topological descriptors of pore networks for material science. IEEE Trans. Vis. Comput. Graph. **18**(12), 2041–2050 (2012)
22. M. van Kreveld, R. van Oostrum, C. Bajaj, V. Pascucci, D. Schikore, Contour trees and small seed sets for isosurface traversal, in *Proceedings of the Annual Symposium Computational Geometry*, New York, 1997, pp. 212–220
23. G.H. Weber, P.-T. Bremer, M.S. Day, J.B. Bell, V. Pascucci, Feature tracking using reeb graphs, in *Topological Methods in Data Analysis and Visualization: Theory, Algorithms, and Applications* (Springer, Berlin/Heidelberg, 2011) pp. 241–253
24. G.H. Weber, S.E. Dillard, H. Carr, V. Pascucci, B. Hamann, Topology-controlled volume rendering. IEEE Trans. Vis. Comput. Graph. **13**(2), 330–341 (2007)

Clear and Compress: Computing Persistent Homology in Chunks

Ulrich Bauer, Michael Kerber, and Jan Reininghaus

Abstract We present a parallel algorithm for computing the persistent homology of a filtered chain complex. Our approach differs from the commonly used reduction algorithm by first computing persistence pairs within local chunks, then simplifying the unpaired columns, and finally applying standard reduction on the simplified matrix. The approach generalizes a technique by Günther et al., which uses discrete Morse Theory to compute persistence; we derive the same worst-case complexity bound in a more general context. The algorithm employs several practical optimization techniques, which are of independent interest. Our sequential implementation of the algorithm is competitive with state-of-the-art methods, and we further improve the performance through parallel computation.

1 Introduction

Persistent homology has developed from a theoretical idea to an entire research area within the field of computational topology. One of its core features is its multi-scale approach to analyzing and quantifying topological features in data. Recent examples of application areas are shape classification [5], topological denoising [3], or developmental biology [10].

Another major feature of persistent homology is the existence of a simple yet efficient computation method: The standard reduction algorithm as described in [11, 20] computes the persistence pairs by a simple sequence of column operations;

U. Bauer (✉) • J. Reininghaus
Institute of Science and Technology Austria, Klosterneuburg, Austria
e-mail: mail@ulrich-bauer.org; jan.reininghaus@ist.ac.at

M. Kerber
Stanford University, Stanford, CA, USA

Max-Planck-Institut für Informatik, Saarbrücken, Germany
e-mail: mkerber@mpi-inf.mpg.de

P.-T. Bremer et al. (eds.), *Topological Methods in Data Analysis and Visualization III*, Mathematics and Visualization, DOI 10.1007/978-3-319-04099-8_7,

Algorithm 1 gives a complete description in just 10 lines of pseudo-code. The worst-case complexity is cubic in the input size of the complex, but the practical behavior has been observed to be closer to linear on average. Various variants have been proposed in order to improve the theoretical bounds [6, 17] or the practical behavior [8].

Our first contribution consists of two simple optimization techniques of the standard reduction algorithm, called *clearing* and *compression*. Both approaches exploit the special structure of a filtered chain complex in order to significantly reduce the number of operations on real-world instances. However, the two methods cannot be easily combined because they require the columns of the boundary matrix to be processed in different orders.

Our second contribution is a novel algorithm that incorporates both of the above optimization techniques, and is also suitable for parallelization. It proceeds in three steps: In the first step, the columns of the matrix are partitioned into consecutive *chunks*. Each chunk is reduced independently, applying the clearing optimization mentioned above. In this step, the algorithm finds at least all persistence pairs with (index) persistence less than the size of the smallest chunk; let g be the number of columns not paired within this first step. In the second step, the g unpaired columns are compressed using the method mentioned above. After compression, each column has at most g non-zero entries and the unpaired columns form a nested $(g \times g)$-matrix. In the third and final step, this nested matrix is reduced, again applying the clearing optimization.

The chunk algorithm is closely related to two other methods for computing persistence. The first is the *spectral sequence algorithm* [9, Sect. VII.4], which decomposes the matrix into blocks and proceeds in several phases, computing in phase r the persistence pairs spanning r consecutive blocks. The first step of the chunk algorithm consists in applying the first two phases of the spectral sequence algorithm. Note Second, the three step approach of the chunk algorithm is inspired by an algorithm of Günther et al. [13], which uses discrete Morse theory for persistence computation of 3D image data. The first step of that algorithm consists in constructing a discrete gradient field consistent with the input function; such a gradient field can be interpreted [1] as a set of persistence pairs that are incident in the complex and have persistence 0. We replace this method by local persistence computations, allowing us to find pairs with are not incident in the complex.

We analyze the chunk algorithm in terms of time complexity. Let n be the number of generators (simplices, cells) of the chain complex, m the number of chunks, ℓ the maximal size of a chunk, and g as above. We obtain a worst-case bound of

$$O(m\ell^3 + g\ell n + g^3),$$

where the three terms reflect the worst-case running times of the three steps. For the filtration of a cubical complex induced by a d-dimensional grayscale image (with d some fixed constant), this bound simplifies to

$$O(gn + g^3),$$

assuming that the chunks are given by the cells appearing simultaneously in the filtration. This bound improves on the previous general bound of $O(g^2 n \log n)$ from [7]. Moreover, it matches the bound in [13], but applies to arbitrary dimensions. Of course, the bound is still cubic if expressed only in terms of n, since $g \in \Omega(n)$ in the worst case.

We implemented a sequential and a parallel version of the chunk algorithm. The sequential code already outperforms the standard reduction algorithm and is competitive to other variants with a good practical behavior. The parallel version using 12 cores yields a speed-up factor between 3 and 11 (depending on the example) in our tests, making the implementation the fastest among the considered choices. This is the first result where the usefulness of parallelization is shown for the problem of persistence computation through practical experiments.

2 Background

This section summarizes the theoretical foundations of persistent homology as needed in this work. We limit our scope to simplicial homology over $\mathbb{Z}_2$ just for the sake of simplicity in the description; our methods generalize to chain complexes over arbitrary fields.

2.1 Homology

Homology is an algebraic tool for analyzing the connectivity of topological spaces. Let K be a simplicial complex of dimension d. In any dimension p, we call a *p-chain* a formal sum of the p-simplices of K with $\mathbb{Z}_2$ coefficients. The p-chains form a group called the *pth chain group* C_p. The *boundary* of a p-simplex σ is the $(p-1)$-chain formed by the sum of all faces of σ of codimension 1. This operation extends linearly to a *boundary operator* $\delta : C_p \to C_{p-1}$. A p-chain γ is a *p-cycle* if $\delta(\gamma) = 0$. The p-cycles form a subgroup of the p-chains, which we call the *pth cycle group* Z_p. A p-chain γ is called a *p-boundary* if $\gamma = \delta(\xi)$ for some $(p+1)$-chain ξ. Again, the p-boundaries form a group B_p, and since $\delta(\delta(\xi)) = 0$ for any chain ξ, p-boundaries are p-cycles, and so B_p is a subgroup of Z_p. The *pth homology group* H_p is defined as the quotient group Z_p / B_p. The rank of H_p is denoted by β_p and is called the *pth Betti number*. In our case of $\mathbb{Z}_2$ coefficients, the homology group is a vector space isomorphic to $\mathbb{Z}_2^{\beta_p}$, hence it is completely determined by the Betti number. Roughly speaking, the Betti numbers in dimension 0, 1, and 2 yield the number of connected components, tunnels, and voids of K, respectively.

2.2 Persistence

Let $\{\sigma_1, \ldots, \sigma_n\}$ denote the simplices of K. We assume that for each $i \leqslant n$, $K_i := \{\sigma_1, \ldots, \sigma_i\}$ is a simplicial complex again. The sequence of inclusions $\emptyset = K_0 \subset \ldots \subset K_i \ldots \subset K_n = K$ is called a *simplexwise filtration* of K. For every dimension p and every K_i, we have a homology group $H_p(K_i)$; we usually consider all dimensions at once and write $H(K_i)$ for the direct sum of the homology groups of K_i in all dimensions. The inclusion $K_i \hookrightarrow K_{i+1}$ induces a homomorphism $g_i^{i+1} : H(K_i) \to H(K_{i+1})$ on the homology groups. These homomorphisms compose and we can define $g_i^j : H(K_i) \to H(K_j)$ for any $i \leqslant j$. We say that a class $\alpha \in H(K_i)$ is *born at (index) i* if

$$\alpha \notin \operatorname{im} g_{i-1}^i.$$

A class α born at index i *dies entering (index) j* if

$$g_i^j(\alpha) \in \operatorname{im} g_{i-1}^j \text{ but } g_i^{j-1}(\alpha) \notin \operatorname{im} g_{i-1}^{j-1}.$$

In this case, the index pair (i, j) is called a *persistence pair*, and the difference $j - i$ is the *(index) persistence* of the pair. The transition from K_{i-1} to K_i either causes the birth or the death of an homology class. We call the added simplex σ_i *positive* if it causes a birth and *negative* if it causes a death. Note that homology classes of the full complex K do not die during the filtration. We call a simplex σ_i that gives birth to such a class *essential*. All other simplices are called *inessential*.

2.3 Boundary Matrix

For a matrix $M \in \mathbb{Z}_2^{n \times n}$, we let M_j denote its j-th column, M^i its i-th row, and $M_j^i \in \mathbb{Z}_2$ its entry in row i and column j. For a non-zero column $0 \neq M_j = (m_1, \ldots, m_n) \in \mathbb{Z}_2^n$, we set $\operatorname{pivot}(M_j) := \max\{i = 1, \ldots, n \mid m_i = 1\}$ and call it the *pivot index* of that column.

The *boundary matrix* $D \in (\mathbb{Z}_2)^{n \times n}$ of a simplexwise filtration $(K_i)_i$ is a $n \times n$ matrix with $D_j^i = 1$ if and only if σ_i is a face of σ_j of codimension 1. In other words, the jth column of D encodes the boundary of σ_j. D is an upper-triangular matrix because any face of σ_j must precede σ_j in the filtration. Since the jth row and column of D corresponds to the jth simplex σ_j of the filtration, we can talk about *positive columns*, *negative columns*, and *essential columns* in a natural way, and similarly for rows.

Algorithm 1 Left-to-right persistence computation

procedure PERSISTENCE_LEFT_RIGHT(D)
  $R \leftarrow D$; $L \leftarrow [0, \dots, 0]$; $P \leftarrow \emptyset$ ▷ $L \in \mathbb{Z}^n$
  for $j = 1, \dots, n$ **do**
    while $R_j \neq 0$ and $L[\text{pivot}(R_j)] \neq 0$ **do**
      $R_j \leftarrow R_j + R_{L[\text{pivot}(R_j)]}$
    if $R_j \neq 0$ **then**
      $i \leftarrow \text{pivot}(R_j)$
      $L[i] \leftarrow j$
      Mark columns i and j as paired and add (i, j) to P
  return P

2.4 The Reduction Algorithm

A column operation of the form $M_j \leftarrow M_j + M_k$ is called *left-to-right* if $k < j$. We call a matrix M' *derived from* M if M can be transformed into M' by left-to-right operations. Note that in a derivation M' of M, the jth column can be expressed as a linear combination of the columns $1, \dots, j$ of M, and this linear combination includes M_j. We call a matrix R *reduced* if no two non-zero columns have the same pivot index. If R is derived from M, we call it a *reduction of* M. In this case, we define

$$P_R := \{(i, j) \mid R_j \neq 0 \wedge i = \text{pivot}(R_j)\}$$
$$E_R := \{i \mid R_i = 0 \wedge \text{pivot}(R_j) \neq i \,\forall j = 1, \dots, n\}.$$

Although the reduction matrix R is not unique, the sets P_R and E_R are the same for any choice of reduction; therefore, we can define P_M and E_M to be equal to P_R and E_R for any reduction R of M. We call the set P the *persistence pairs* of M. When obvious from the context, we omit the subscripts and simply write P for the persistence pairs. For the boundary matrix D of K, the pairs $(i, j) \in P$ are the persistence pairs of the filtration $(K_i)_{0 \leqslant i \leqslant n}$, and the indices in E correspond to the essential simplices of the complex. Note that E is uniquely determined by P and n as the indices between 1 and n that do not appear in any pair of P.

The simplest way of reducing D is to process columns from left to right; for every column, other columns are added from the left until the pivot index is unique (Algorithm 1). A lookup table can be used to identify the next column to be added in constant time. A flag is used for every column denoting whether a persistence pair with the column index has already been found. After termination, the unpaired columns correspond to the essential columns. The running time is at most cubic in n, and this bound is actually tight for certain input filtrations, as demonstrated in [19].

Let M be derived from D. A column M_j of M is called *reduced* if either it is zero, or if $(i, j) \in P$ with $i = \text{pivot}(M_j)$. With this definition, a matrix M is a reduction of D if and only if every column is reduced.

3 Speed-Ups

Algorithm 1 describes the simplest way of reducing the boundary matrix, but it performs more operations than actually necessary to compute the persistence pairs. We now present two simple techniques, which both lead to a significant decrease in the number of required operations.

3.1 Clearing Positive Columns

The key insight behind our first optimization is the following fact: if i appears as the pivot in a reduced column of M, the index i is positive and hence there exists a sequence of left-to-right operations on M_i that turn it to zero. Instead of explicitly executing this sequence of operations, we define the *clear* operation by setting column M_i to zero directly. Informally speaking, a clear is a shortcut to avoid some column operations in the reduction when the result is evident.

In order to apply this optimization, we change the traversal order in the reduction by first reducing the columns corresponding to simplices with dimension d (from left to right), then all columns with dimension $d-1$, and so on. After having reduced all columns with dimension δ, we have found all positive inessential columns with dimension $\delta - 1$ and clear them before continuing with $\delta - 1$. This way all positive inessential columns of the complex are cleared without performing any column additions on them. See [7] for a more detailed description.

3.2 Compression

Alternatively, we can try to save arithmetic operations by reducing the number of non-zero rows among the unpaired columns. A useful observation in this context is given next.

Lemma 1. *Let M_j be a non-zero column of M with $i = \text{pivot}(M_j)$. Then M_i is a positive and inessential column.*

Proof. The statement is clearly true if M_j is reduced, because in this case (i, j) is a persistence pair. If M_j is not reduced, this means that after applying some sequence of left-to-right column operations, some reduced column has i as pivot index.

Corollary 1. *Let M_i be a negative column of M. Then i is not the pivot index of any column in M.*

As a consequence, whenever a negative column with index j has been reduced, row j can be set to zero before further reducing, as suggested in [20].

Corollary 2. *Let M_i be a negative column and let M_j be a column with $M_j^i = 1$. Then setting M_j^i to zero does not affect the pairs.*

But we can even do more: let i be the pivot index of the reduced column M_j and assume that the submatrix of M with column indices $\{1, \ldots, j\}$ and row indices $\{i, \ldots, n\}$ is reduced, i.e., the pivot indices are unique in this submatrix. By adding column j to each unreduced column in the matrix that has a non-zero entry at row i, we can eliminate all non-zero entries in row i from the unreduced columns. Note that if $k < j$ and $M_k^i \neq 0$, then $\text{pivot}(M_k) \geqslant i$ and thus, by assumption, M_k must be a reduced negative column. Therefore, for each unreduced column M_k, the operation $M_k \leftarrow M_k + M_j$ is a left-to-right addition and thus does not affect the pairs.

4 Reduction in Chunks

The two optimization techniques from Sect. 3 both yield significant speed-ups, but they are not easily combinable, because clearing requires to process a simplex before its faces, whereas compression works in the opposite direction. In this section, we present an algorithm that combines both optimization techniques.

Let $m \in \mathbb{N}$. Fix $m + 1$ numbers $0 = t_0 < t_1 < \ldots < t_{m-1} < t_m = n$ and define the ith *chunk* of D to be the columns of D with indices $\{t_{i-1} + 1, \ldots, t_i\}$. We call a column D_j *local* if it forms a persistence pair with another column in the same chunk or in one of the adjacent chunks. In this case, we also call the persistence pair local. Non-local columns (and pairs) are called *global*. If ℓ is a lower bound on the size of each chuck, then every global persistence pair has index persistence at least ℓ. We also call an index j local if the jth column of D is local, and the same for global. We denote the number of global columns in D by g. The high-level description of our new algorithm consists of three steps:

1. Partially reduce every chunk independently, applying the clearing optimization, so that all local columns are completely reduced.
2. Independently compress every global column such that all its non-zero entries are global.
3. Reduce the submatrix consisting only of the global rows and columns.

We give details about the three steps in the rest of this section. The first two steps can be performed in parallel, whereas the third step only needs to reduce a matrix of size $g \times g$ instead of $n \times n$. In many situations, g is significantly smaller than n.

4.1 Local Chunk Reduction

The first step of our algorithm computes the local pairs by performing two phases of the spectral sequence algorithm [9, Sect. VII.4]. Concretely, we apply left-to-right operations as usual, but in the first phase we only add columns from the same chunk, and in the second phase we only add columns from both the same chunk and its left neighbor. After phase r, for each $b \in \{r, \ldots, m\}$ the submatrix with column indices $\{1, \ldots, t_b\}$ and row indices $\{t_{b-2} + 1, \ldots, n\}$ is reduced. If the reduction

Algorithm 2 Local chunk reduction

procedure LOCAL_REDUCTION($M, t_0, \ldots, t_m$)
  $R \leftarrow M; L \leftarrow [0, \ldots, 0]; P \leftarrow \emptyset$ ▷ $L \in \mathbb{Z}^n$
  for $\delta = d, \ldots, 0$ **do**
    for $r = 1, 2$ **do** ▷ Perform two phases of the spectral sequence algorithm
      for $b = r, \ldots, m$ **do** ▷ Loop is parallelizable
        for $j = t_{b-1} + 1, \ldots, t_b$ with $\dim \sigma_j = \delta$ **do**
          if j is not marked as paired **then**
            while $R_j \neq 0 \wedge L[\mathrm{pivot}(R_j)] \neq 0 \wedge \mathrm{pivot}(R_j) > t_{b-r}$ **do**
              $R_j \leftarrow R_j + R_{L[\mathrm{pivot}(j)]}$
            if $R_j \neq 0$ **then**
              $i \leftarrow \mathrm{pivot}(R_j)$
              if $i > t_{b-r}$ **then**
                $L[i] \leftarrow j$
                $R_j \leftarrow 0$ ▷ Clear column i
                Mark i and j as paired and add (i, j) to P
  return (R, L, P)

of column j stops at a pivot index $i > t_{b-r}$, row j cannot be reduced any further by adding any column, so we identify (i, j) as a local persistence pair. Conversely, any local pair (i, j) is detected by this method after two phases. We incorporate the clearing operation for efficiency, that is, we proceed in decreasing dimension and set detected local positive columns to zero; see Algorithm 2. After its execution, $L[i]$ contains the index of the local negative column with pivot index i for any local positive column i, and the resulting matrix R is a derivation of D in which all local columns are reduced.

4.2 *Global Column Compression*

Let R be the matrix returned by Algorithm 2. Before computing the global persistence pairs, we first compress the global columns, using the ideas from Sect. 3; recall that negative rows can simply be set to zero, while entries in positive rows can be eliminated by an appropriate column addition. Note, however, that a full column addition might actually be unnecessary: for instance, if all non-zero row indices in the added column belong to negative columns (except for the pivot), the entry in the local positive row could just have been zeroed out in the same way as in Corollary 2. Speaking more generally, it is more efficient to avoid column additions that have no consequences for global indices, neither directly nor indirectly.

In the spirit of this observation, we call an index i *inactive* if either it is a local negative index or if (i, j) is a local pair and all indices of non-zero entries in column R_j apart from i are inactive. Otherwise, the index is called *active*. By induction and Corollary 2, we can show:

Algorithm 3 Determining active entries

```
procedure MARK_ACTIVE_ENTRIES(R)
    for each unpaired column k do                          ▷ Loop is parallelizable
        MARK_COLUMN(R, k)
function MARK_COLUMN(R, k)
    if k is marked as active/inactive then return true/false
    for each non-zero entry index i of R_k do
        if ℓ is unpaired then
            mark k as active and return true
        else if i is positive then
            j ← L[i]                                       ▷ (i, j) is persistence pair
            if j ≠ k and MARK_COLUMN(R, j) then
                mark k as active and return true
    mark k as inactive and return false
```

Algorithm 4 Global column compression

```
procedure COMPRESS(R, k)
    Uses variables: L
    for each non-zero entry index ℓ of R_k in decreasing order do
        if ℓ is paired then
            if ℓ is inactive then
                R^k_ℓ ← 0
            else
                j ← L[ℓ]                                   ▷ (ℓ, j) is persistence pair
                R_k ← R_k + R_j
```

Lemma 2. *Let i be an inactive index and let M_j be any column with $M_j^i = 1$. Then setting M_j^i to zero does not affect the persistence pairs.*

The compression proceeds in two steps: first, every non-zero entry of a global column is classified as active or inactive (using depth-first search; see Algorithm 3). Then, we iterate over the global columns, set all entries with inactive index to zero, and eliminate any non-zero entry with a local positive index ℓ by column addition with $L[\ell]$ (see Algorithm 4). After this process, we obtain a matrix R' with the same persistence pairs as R, such that the global columns of R' have non-zero entries only in the global rows.

4.3 Submatrix Reduction

After having compressed all global columns, these form a $g \times g$ matrix "nested" in R (recall that g is the number of global columns). To complete the computation of the persistence pairs, we simply perform standard reduction on the remaining matrix. For efficiency, we perform steps 2 and 3 alternatingly for all dimensions in decreasing order and apply the clearing optimization; this way, we avoid the compression of positive global columns. Algorithm 5 summarizes the whole method.

Algorithm 5 Persistence in chunks

```
procedure PERSISTENCE_IN_CHUNKS(D, t_0, ..., t_m)
    (R, L, P) ←LOCAL_REDUCTION(D, t_0, ..., t_m)        ▷ step 1: reduce local columns
    MARK_ACTIVE_ENTRIES(R)
    for δ = d, ..., 0 do
                                                        ▷ step 2: compress global columns
        for j = 1, ..., n with dim σ_j = δ do           ▷ Loop is parallelizable
            if column j is not paired then
                COMPRESS(R, j)
        for j = 1, ..., n with dim σ_j = δ do           ▷ step 3: reduce global columns
            while R_j ≠ 0 ∧ L[pivot(R_j)] ≠ 0 do
                R_j ← R_j + R_{L[pivot(j)]}
            if R_j ≠ 0 then
                i ← pivot(R_j)
                L[i] ← j
                R_i ← 0                                 ▷ Clear column i
                Mark i and j as paired and add (i, j) to P
    return P
```

5 Analysis

Algorithm 5 permits a complexity analysis depending on the following parameters: n, the number of simplices; m, the number of chunks; ℓ, the maximal size of a chunk; and g, the number of global columns. We assume that for any simplex, the number of non-zero faces of codimension 1 is bounded by a constant (this is equivalent to assuming that the dimension of the complex is a constant).

5.1 General Complexity

We show that the complexity of Algorithm 5 is bounded by

$$O(m\ell^3 + g\ell n + g^3). \tag{1}$$

The three summands correspond to the running times of the three steps.[1] Note that $g \in O(n)$ in the worst case.

For the complexity of Algorithm 2, we consider the complexity of reducing one chunk, which consists of up to ℓ columns. Within the local chunk reduction, every column is only added with columns of the same or the previous chunk, so there are only up to 2ℓ column additions per column. Moreover, since the number of non-zero entries per column in D is assumed to be constant, there are only $O(\ell)$

[1]The running time of the third step could be lowered to g^ω, where ω is the matrix-multiplication exponent, using the method of [17].

many entries that can possibly become non-zero during the local chunk reduction. It follows that the local chunk reduction can be considered as a reduction on a matrix with ℓ columns and $O(\ell)$ rows. If we represent columns by linked lists (containing the non-zero indices in sorted order), one column operation can be done in $O(\ell)$ primitive operations, which leads to a total complexity of $O(\ell^3)$ per chunk.

The computation of active columns in Algorithm 3 is done by depth-first search on a graph whose vertices are given by the columns and whose edges correspond to their non-zero entries. The number of edges is $O(n\ell)$, so we obtain a running time of $O(n\ell)$.

Next, we consider the cost of compressing a global column with index j. After the previous step, the column has at most $O(\ell)$ non-zero entries. We transform the presentation of the column from a linked list into a bit vector of size n. In this representation, adding another column in list representation with v entries to column j takes time proportional to v. In the worst case, we need to add all columns with indices $1, \ldots, j-1$ to j. Each such column has $O(\ell)$ entries. At the end, we transform the bit vector back into a linked list representation. The total cost is $O(n + (j-1)\ell + n) = O(n\ell)$ per global column.

Finally, the complexity of the global reduction is $O(g^3)$, as in the standard reduction.

5.2 *Choosing Chunks*

We discuss different choices of chunk size and their complexities. A generic choice for an arbitrary complex is to choose $O(\sqrt{n})$ chunks of size $O(\sqrt{n})$ each. With that, the complexity of (1) becomes

$$O(n^2 + g_1 n\sqrt{n} + g_1^3).$$

Alternatively, choosing $O(\frac{n}{\log n})$ chunks of size $O(\log n)$, the complexity becomes

$$O(n\log^2 n + g_2 n\log n + g_2^3).$$

We replaced g by g_1 and g_2 to express that the number of global columns is different in both variants. In general, choosing larger chunks is likely to produce less global columns, since every global persistence pair has index persistence at least ℓ (the size of the smallest chunk).

5.3 *Cubical Complexes*

We consider an important special case of boundary matrices: consider a d-dimensional image with p hypercubes, where each vertex contains a grayscale value. We assume that the cubes are triangulated conformally in order to get simplicial input – the argument also works, however, for the case of cubical cells. We assign function values inductively, assigning to each simplex the maximal value of its faces. Assuming that all vertex values are distinct, the *lower star* of vertex v is the set of all simplices that have the same function value as v. Filtering the simplices in a way that respects the order of the function values, we get a *lower star filtration* of the image. Now choose the lower stars as the chunks in our reduction algorithm. Note that the lower star is a subset of the star of the corresponding vertex, which is of constant size (assuming that the dimension d is constant). Therefore, the complexity bound (1) reduces to

$$O(n + gn + g^3) = O(gn + g^3).$$

Note that global columns with large index persistence might still have very small, or even zero, persistence with respect to the function values, for instance in the presence of a flat region in the image where many vertices have similar values.

6 Experiments

We implemented two versions of the algorithm presented in Sect. 4: a sequential and a parallel version (using OPENMP), in which the first two steps of the algorithm are performed simultaneously on each chunk and on each global column, respectively. In both cases, we use $\lfloor\sqrt{n}\rfloor$ as the chunk size. For a fair comparison, we also re-implemented the *standard* reduction algorithm introduced in [11] and the *twist* algorithm [7] (clearing optimization applied to standard reduction) in the same framework, i.e., using the same data representations and low-level operations such as column additions. All implementations are available as part of the open source library PHAT [2]. Additionally, we compare with the memory efficient algorithm [13] based on discrete Morse theory [12] and to the implementation of the persistent cohomology algorithm [8] found in the DIONYSUS library [18]. An optimized implementation of the latter algorithm has been announced in [4].

To find out how these algorithms behave in practice, we apply them to five representative data sets. The first three are 3D image data sets with a resolution of 128^3. The first of these is given by a Fourier sum with random coefficients and is representative of smooth data. The second is uniform noise. The third is the sum of the first two and represents large-scale structures with some small-scale noise. These data sets are illustrated in Fig. 1 by an isosurface.

In addition to the lower star filtrations of these image data sets, we also consider an alpha shape filtration defined by 10,000 samples of a torus embedded in $\mathbb{R}^3$, and the 4-skeleton of the Rips filtration given by 50 points randomly chosen from the Mumford data set [14].

As pointed out in [8], the pairs of persistent cohomology are the same as those of persistent homology. We therefore also applied all algorithms to the corresponding cochain filtration, given in matrix form by transposing the boundary

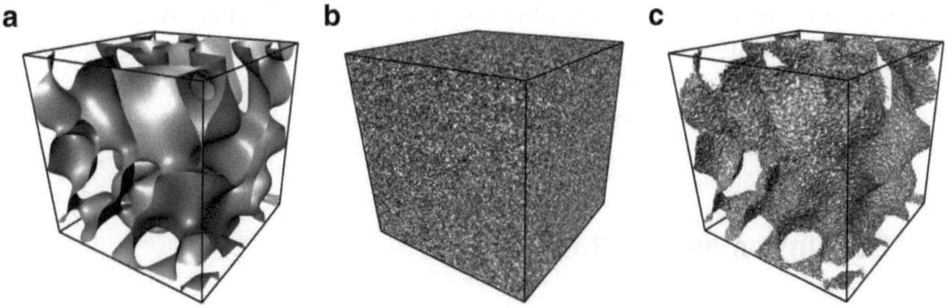

Fig. 1 A single isosurface of the representative data sets used to benchmark and compare our algorithm: (**a**) a smooth data set, (**b**) uniformly distributed noise, (**c**) the sum of (**a**) and (**b**)

Table 1 Running time comparison of various persistent homology algorithms applied to the data sets described in Sect. 6. The last three columns contain information relevant to the algorithm presented in this paper: the fraction $\tilde{g}/n$, where $\tilde{g} \leqslant g$ is the number of non-zero global columns after global column compression, and the running times using 1 and 12 cores, respectively

Dataset	$n \cdot 10^{-6}$	std. [11] (s)	twist [7] (s)	cohom. [8] (s)	DMT [13] (s)	$\tilde{g}/n$ (%)	chunk (1×) (s)	chunk (12×) (s)
Smooth	16.6	288.5	0.8	65.8	2.0	0	1.6	0.6
Smooth$^{\perp}$	16.6	299.9	5.3	20.8	–	0	1.9	0.6
Noise	16.6	279.2	11.5	15,971	13.0	7	14.7	4.4
Noise$^{\perp}$	16.6	940.2	24.0	190.1	–	9	14.9	5.0
Mixed	16.6	206.4	2.6	50,927	12.3	4	14.1	1.9
Mixed$^{\perp}$	16.6	299.2	7.6	32.7	–	5	14.3	2.3
Torus	0.6	40.4	0.1	1.6	–	5	0.2	0.1
Torus$^{\perp}$	0.6	16.7	0.3	1.4	–	6	0.6	0.2
Mumford	2.4	29.8	27.7	2.8	–	82	11.5	1.9
Mumford$^{\perp}$	2.4	48.8	0.1	184.1	–	1	1.1	0.4

matrix and reversing the order of the columns; this operation is denoted by $(\cdot)^{\perp}$. When reducing such a coboundary matrix with the clearing optimization, columns are processed in order of increasing instead of decreasing dimension. Note that the persistent cohomology algorithm in [8] is a *dual* algorithm, i.e., it computes persistent cohomology from a boundary matrix and persistent homology from a coboundary matrix. The other algorithms are *primal* algorithms, i.e., they compute persistent (co)homology from a (co)boundary matrix.

Table 1 contains the running times of the above algorithms applied to filtrations of these five data sets run on a PC with two Intel Xeon E5645 CPUs. All results (except for the DIONYSUS and DMT codes) are obtained with PHAT 1.2.1 using the included `benchmark` tool with the option `--sparse_pivot_column`. The data sets are available at the PHAT website [2].

We observe a significant speed-up caused by the clearing optimization, as already reported in [7]. We point out that a similar optimization is also used in the cohomology algorithm of [8]. We can see that our *chunk* algorithm performs slightly worse than the one of [7] when executed sequentially, but often faster when parallelized. We also observe that the algorithms *twist* and *chunk* generally behave worse when computing persistent cohomology, except for the Rips filtration of the Mumford data set, where the converse is true.

7 Conclusion and Outlook

We have presented an algorithm for persistent homology that includes two simple optimization techniques into the reduction algorithm. It can be fully parallelized, except for the reduction of compressed global columns, whose number is often small. Besides our asymptotic complexity bounds, which give a detailed dependence on the parameters of the algorithm, our experiments show that significant speed-ups can be achieved through parallel persistence computation. Similar observations have been made recently by Lewis and Zomorodian [15] for the computation of (non-persistent) homology; see also [16]. Our experiments also reveal that the running times for computing persistent homology versus cohomology can be very different, with a preference for homology or cohomology depending on the data but not so much on the type of algorithm (primal or dual) used. We plan a more extensive discussion of these effects in future work.

Acknowledgements The authors thank Chao Chen, Herbert Edelsbrunner, and Hubert Wagner for helpful discussions. This research is partially supported by the TOPOSYS project FP7-ICT-318493-STREP and the Max Planck Center for Visual Computing and Communication.

References

1. U. Bauer, Persistence in discrete Morse theory. PhD thesis, Georg-August-Universität Göttingen, 2011
2. U. Bauer, M. Kerber, J. Reininghaus, PHAT: persistent homology algorithm toolbox. http://phat.googlecode.com/
3. U. Bauer, C. Lange, M. Wardetzky, Optimal topological simplification of discrete functions on surfaces. Discret. Comput. Geom. **47**, 347–377 (2012)
4. J.-D. Boissonnat, T.K. Dey, C. Maria, The compressed annotation matrix: an efficient data structure for computing persistent cohomology, in *Algorithms – ESA 2013*, Sophia Antipolis, ed. by H.L. Bodlaender, G.F. Italiano. Volume 8125 of Lecture Notes in Computer Science (Springer, Berlin/Heidelberg, 2013), pp. 695–706
5. F. Chazal, D. Cohen-Steiner, L. Guibas, F. Memoli, S. Oudot, Gromov–Hausdorff stable signatures for shapes using persistence, in *Eurographics Symposium on Geometry Processing*, Berlin, 2009, pp. 1393–1403
6. C. Chen, M. Kerber, An output-sensitive algorithm for persistent homology, in *Proceedings of the 27th Annual Symposium on Computational Geometry*, Paris, 2011, pp. 207–215
7. C. Chen, M. Kerber, Persistent homology computation with a twist, in *27th European Workshop on Computational Geometry (EuroCG)*, Morschach, 2011, pp. 197–200. Extended abstract

8. V. de Silva, D. Morozov, M. Vejdemo-Johansson, Dualities in persistent (co)homology. Inverse Probl. **27**, 124003 (2011)
9. H. Edelsbrunner, J. Harer, *Computational Topology, An Introduction* (American Mathematical Society, Providence, 2010)
10. H. Edelsbrunner, C.-P. Heisenberg, M. Kerber, G. Krens, The medusa of spatial sorting: topological construction (2012). arXiv:1207.6474
11. H. Edelsbrunner, D. Letscher, A. Zomorodian, Topological persistence and simplification. Discret. Comput. Geom. **28**, 511–533 (2002)
12. R. Forman, Morse theory for cell complexes. Adv. Math. **134**, 90–145 (1998)
13. D. Günther, J. Reininghaus, H. Wagner, I. Hotz, Efficient computation of 3D Morse–Smale complexes and persistent homology using discrete Morse theory. Vis. Comput. 1–11 (2012)
14. A.B. Lee, K.S. Pedersen, D. Mumford, The nonlinear statistics of high-contrast patches in natural images. Int. J. Comput. Vis. **54**, 83–103 (2003)
15. R.H. Lewis, A. Zomorodian, Multicore homology (2012). Unpublished manuscript
16. D. Lipsky, P. Skraba, M. Vejdemo-Johansson, A spectral sequence for parallelized persistence (2011). arXiv:1112.1245
17. N. Milosavljević, D. Morozov, P. Škraba, Zigzag persistent homology in matrix multiplication time, in *Proceedings of the 27th Annual Symposium on Computational Geometry*, Paris, 2011, pp. 216–225
18. D. Morozov, Dionysus: a C++ library for computing persistent homology. http://www.mrzv.org/software/dionysus/
19. D. Morozov, Persistence algorithm takes cubic time in the worst case, in *BioGeometry News* (Duke Computer Science, Durham, 2005)
20. A. Zomorodian, G. Carlsson, Computing persistent homology. Discret. Comput. Geom. **33**, 249–274 (2005)

Parallel Computation of Nearly Recurrent Components of Piecewise Constant Vector Fields

Nicholas Brunhart-Lupo and Andrzej Szymczak

Abstract We describe a simple way to parallelize the algorithm for computing Morse decompositions and nearly recurrent sets for piecewise constant vector fields. In this chapter, we focus on the most computationally expensive component of the algorithm, namely refinement of the transition graph, which often takes up over 90 % of the running time.

The original implementation of the refinement algorithm refines nodes of the graph one at a time, in an arbitrary order. As a node is refined, arcs out of and into the resulting node are updated to preserve the key property of the graph, i.e. its ability to represent every trajectory of the vector field as a path in the graph. A significant portion of time is spent on updating and maintaining the integrity of the graph for each node that undergoes refinement. This approach is attractive and elegant since it allows one to refine the nodes in arbitrary order and to an arbitrary depth while guaranteeing the correctness of the graph. However, its performance is suboptimal for most transition graph refinement scenarios, not only because of the cost of maintaining the integrity of the graph for each node refinement operation, but also because of uncontrollable memory access patterns.

The method described in this chapter is tailored for the special case of nodes of the graph refined to the same depth, which is sufficient for most applications described in prior work. The idea is to stream through the set of arcs of the coarse graph to generate a compact bitmap based representation of the set of arcs of the refined graph. In the subsequent pass over the bitmap, the refined transition graph's arcs are constructed from scratch, using only arc additions. The simplicity of the two passes makes them easy to parallelize and results in better cache performance. The bitmap building pass can be offloaded to the GPU to further increase the performance.

N. Brunhart-Lupo (✉) • A. Szymczak
Department of Electrical Engineering & Computer Science, Colorado School of Mines, Golden, CO, USA
e-mail: nbrunhar@mymail.mines.edu

P.-T. Bremer et al. (eds.), *Topological Methods in Data Analysis and Visualization III*, Mathematics and Visualization, DOI 10.1007/978-3-319-04099-8_8,

1 Introduction

Because of applications in fluid dynamics, aerodynamics, electromagnetism, biology and computer vision, there has been an increasing interest in vector field analysis. One of the tools developed for this task is *vector field topology*, which aims to describe the structure of a vector field in terms of features such as sources, sinks, saddle points, periodic trajectories and separatrices.

A robust and efficient algorithm for analyzing and extracting close to recurrent (i.e. stagnant or circulating) flow in a vector field has been developed in [16] for piecewise constant (PC) 2D vector fields. More recently, this algorithm was extended to 3D in [15]. While these algorithms are relatively efficient, the implementations available so far are either serial or rather naively parallelized, with a modest speedup on multicore CPUs. By reworking these algorithms into a universal, parallel and cache aware approach, higher performance can be achieved, allowing the exploration of larger datasets that are common in today's research.

In this chapter, we put forth an alternate formulation of the algorithm as found in [15]. The refinement process is replaced with a multi-pass procedure that is friendly to concurrent implementations, and improves memory locality. The ideas presented in this chapter apply to both 2D and 3D PC vector fields, however, we focus on the 3D case.

This chapter is organized into the following sections. Section 2 contains a brief overview of related vector field topology algorithms, and Sect. 3 discusses background on the original specification and implementation of the algorithm. Section 4 discusses the improved procedure. Section 5 details our implementation and results we have obtained thus far. The conclusion in Sect. 6 closes the chapter with possible future work.

2 Related Work

In this section we discuss prior work on graph based analysis of general (not necessarily gradient) vector field topology. More in-depth discussion of vector field topology can be found in [10].

In [3], *edge maps* are introduced. Edge maps are a mapping of line segments obtained by subdividing the mesh edges based on vector field dynamics. The goal of this approach is to provide an approximation of arbitrary quality of the piecewise linear vector field flow, while at the same time guaranteeing consistency of the results. In this scheme on a triangulated 2D mesh, we can map pieces of edges of triangles that are connected by flow inside that triangle, creating an origin-destination pair. The exit point of a trajectory can be approximated as a linear function mapping the origin edge pieces into the destination edge pieces. If higher precision is needed, edges can be subdivided multiple times and reattached to build a finer set of origin-destination pairs, resulting in a better approximation of the trajectories of the piecewise linear vector field. Following a trajectory through a triangle requires one linear interpolation operation based on an edge map, instead of using numerical integration. As a drawback, this approach has not yet been extended into 3D, and it is not clear how to accomplish this task.

Recent work on reliable extraction of vector field topology is built off of discrete Morse Theory [8]. An example of this is in [13]. Here, a purely combinatorial approach to vector field analysis is presented; the algorithm attempts to build a graph representation of the flow. This graph representation is a matching between 2D simplices and incident mesh edges. The matching is computed using a maximum weight matching algorithm. The objective is to make the edges of the matching follow the direction of the flow throughout the domain. Discrete trajectories (approximating the trajectories of the underlying vector field) are sequences of vertices in which every second pair is connected by a matching edge and all other pairs are not. They can be analyzed using simple graph search algorithms. This leads to a robust algorithm for vector field topology which in this case consists of discrete stationary points, periodic trajectories and separatrices. In [14], this matching algorithm was furnished with a GPU/multi-threaded algorithm to reduce the penalty of the original quadratic runtime. Unfortunately, the basic approach suffers from a significant issue; discrete trajectories are a poor approximation, and even if the mesh size (i.e. maximum diameter of a triangle) trends to zero, the trajectories emitted will not converge to the their continuous counterparts. This is remedied somewhat in [9], wherein they increase the expected accuracy of combinatorial gradient vector fields by employing a randomized algorithm and subdivision.

In [5], Morse decomposition was used as a way to robustly detect features in a general (not necessarily gradient) vector field. A piecewise linear input vector field (on a triangle mesh) is converted into a directed graph by inserting a node for every triangle. The nodes are connected based on flow along the edges of their representative triangle. If a particle can flow from one adjacent triangle to another, they will be connected in the graph. Once all the proper arcs have been inserted, Morse sets can be extracted by using a strongly connected component algorithm.

In [6], an algorithm for computing Morse decompositions based on a different graph representation, called the τ-map, is introduced. To build the τ-map graph (whose nodes also correspond to the mesh triangles), a triangle T is pushed along the flow by τ forward in time to yield a set of points T'. The arc $T \rightarrow S$ is inserted into the τ-map graph whenever the triangle S intersects T'. The resulting Morse decompositions are typically much finer than the ones obtained using the original approach in [5], but in both cases the algorithms are still limited by the mesh resolution.

Piecewise constant vector field algorithms (discussed in detail in the next section) offer an attractive alternative [15, 16]. Firstly, they are based on a graph representation whose construction does not rely on numerical integration. Precision of the output is not restricted to the mesh itself; arbitrary sub-triangle/sub-cube accuracy of features is achieved through a refinement process. Extension of the algorithms to 2D and 3D have been accomplished, and higher dimensional specializations are straightforward to describe. A drawback of this approach is that the transition graph can grow to very large sizes as higher levels of refinement are sought or very large data sets are explored. Merely processing millions of nodes in the graph can take a non-trivial amount of time, even if the individual refinement operation is simple.

3 Piecewise Constant Vector Fields: Background

In this section we provide a review of the Morse decomposition algorithm for PC vector fields that this contribution improves. A piecewise constant (PC) 3D vector field provides a non-zero vector $f(D)$ for every 3D cell (cube) of a regular grid M. In order to build a graph representation of the flow, f is extended to lower dimensional grid cells (details are provided in [15, 16]). In general, for a grid cell D, $f(D)$ is defined as the intersection of the convex hull of vectors assigned to 3D cells containing D and the linear space of D (note that this means that every vector in $f(D)$ points along D).

In a PC vector field, trajectories are greatly simplified as opposed to a PL vector field, and appear as concatenated *simple segments*. Simple segments are continuous curves defined on a closed interval and contained fully in a mesh cell, referred to as the segment's *carrier*. For 3D, 2D, or 0D carrier C, simple segments move through C with constant velocity, given by the only vector in $f(C)$, if such a vector exists. For 1D carriers (mesh edges), simple segments may be more complex; they move along C with average velocity in $f(C)$ over any nonzero length interval.

To represent these trajectories, we use a transition graph. The transition graph is a finite directed graph which records all trajectories in the PC vector field as paths. The nodes of this graph correspond to n-sets; vertices, edge pieces (either whole or from subdivided edges), and face pieces (obtained from 2D faces by means of quad-tree-like subdivision). An arc $A \to B$ in the graph indicates that there is a trajectory in the vector field starting from node A and ending in node B, and is inserted into the graph pending a connectivity test. Details on how this connectivity check is performed can be found in [15] as it remains unchanged in this chapter; in general, it requires projection of the two n-sets and an intersection test. Arcs possess an *owner* cell, which is the carrier of the simple segments represented by that arc.

In Fig. 1, we demonstrate the transition graph contributions for each dimension of cell in a 3D grid. Note that these are contributions; all cells must be considered and all their derived arcs added to the transition graph. Using the 3D cell case, the strength of this approach becomes apparent. By projecting the lower dimensional subcells of the cell along f into a 2D space (as seen in the figure) we know instantly that a trajectory entering the 3D cell by face $\square_{cdgh}$ can leave the 3D cell by either $\square_{efgh}$ or $\square_{bcgf}$ as these three 2D cells intersect in this projected space. This is repeated for 2D cells; here we project 1D and 0D subcells to a 1D space, and determine graph connectivity by projections. For 1D cells, the convex hull of adjacent 3D cells' f, and its intersection with the linear space of the cell result in a set of vectors with varying magnitude. We only need to know the direction (either

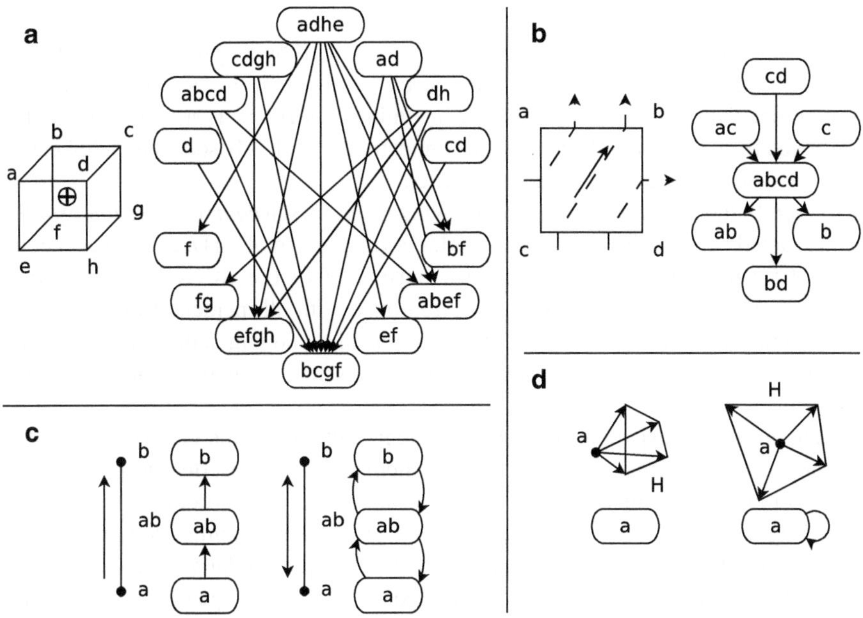

Fig. 1 Example transition graph contributions for each grid cell dimension, arranged by 3D to 0D starting from the *upper left*. In the 3D case (**a**), the cell is projected by f, which points into the paper (*circled-plus*). In (**b**), the 2D case, f is denoted by the *bold arrow* in the center. *Dashed lines* illustrate simple segments within the face. If there is no flow along a 2D cell, it will contribute no arcs to the graph. The 1D edge subfigure in (**c**) shows the single and bidirectional case: f for this cell may actually be an interval of vectors, but we only need to know the direction, not magnitudes. It is also possible that there will be no flow in this cell; in this situation, it contributes no arcs to the graph. The 0D figure (**d**) shows that a vertex cell can add but a single arc to the transition graph

one-way, or bidirectional) to determine its contribution to the transition graph. 0D cells require no projection steps; we check to see if the zero vector is within the convex hull of adjacent 3D cells' f vectors. If so, we add an arc to itself in the transition graph.

The non-trivial strongly connected components (SCCs) of this transition graph represent the regions of nearly recurrent flow in the vector field. Here, non-trivial refers to SCCs that either contain more than one node, or components with a single node only if that node has an arc to itself.

By refining this transition graph, the upper bound on nearly recurrent sets can be made tighter. The refinement operation refers to the subdivision of an n-set into smaller n-sets. In the algorithm, edge pieces are subdivided into two equal sized smaller edge pieces and face pieces are split into four, in a quad-tree-like manner. These refined n-sets replace the original in the graph; to maintain the ability for the graph to represent all trajectories, we are required to update arcs after refinement. This update is designed to ensure that any two n-sets connected by a simple segment are connected by an arc in the refined graph.

If A is a node in the transition graph, and A' is one of the n-sets resulting from the refinement operation on A, we construct arcs out of A' by looping over all outgoing arcs $A \to B$ for A. If the connectivity check returns true, $A' \to B$ is inserted into the graph. This process is repeated for incoming arcs $B \to A$. A is then deleted or unlinked from the graph. Note that the connectivity check is based on an intersection test for n-sets projected as described earlier.

Overall, the algorithm takes the following steps. First a *coarse transition graph* is constructed from the input mesh. Its nodes correspond to 0D, 1D, and 2D grid cells. The SCCs are then computed. All nodes in trivial SCCs (i.e. SCCs that contain only a single node with no arc to itself) are removed with all incident arcs. If the user desires further refinement of the representation, all remaining nodes are then refined (only one at a time), and the same process is repeated by again computing SCCs, etc.

The most expensive part of the refinement operation is the connectivity test of A' and B to determine if an arc between them should be inserted in the graph. In the 3D implementation of [15], this requires a number of dot product comparisons. Deleting arcs is also an expensive operation, as we must maintain a doubly linked arc data structure. Inserting the arcs into the graph requires relatively less processing time, but is still considerable. We would like to take advantage of the fact that the connectivity tests can all be computed in parallel.

3.1 Naive Parallelization

As an initial attempt to gain a performance enhancement, the refinement process was naively parallelized. This approach added multi-threading to the refinement pass, with a large number of critical sections (forced serialization) to prevent consistency issues. The results of [15] are based off of this first effort.

Here, t threads (usually equal to the available logical cores of the host CPU) were spawned, each with an allotment of nodes to refine. Since the refinement process requires the addition and removal of arcs in the graph, it is disadvantageous to have two different threads operate on adjacent nodes. Thus, a preprocessing stage was run in an attempt to create a list of independent nodes, where no node n_i shares an arc with node n_j in the list. This list is then consumed by the threads, each taking a node, refining it, adding the subdivided nodes to the graph, updating the arcs, and then deleting/unlinking the original node.

There are many problems with this approach. A large amount of time is spent finding these independent nodes to process in parallel. We must also run the independence algorithm in a loop (as further independent sets can be extracted after visiting one set), exacerbating the problem. While the number of nodes that can eventually be processed in these independent sets is usually a large fraction of the overall graph, there will be a number of nodes left over that must be done in serial. A more optimal formulation would be able to process all nodes with no such expensive preparatory step.

While the situation varies from dataset to dataset, it is not uncommon to experience only an approximate 150 % speedup on eight core processors. To solve these problems, a reformulation of the algorithm is required.

3.2 Contribution

Our contribution is a revision of the algorithm in [15], allowing a better provisioning of processing resources to accelerate the refinement process. This new algorithm is not limited to dimension; it applies to both 2D and 3D cases.

4 Algorithm

In this section, we describe our parallel implementation of transition graph subdivision. The basic idea is to construct a new graph for each refinement round instead of executing refinements for one node at a time. In our implementation we describe the 3D case, but the process readily applies to 2D as well.

4.1 Transition Graph Representation

Before we discuss the new algorithm, we first describe an alteration to the underlying graph data structures used. The graph consists of many lists of nodes; each list consists of nodes that were obtained by subdividing the same grid cell. More precisely, with nodes organized into multiple lists, addition of nodes to the graph no longer requires a single graph global mutex as in previous implementations; we use a separate mutex for each list. The objective of this representation is to maximize concurrent use by taking advantage of our problem context.

While this is not fully optimal, adding a mutex for each node to maximize concurrency may cause memory problems as the graphs can grow to several millions of nodes with ease.

4.2 Transition Graph Construction

4.2.1 Coarse Graph

First, we construct a new empty transition graph G, whose nodes correspond to grid cells of dimension no more than 2 as described in [15]. The compute device will then receive a list of projection matrices to be used in the refinement operation. For

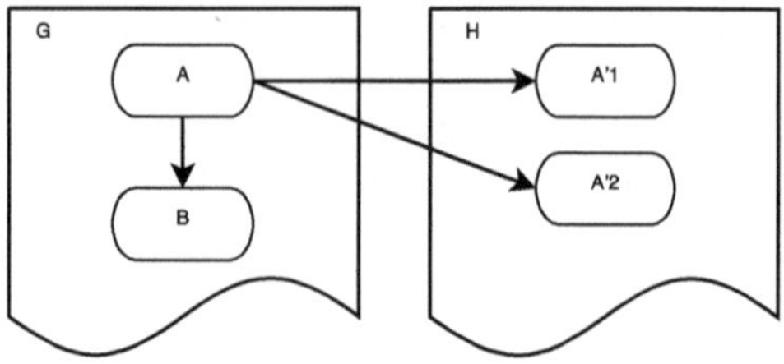

Fig. 2 Multiple graphs during subdivision. Node A is subdivided into $A'1$ and $A'2$, with pointers to these new child nodes. The parent node A is not removed; G remains a consistent transition graph while H is under construction

i rounds of refinement (specified by the user) we cycle through the main loop of our algorithm.

4.2.2 Refinement

To complete a single iteration, we first create an empty transition graph H, containing no nodes and no arcs. For each node A in the SCCs of G, we subdivide A and insert all its children into H. Again, we use the data structure outlined in Sect. 4.1, using a node list per grid cell. In particular, nodes belonging to different lists can be subdivided in parallel.

To populate H, we process the nodes in the strongly connected components of G. For each node A in an SCC of G, we subdivide A into child nodes $A'_1, \ldots, A'_n$. In 3D, this means edge n-sets being split in two and face n-sets being split into four pieces. These child nodes are inserted in H. The parent node A still possesses its connectivity information to other nodes in G. This is presented in Fig. 2.

Because of the change to graph implementation, node insertions can be done in parallel without the need for any critical sections by having threads operate on a cell level. A single thread subdivides all nodes that are owned by cell C, before moving to the next cell. In this manner, no thread will conflict with another over a cell's node list.

With H populated with nodes, we now determine which are to be connected by an arc. To do this, we examine the parent nodes. For each parent node A, we loop over all outgoing arcs $A \to B$ (with owner cell C). Our goal is to insert an arc from child node A' to B' in H if the connectivity check returns true. To do this, we add all A, B as well as C to the parallel processing list L. The goal of L is to provide a concise representation of all arcs along with arc owner information in a parallel-ready structure.

Each record in L is then processed in parallel, for example, by a kernel function. Since L is the only information sent to the device, before connectivity can be computed, the children of A and B are reconstructed on the compute device (this is not required if the compute device shares memory address space with the graph data structure).

While this duplicates the subdivision work of the previous step of the algorithm, the alternative is to fill the list with children nodes and their projection contexts. In the case of two 2D nodes (faces), this means 16 records (along with redundant

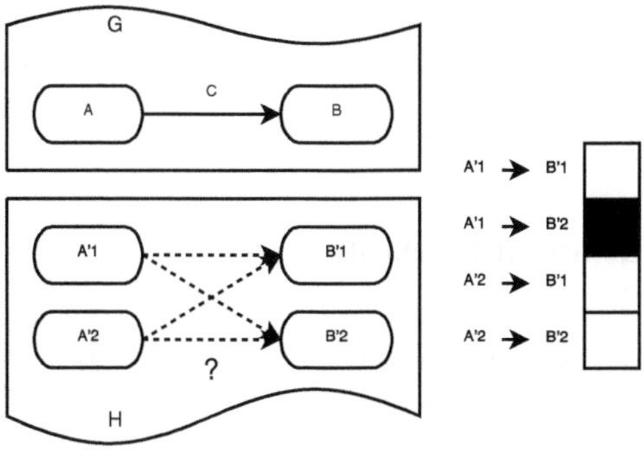

Fig. 3 Example of children arc testing and encoding. Nodes A, B and arc/owner cell C (which provides projection information) are the input elements, and the bitmap is returned as the connectivity output. In this example, there are two children each for nodes A and B. The children nodes here are temporarily constructed on the compute device, but their connectivity will be provided to the child nodes in H afterwards. The output bitmap shows one arc is to be constructed between the child nodes A'_1 and B'_2

projection information) must be added to the list and possibly transferred to the compute device. By only sending nodes and duplicating the child computation, we limit the effect of the memory transfer bottleneck between host and device, which is the commonly the slowest portion of the compute process. Further, the children of a node are relatively inexpensive to compute.

The kernel/thread will look up the associated projection information associated with C, project the children of A and B, and compute intersection tests to determine which children of A should be connected to which children of B. It then encodes this information in a bitmap O (each tuple owns a portion of O). An example of processing a record is in Fig. 3.

The device takes each child of A and computes the connectivity to each child of B, using the projection information of C. If the connectivity test between the two n-sets is true, we encode a 1 into the bitmap. In Fig. 3, we can see a single arc that must be constructed.

The bitmap O is returned from the compute device. We now insert arcs into H. For each pair of nodes $A \to B$ in the upload list, we will connect the children $A'_i \to B'_j$ if their bit is true in the bitmap. This operation can also be done roughly in parallel. Adding an arc requires the exclusive access of each nodes' arc lists. When processing a single bitmap slot in serial, this is a non-issue. For parallelization of the entirety of the bitmap, the thread will have to lock the mutex of A's owning grid cell, as well as the mutex of B's owning grid cell.

Next, O is returned from the device, where the bitmap is decoded. Whenever the bitmap indicates that the projections of two refined nodes intersect, arcs connecting them are inserted into H. G is now destroyed.

We compute the strongly connected components of H. As in the original algorithm [15], we use Tarjan's algorithm [17] to compute SCCs. Although this step could potentially be parallelized [2, 7], this is not the focus of this chapter.

Finally the contents of G are replaced with H (at most this requires swapping a pointer), to prepare for the next round of refinement. This process can be repeated as many times as the user desires.

5 Implementation and Results

In this section, we present two implementations of this algorithm, and their results on selected data sets.

5.1 Implementations

Our first implementation has been applied in part to run on a multi-core computer for 3D input data using multi-threading in a shared memory environment. The implementation has dual graph support, and fully parallelized connectivity testing. A buffer system has been added in preparation for OpenCL powered testing. Arc addition has been partially parallelized through the use of threading and mutexes.

In our second, experimental, implementation, we have added support for a CPU OpenCL compute device to build the arc connectivity bitmap. It shares the same technology of the first implementation, with the addition of pooling of nodes in memory. Because of the current design of OpenCL, this implementation will always use the maximum number of logical cores available on the platform to build the bitmap. As our graph implementation is contained in CPU addressable memory and is in a separate address space from OpenCL, we interpret and connect the arcs using threading and mutex exclusion.

5.2 Preliminary Results

To show improvements in the previous version of this algorithm, we test our approach with the same 3D datasets as from [15].

Lorenz: This dataset is a $24 \times 24 \times 24$ grid vector field originally presented in [11], where the vector field is described by $\dot{x} = \sigma(x - y)$, $\dot{y} = \rho x - y - xz$, $\dot{z} = xy - bz$. Using the common parameters $\sigma = 10$, and $b = \frac{8}{3}$, we use $\rho = 350$, which, by Mrozek and Piłarczyk [12], will provide a periodic orbit within the domain $[-150, 150] \times [-150, 150] \times [200, 500]$.

Bénard-Rayleigh Convection: This set is a simulation of convection by Daniel Weiskopf and appears in [18]. We have subsampled the dataset to $128 \times 32 \times 64$. Further, to reveal more detail in the structure of this dataset, we have striped one and two voxels from each side to create three datasets in total.

Table 1 Timing results in seconds

Dataset	Refinement	Old (s)	P-1 (s)	P-2 (s)	P-4 (s)	P-8 (s)	P-24 (s)	PCL-24-4 (s)
Lorenz ($\rho = 350$)	7	437	270	217	153	156	174	96
Bénard-Rayleigh	4	419	241	189	132	153	185	77
Bénard-Rayleigh (-1)	4	287	154	120	87	110	118	50
Bénard-Rayleigh (-2)	4	158	91	69	50	62	72	30
Square Cylinder	4	446	263	201	142	172	207	82

Square Cylinder: Prepared by Camarri et al. [4], this simulation places a square cylinder in a fluid flow. By subtracting the average velocity from the flow, we extract interesting swirling structures. This set has subsampled dimensions of $96 \times 32 \times 24$.

Table 1 provides timing of processing these datasets. The number of refinement iterations is listed, as well as the original processing times offered in [15] (the 'Old' algorithm, with naive parallelization, and the time needed to write the output excluded). Our improved algorithm was run several times on a compute node with differing numbers of assigned cores ('P-n', where n is the number of available cores for that run). We also show timings for our experimental OpenCL powered algorithm as 'PCL-24-4', using 24 cores to compute the arc connectivity bitmap, and 4 cores to connect arcs.

The algorithm was tested on a 24-core 2.4 GHz Intel® Xeon® E5645 node with 48 GB of RAM. Overall, the CPU used here and in [15] are comparable.

5.3 Discussion

Table 1 demonstrates improved performance for all input data sets, even in the case where only a single core was used, indicating better use of system resources when in serial mode. In the case of the Lorenz dataset, we reduced execution time from almost 8 min to under 3 min for our first implementation, and less than 2 min for the OpenCL implementation. This is promising, as there are still a number of improvements to make to the implementations to bring them fully in line with the described algorithm. Memory usage of these algorithms is roughly comparable with the algorithm in [15]; our algorithm adds a statically sized buffer for computation, but also uses a more RAM efficient graph representation.

While there are significant gains, the results show that there is not a linear improvement as we increase the number of cores used. This is ultimately the effect of Amdahl's Law [1]; with only the bitmap generation being perfectly parallelized, the weight of a highly contentious arc insertion procedure and a serial SCC algorithm become nontrivial. The OpenCL version is able to achieve better times by being able to use the full 24 cores available to compute connectivity, while only

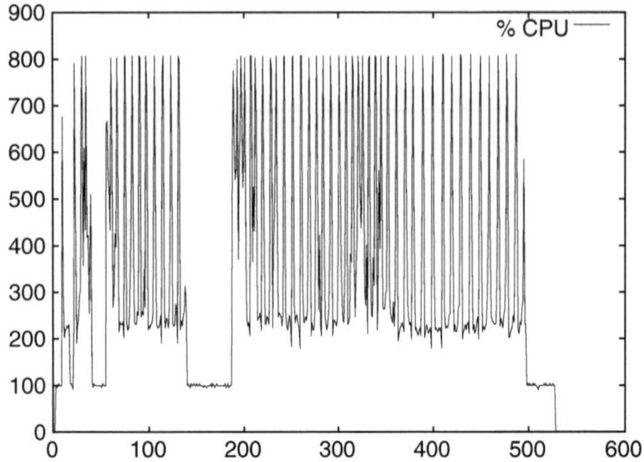

Fig. 4 Example runtime of the Bénard-Rayleigh (4R) dataset on eight processors ('P-4'). The x-axis represents the sample number (samples taken around four times a second), while the y-axis denotes the percent of system CPU activity

having 4 contentious threads to insert arcs, as opposed to the first implementation, which would use the same number of cores for both tasks.

Figure 4 shows an example CPU profile of an eight core processor node running the first parallel algorithm on the Bénard-Rayleigh dataset over four refinement iterations. The features we note are blocks of large spikes of parallelism of 800 % usage, separated by small troughs that extend to ≈200 % usage. These blocks are separated by larger troughs at 100 % usage (single core saturation). This last feature represents the strongly connected component algorithm. As this is not yet parallelized on our implementation, it runs on a single core. The time spent in this serial procedure (up to 20 % in some cases) is large enough to justify a parallel SCC addition to the software. Of greater interest are the troughs at 200 %. In our current implementation, we repeatedly fill a buffer to be processed and then compute the arc connections, repeating this every time the buffer is full. Here we see the buffer being processed (the large 800 % spikes) and the arcs being connected (200 % troughs). The arc connection portion is threaded, but uses mutexes to prevent concurrency related issues; low processor usage here denotes that the program is waiting for these mutexes to be unlocked by other threads.

Thus, while the larger issue of computing intersections in parallel has been solved, these two smaller issues (SCC computation and parallel arc insertion) must be strongly considered for revision in future implementations of this algorithm. As mentioned, parallel SCC algorithms have already been developed and are readily available. The problem of parallel arc insertion is more involved; double buffering, queues, and other methods could be used as a remedy. A possible solution involves the use of strong partitioning, where each thread/process owns a portion of the graph outright, so that no mutexes are used at any stage of the algorithm. This approach is specifically being explored as we consider out-of-core applications and distributed systems.

6 Conclusion

In this document, we have presented a parallel reformulation of the refinement process for piecewise constant transition graphs. The new approach separates connectivity testing from the arc insertion and node insertion procedures, showing improvements in computation performance even when run in serial. When run in parallel, run times are greatly reduced, facilitating study of larger datasets.

Further work on this subject includes parallel strongly connected component computation. As a larger goal, a reformulation of the algorithm intended for distributed systems (i.e. MPI) with out-of-core capability is also planned, using partitioned graphs. It is anticipated that this new scheme will eliminate issues from arc insertion, boosting performance while scaling to hundreds of cores.

References

1. G. Amdahl, Validity of the single-processor approach to achieving large scale computing capabilities, in *Proceedings of AFIPS Conference*, Atlantic City, 1967, pp. 483–485
2. J. Barnat, P. Bauch, L. Brim, M. Ceska, Computing strongly connected components in parallel on CUDA, in *Proceedings of the 2011 IEEE International Parallel & Distributed Processing Symposium*, IPDPS'11, Washington, DC (IEEE Computer Society, 2011), pp. 544–555
3. H. Bhatia, S. Jadhav, P.-T. Bremer, G. Chen, J.A. Levine, L.G. Nonato, V. Pascucci, Edge maps: representing flow with bounded error, in *Proceedings of IEEE Pacific Visualization Symposium*, Hong Kong, 2011
4. S. Camarri, M.-V. Salvetti, M. Buffoni, A. Iollo, Simulation of the three-dimensional flow around a square cylinder between parallel walls at moderate reynolds numbers, in *XVII Congresso di Meccanica Teorica ed Applicata*, Florence, Sept 2005
5. G. Chen, K. Mischaikow, R.S. Laramee, P. Pilarczyk, E. Zhang, Vector field editing and periodic orbit extraction using Morse decomposition. IEEE Trans. Vis. Comput. Graph. **13**, 769–785 (2007)
6. G. Chen, K. Mischaikow, R.S. Laramee, E. Zhang, Efficient Morse decompositions of vector fields. IEEE Trans. Vis. Comput. Graph. **14**, 848–862 (2008)
7. L. Fleischer, B. Hendrickson, A. Pinar, On identifying strongly connected components in parallel, in *Proceedings of the 15 IPDPS 2000 Workshops on Parallel and Distributed Processing*, IPDPS'00, London (Springer, 2000), pp. 505–511
8. R. Forman, A user's guide to discrete Morse theory, in *Proceedings of the 2001 International Conference on Formal Power Series and Algebraic Combinatorics*, Arizona. Advances in Applied Mathematics (2001)
9. A. Gyulassy, P.-T. Bremer, V. Pascucci, Computing Morse-Smale complexes with accurate geometry. IEEE Trans. Vis. Comput. Graph. **18**(12), 2014–2022 (2012)
10. R.S. Laramee, H. Hauser, L. Zhao, F.H. Post, Topology-based flow visualization, the state of the art, in *Topology-Based Methods in Visualization (Proceedings of TopoInVis 2005)*, Budmerice, ed. by H. Hauser, H. Hagen, H. Theisel (Springer, 2007), pp. 1–19
11. E.N. Lorenz, Deterministic nonperiodic flow. J. Atmos. Sci. **20**(2), 130–148 (1963)

12. M. Mrozek, P. Piłarczyk, The conley index and rigorous numerics of attracting periodic orbits, in *Conference on Variational and Topological Methods in the Study of Nonlinear Phenomena*, Pisa, 2000
13. J. Reininghaus, I. Hotz, Combinatorial 2D vector field topology extraction and simplification, in *Mathematics in Visualization (TopoInVis)*, Snowbird, 2009
14. J. Reininghaus, C. Lowen, I. Hotz, Fast combinatorial vector field topology. IEEE Trans. Vis. Comput. Graph. **99**, 1433–1443 (2010)
15. A. Szymczak, N. Brunhart-Lupo, Nearly recurrent components in 3D piecewise constant vector fields. Comput. Graph. Forum **31**(3), 1115–1124 (2012)
16. A. Szymczak, E. Zhang, Robust Morse decompositions of piecewise constant vector fields. IEEE Trans. Vis. Comput. Graph. **18**(6), 938–951 (2012)
17. R. Tarjan, Depth-first search and linear graph algorithms. SIAM J. Comput. **1**(2), 146–160 (1972)
18. D. Weiskopf, T. Schafhitzel, T. Ertl, Texture-based visualization of unsteady 3D flow by real-time advection and volumetric illumination. IEEE Trans. Vis. Comput. Graph. **13**(3), 569–582 (2007)

Part III
Simplification, Approximation, and Distance Measures

Notes on the Simplification of the Morse-Smale Complex

David Günther, Jan Reininghaus, Hans-Peter Seidel, and Tino Weinkauf

Abstract The Morse-Smale complex can be either explicitly or implicitly represented. Depending on the type of representation, the simplification of the Morse-Smale complex works differently. In the explicit representation, the Morse-Smale complex is directly simplified by explicitly reconnecting the critical points during the simplification. In the implicit representation, on the other hand, the Morse-Smale complex is given by a combinatorial gradient field. In this setting, the simplification changes the combinatorial flow, which yields an indirect simplification of the Morse-Smale complex. The topological complexity of the Morse-Smale complex is reduced in both representations. However, the simplifications generally yield different results. In this chapter, we emphasize properties of the two representations that cause these differences. We also provide a complexity analysis of the two schemes with respect to running time and memory consumption.

1 Introduction

The Morse-Smale (MS) complex [17, 21] has been proven useful in many applications due to its compact representation of the input data. However, a simplification of this complex is mandatory to determine the dominant features within the data.

D. Günther (✉)
CNRS LTCI, Institut Mines-Télécom, Télécom ParisTech, Paris, France
e-mail: gunther@telecom-paristech.fr

J. Reininghaus
IST Austria, Klosterneuburg, Austria
e-mail: jan.reininghaus@ist.ac.at

H.-P. Seidel • T. Weinkauf
MPI for Informatics, Saarbrücken, Germany
e-mail: hpseidel@mpi-inf.mpg.de; weinkauf@mpi-inf.mpg.de

P.-T. Bremer et al. (eds.), *Topological Methods in Data Analysis and Visualization III*, Mathematics and Visualization, DOI 10.1007/978-3-319-04099-8_9,

Over the last few years, two concepts have been established to represent the MS-complex. The first concept goes back to Edelsbrunner et al. [4] and was originally proposed for piecewise linear input data given on a triangulated domain. In this setting, the MS-complex is explicitly represented as a graph called the 1-skeleton. The graph consists of nodes and edges – called *links* in this chapter to avoid confusion with geometric edges. The nodes are the critical points of the data and the links describe their neighborhood relation given by the separatrices. The simplification of the MS-complex works iteratively on this graph. Two critical points can be removed from the graph if they are connected by a single link. After a removal, the neighborhood information of their respective adjacent critical points needs to be updated, i.e., the nodes of the graph are newly connected.

The second concept follows a formulation proposed by Forman [6] to describe the MS-complex. In contrast to the first approach, the MS-complex is only implicitly represented by a combinatorial gradient field. A simplification is now applied to the gradient field by changing the direction of the combinatorial flow along a separatrix; see right column of Fig. 1. This change in direction implicitly simplifies the MS-complex. Its explicit representation is computed from a simplified gradient field.

The motivation behind these two simplification strategies is different. While the explicit procedure refers to an explicit computation of a true simplified scalar field, the implicit case directly addresses the (combinatorial) gradient flow. However, both simplification procedures follow the same rules to determine which simplification step is admissible and they may also follow both the same principle to determine the next removal. But the results of the explicit and implicit simplification procedures differ, in general.

For 2D data sets, the difference is limited to the geometric embedding of the separation lines. For 3D data sets, on the other hand, the simplification procedures may even lead to different hierarchies: the order of the removal as well as the connectivity of critical points might differ. In this work, we investigate specific properties of the two representations and show that these properties can cause these differences. The different representations also affect the algorithmic side. Therefore, we also provide a complexity analysis of the two simplification schemes.

In Sect. 2, we provide a brief overview of prior work. Both simplification strategies are presented in Sect. 3, and we discuss specific properties in Sect. 4. We conclude this chapter in Sect. 5 by discussing the effects of these properties.

2 Related Work

In the following, we give a brief overview about the computation and simplification of the MS-complex in the discrete setting. The MS-complex [17, 21] consists of critical points (minima, saddles, maxima) and separatrices – the gradient lines that connect the critical points.

The MS-complex for 2D or 3D piecewise linear data can be computed using an approach proposed by Edelsbrunner et al. [3, 4]. The critical points are given by an

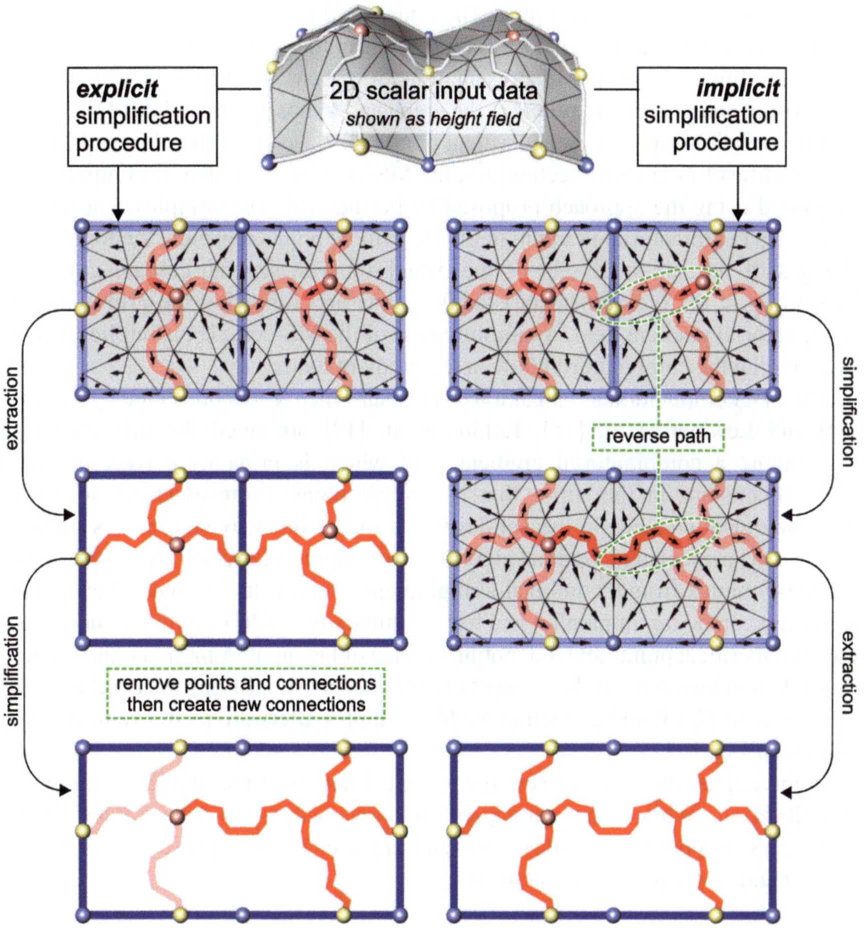

Fig. 1 Illustration of the explicit and implicit simplification of the MS-complex. The scalar input data is given on a discretization. From this input, the explicit and implicit simplification procedure computes a combinatorial gradient field that represents the MS-complex (first row). The *left column* shows the explicit procedure. From the gradient field, the MS-complex is extracted and represented as a graph (second row). The simplification now removes pairs of critical points in the graph (third row). This involves a local change of the connectivity of the affected critical points: incident connections are deleted and new connections are added. The *right column* shows the implicit procedure. The MS-complex is implicitly simplified by reversing the direction of the flow along a combinatorial gradient line that connects two critical points (second row). An explicit representation of the MS-complex is computed after the gradient field was simplified (third row)

analysis of the lower star of each vertex. The separatrices are approximated as a sequence of steepest edges in the triangulation. An extension to more general input was proposed by Gyulassy et al. [11] using a region-growing approach based on the definition [21] of the MS-complex. These approaches result in an explicit graph representation of the MS-complex. The nodes of this graph are the critical points and its links represent the separatrices that connect the critical points.

A simplification of this representation of the MS-complex is obtained by eliminating pairs of critical points in the graph and a subsequent update of the neighborhood relationship of adjacent critical points [4, 9]; see the left column of Fig. 1.

In contrast to the above techniques, the MS-complex can also be computed and simplified using the approach proposed by Forman [6]. The complex is implicitly given by a combinatorial gradient field. The critical points represent the topological changes of the sub-level sets of the data [16]. The separatrices are computed by starting at the (combinatorial) saddle points and following the grid along the combinatorial gradient field. A simplification of the MS-complex is implicitly done by changing the combinatorial flow in the gradient field.

The first computational realization of Forman's theory was presented by Lewiner [14] and Lewiner et al. [15]. Robins et al. [19] presented the first algorithm to compute a combinatorial gradient field which is provably correct in up to three dimensions in the sense that its critical points correspond one-to-one to the topological changes in the sub-level sets of the input. Based on this method, Günther [7] presented an optimal algorithm to compute an explicit representation of the MS-complex from a combinatorial gradient field for the 2D and 3D case. This algorithm can be run in parallel and has a complexity of $O(cn)$ with c denoting the number of critical points and n denoting the size of the input. A memory and running time efficient hierarchy of this complex is obtained by simplifying the combinatorial gradient field [7, 18] and extracting the MS-complex afterward as shown in the right column of Fig. 1.

In the next section, we present the technical details of the explicit and implicit simplification. While other strategies are conceivable, we concentrate on the techniques proposed by Gyulassy [9] and Günther [7] as representatives for the explicit and implicit case, respectively.

3 Simplification of the Morse-Smale Complex

In this section, we briefly present the explicit and implicit simplification scheme. The basis for both approaches is an initial combinatorial gradient field that describes the combinatorial flow of the input data. Although other settings are conceivable, we concentrate on the setting of discrete Morse theory [6].

3.1 Combinatorial Gradient Field and Morse-Smale Complex

A combinatorial gradient field represents the gradient flow of a given input but restricted to the discretization of the underlying domain. The discretization decomposes into cells of different dimensions. Assuming that the domain is given

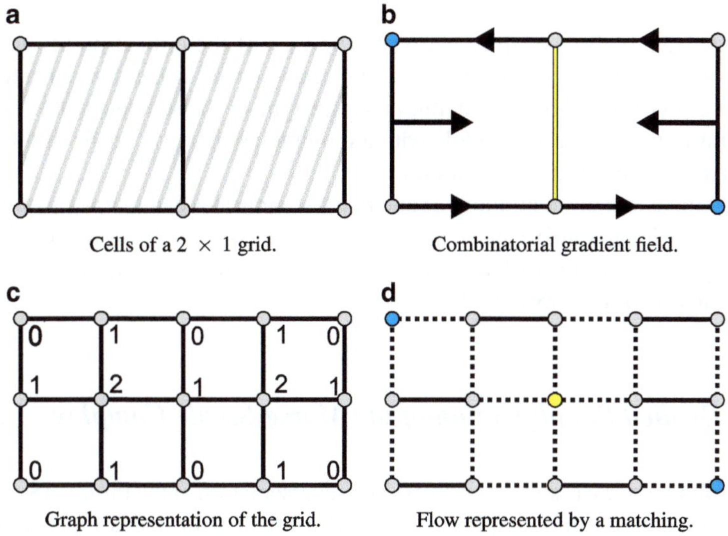

Fig. 2 Illustration of a combinatorial flow. (**a**) shows the cells of 2D grid: vertices (*circles*), edges (*black lines*), quads (*shaded area*). (**b**) shows a combinatorial gradient field: cells of consecutive dimension are paired, unpaired cells (*blue* and *yellow*) are critical cells. (**c**) shows a graph representation of the grid shown in (**a**): each node (sphere) is labeled by the dimension of the cell it represents, the adjacency of the cells is represented by links (*black lines*). (**d**) shows a matching representation of the combinatorial flow shown in (**b**): paired cells are represented by links (*solid black lines*), nodes with no incident matched links are critical (*blue* and *yellow*)

as a 2D grid as shown in Fig. 2a, the cells are the k-dimensional entities of the grid: vertices, edges, quads. Similarly, higher dimensional grids or triangulations also decomposes into cells. A combinatorial gradient field now pairs cells of consecutive dimension yielding a combinatorial flow restricted to the entities of the discretization, see Fig. 2b. Unpaired cells are the combinatorial analogue of the continuous critical points [6].

Recently, several approaches [9,13,20] to construct such a combinatorial gradient field based on an input field are proposed. The first provably correct algorithm, however, is proposed by Robins et al. [19]. While all of the approaches work in practice, it is beneficial to use this algorithm since the topological structures in the gradient field are reduced to a minimum, i.e., all structures correspond to evolution of the sub-level sets and no falsely identified structures are present.

Given a combinatorial gradient field, an explicit representation of the MS-complex C^{MS} can be computed using the intersection of the underlying ascending and descending manifolds [21]. The MS-complex consists of the critical points and separatrices. The critical points are the unpaired cells in the combinatorial flow (Fig. 2b). The dimension k of the cells defines the type of the critical point. For a 2D input, we call a critical point a minimum ($k = 0$), a saddle ($k = 1$), or a maximum ($k = 2$). In case of a 3D input, the critical points are called: minimum ($k = 0$), 1-saddle ($k = 1$), 2-saddle ($k = 2$), or maximum ($k = 3$).

The separatrices, on the other hand, are special combinatorial gradient lines that connect critical points of different type. The type of the critical points define the dimensions of the cells a separatrix is allowed to cover in the discretization. The smaller dimension of the cells defines then the type of the separatrix. For example, a separatrix that connects a saddle with a minimum only covers cells of dimension zero and one; we call it a 0-separatrix.

In contrast to their continuous counterpart, separatrices can merge or split in 2D. In 3D, a single separatrix can even merge *and* split. As we will see in the following, this property causes the differences in the simplification of the explicit and implicit representation of the MS-complex.

3.2 Explicit Simplification of the Morse-Smale Complex

In the following, we assume that an explicit representation of the initial MS-complex C_0^{MS} is given, i.e., the complex is given as a graph with nodes representing the critical points and links representing the separatrices. Topologically simplifying this graph means reducing the number of nodes and consequently the topological complexity of the MS-complex. Based on a simplification guideline, pairs of critical points are iteratively removed from C^{MS} which yields a hierarchy $\mathscr{C}$ of MS-complexes

$$\mathscr{C} = (C_k^{MS})_{k=0,\ldots,m}. \tag{1}$$

A pair of critical points is a valid candidate for a removal if it is connected by a single link. Let p and q denote two critical points of index ℓ and $\ell+1$, respectively. We denote the ℓ-neighborhood of a critical point q by N_q^ℓ, i.e., N_q^ℓ contains all critical points of index ℓ that are connected to q in the MS-complex. Let the pair (p,q) be connected by a single link, i.e., $q \in N_p^{\ell+1}$ and $p \in N_q^\ell$.

The removal of the pair (p,q) now changes the neighborhood of all their adjacent critical points. The removal in the graph consists of two operations:

1. The nodes p and q and all their incident links are deleted from the graph.
2. New connections between each node in $N_p^{\ell+1} \setminus \{q\}$ and $N_q^\ell \setminus \{p\}$ are created.

The simplification process terminates if no valid pair of critical points can be found in the MS-complex. For more details on this simplification, we refer to [9].

3.3 Implicit Simplification of the Morse-Smale Complex

In the following, we assume that an initial combinatorial gradient field V_0 is given. In contrast to the explicit representation, the implicit simplification directly addresses the combinatorial flow. Let a pair of critical points which is uniquely connected by

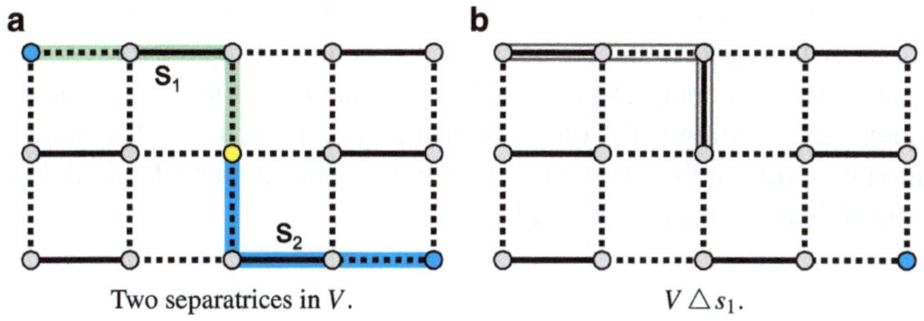

Fig. 3 Illustration of the implicit simplification. (**a**) shows two separatrices (*blue* and *green lines*) that begin at a saddle (*yellow circle*) and end in two minima (*blue circles*). (**b**) shows the change of the combinatorial flow along s_1 (*bounded region*). The incident saddle and minimum of s_1 are matched and no longer critical. The symbol Δ denotes the symmetric difference

a separatrix be given. A simplification is now done by reversing the combinatorial flow along the separatrix. Applying the simplification iteratively yields a sequence $\mathcal{V}$ of combinatorial gradient fields

$$\mathcal{V} = (V_k)_{k=0,\dots,m} \,. \tag{2}$$

From an algorithmic point of view, the simplification can be best expressed in a graph-theoretical notation. A combinatorial gradient field works on the cells of a given discretization, i.e., the vertices, edges, triangles, quads, cubes, etc. Each discretization can be represented as a graph $G = (N, E)$. The nodes N of the graphs are the k-dimensional cells and its links E describe the adjacency of these cells, see Fig. 2c for a 2D example. The combinatorial flow is expressed as a pairing of cells. This pairing can be represented by a subset of links $V \subset E$. Since each cell can only occur in one pair [6], none of the links in V are adjacent; hence, V is a matching. Figure 2d shows a 2D example.

We now interpret the combinatorial gradient field V_0 as a matching. In graph theoretical terms, an combinatorial separatrix $s \subset E$ connecting two critical points is an augmenting path since it is alternating and its start- and end-node are not matched (Fig. 3a). Hence, we can produce a larger matching $V_{k+1} \subset E$ by taking the symmetric difference

$$V_{k+1} = V_k \,\Delta\, s. \tag{3}$$

Equation (3) is called *augmenting* the matching. Since the incident critical nodes of s are matched after the augmentation, the number of critical nodes is decreased by two. Note that Eq. (3) does not depend on the dimension of G, i.e., we can apply the augmentation to 2D as well as 3D data.

The symmetric difference in Eq. (3) reverses the direction of the combinatorial gradient flow. Loosely speaking, the augmentation flips the direction of the arrows representing the combinatorial flow. An illustration of the augmentation is given in

Fig. 3b. The simplification stops if the matching cannot be augmented anymore, i.e., there are no critical points that are connected by a unique separatrix.

The sequence $\mathcal{V}$ is completely defined by the final matching V_m and the (typically rather short) augmenting paths (p_i). An arbitrary element $V_k \in \mathcal{V}$ can be restored by iteratively applying Eq. (3) with respect to (p_i). For a more detailed description of this simplification, we refer to [7, 18].[1]

4 Differences in the Simplifications of the Morse-Smale Complex

In this section, we elucidate the properties of the explicit and the implicit simplification schemes. Both schemes are based on different representations of the MS-complex. The representations influence not only the geometric embedding of the separatrices but also the order of simplifications. In general, both schemes yield different hierarchies of MS-complexes.

4.1 Geometric Embedding of Separatrices

The geometric embedding of the separatrices in the initial MS-complex coincides in the explicit and implicit representation. They are defined by the combinatorial gradient field, and their embedding is given by the cells of the discretization which they cover, see Fig. 4 (left). However, differences occur if the complex is simplified.

In the explicit representation, critical points are newly connected after a simplification. Those new connections involve a merge of separatrices. Figure 4a shows an example. The separatrices (blue lines) emerging at the two saddles (yellow circles) meet each other at a non-critical point. Both separatrices share the same cells from this point on to the central minimum (blue circle). If the saddle and the central minimum are now removed from the MS-complex, the two separatrices merge, which yields a partly overlap of the separatrix with itself.

In the implicit representation, on the other hand, the situation is different, see Fig. 4b. The removal of the right saddle and the central minimum is done by changing the combinatorial flow along the separatrix connecting these two points. The *new* separatrix that emanates from the left saddle directly goes to the right minimum without any self-overlapping.

This situation also occurs in 3D. However, it only affects the geometric embedding of the separatrices and not the simplification process. In the following, we

[1]Note that the idea of reversing the flow along a separation line was also used to modify a scalar field based on a simplified contour tree [23].

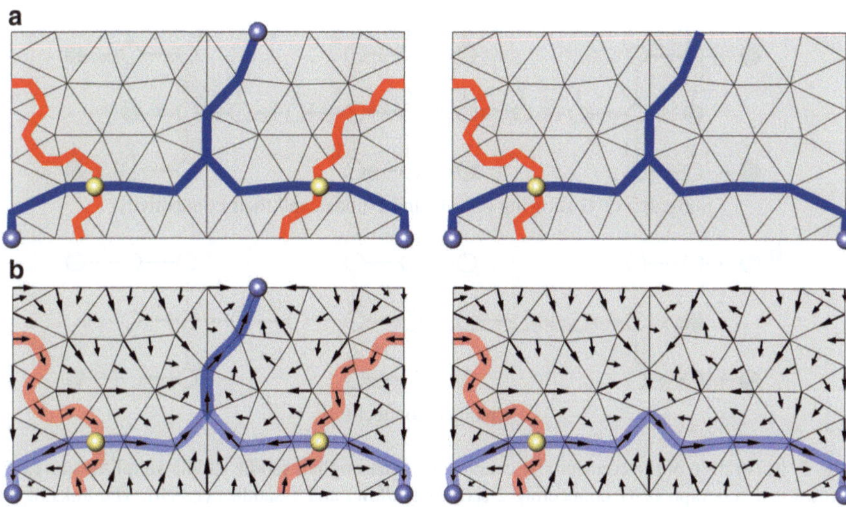

Fig. 4 The explicit and implicit simplification may create different embeddings of the separatrices. (**a**) Before and after the simplification using the *explicit* representation. Two formerly distinct separatrices (*blue*) are stitched together to one new separatrix that partial overlaps with itself. (**b**) Before and after the simplification using the *implicit* representation. Reversing the flow between the two canceled critical points leads to a new embedding, which does not overlap itself

discuss two properties in the explicit and implicit representation which can yield different hierarchies.

4.2 Connectivity of Critical Points

The following situation only occurs when a single separatrix can merge *and* split. Hence, we assume 3D data as input and concentrate on a saddle-saddle simplification, i.e., the simplification of an 1- and 2-saddle and the involved 1-separatrices. For explanatory reasons, we use the above graph-theoretical notation in the following. This allows us to embed the saddle-saddle connections in the plane, see Fig. 5.

A simplification of a saddle-saddle pair can completely change the connectivity of the adjacent saddles in the implicit representation. Figure 5 depicts such a situation. It may happen that the 1-separatrices of multiple 2-saddles merge and share several links before they split again and end in 1-saddles, see Fig. 5a. The shared links describe in some sense a narrow one-way street. All 1-separatrices on the left side of this one-way street must cross it to enter the right side.

However, simplifying the gradient field along one of the 1-separatrices (the blue line in Fig. 5a) changes the connectivity of the saddles. Before the simplification, all 2-saddles (yellow) were connected to the 1-saddles (green) on the right side. Only the central 2-saddle was also connected to the 1-saddles on the left side. After the

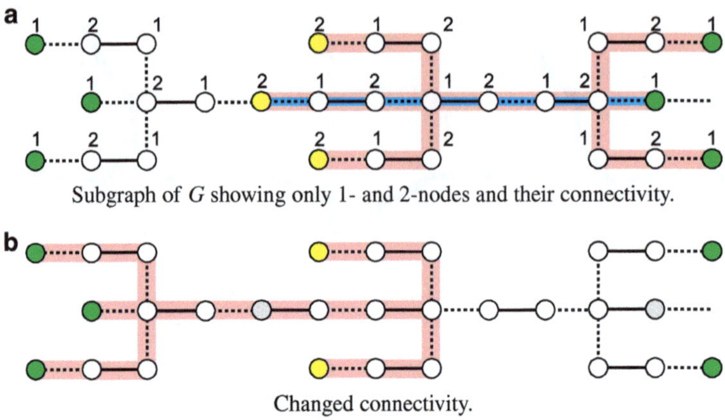

Fig. 5 Illustration of the connectivity change due to an implicit saddle-saddle simplification (3D). Shown is a subgraph of G connecting 2-saddles (*yellow*) and 1-saddles (*green*). The links of the matching are depicted as *black solid lines*. The *blue line* in (**a**) depicts an augmenting path. After the augmentation, the connectivity of the saddles (*red lines*) completely changed (**b**)

simplification, none of the remaining 2-saddles is connected to the right 1-saddles anymore, see Fig. 5b. There is no 1-separation line connecting them. All of them are now connected to the 1-saddles on the left side, which was not the case before.

In the explicit representation, on the other hand, the above situation does not occur. Each saddle-saddle pair and its connectivity is separately stored. Although the geometric embedding of the connections may partly coincide, each connection is independently treated. A simplification only removes the current pair and its connections from the MS-complex, and new connections between all neighboring critical points are created [9]. Therefore, the green 1-saddles on the right are still connected to the two remaining 2-saddles, and they form valid candidates for further simplification in the explicit representation.

Since the neighborhood of a saddle changes differently, the above situation causes different hierarchies in the explicit and implicit representation, in general. However, the differences are caused by the fact that 1-separation lines partly overlap in the discrete setting. Given an initial combinatorial gradient field with non-overlapping separation lines, both schemes would yield the same result.

4.3 Iterative Simplification

A common way to guide the simplification is the use of the height difference of adjacent critical points [7,9,14]. This heuristic assumes that unimportant and noise-induced critical points differ only slightly in their scalar value. Applying an iterative simplification based on this heuristic yields a hierarchy of MS-complexes. The initial MS-complex contains all fine-grained structures in the data. The last level, on the other hand, only contains the large-scale structures of the data.

Pairs of critical points are iteratively removed from the MS-complex such that each pair represents the currently smallest height difference of all possible pairs. This height difference guided removal assigns an importance value to each critical points allowing to distinguish spurious and dominant critical points.

This importance measure is closely related to the concept of persistent homology [5]. In 2D, it was shown that persistent homology can also be used to simplify the MS-complex [4]. The *adjacency lemma* guarantees that two critical points are connected by a unique separatrix at the moment when they should be canceled.

Recently, it was also shown that the pairing generated by the height difference produces the same pairing as by persistent homology [2] for data given on a smooth 2D manifold. However, this is no longer true in 3D. It was shown [1] that there are pairs of critical points generated by persistent homology that can not be obtained by a sequential removal of critical points as described in Sects. 3.2 and 3.3. Therefore, the assessment of importance of critical points based on the height difference differs from persistence, in general.

4.4 Monotonicity of the Simplification

As discussed in Sect. 4.3, the height difference is typically used to guide the simplification assuming that it represents the currently smallest fluctuation. Therefore, a natural assumption is that the height difference ω is monotonically increasing over the simplification process. While this is the case in 2D for the explicit/implicit representation and also for the explicit representation in 3D, it is no longer monotone in the 3D implicit simplification. This stems from the fact that 1-separatices in 3D can merge *and* split. Figure 6 shows such an example together with the combinatorial flow. In the following, we concentrate on saddle-saddle simplifications in an implicit representation and investigate this non-monotonic behavior in detail. For simplicity, we use the above graph-theoretical notation in the following.

Due to the degree of the 1- and 2-nodes, the 1-separation lines can merge *and* split. Consider a subgraph with different saddle-saddle pairs as shown in Fig. 7a. Assume that the central pair represents the smallest height difference ($\omega = 1$) and that this pair is next in line for a simplification. However, this pair cannot be removed since there are three paths connecting them, as shown in Fig. 7b. Since the saddles could still be canceled with their adjacent extrema, they are deferred with a weight defined by these extrema. In the next steps, other saddle pairs are removed with a greater weight ($\omega = 2$), see Fig. 7c. Parts of the corresponding augmenting paths, however, share links with the paths connecting the central pair. Due to the augmentations, the orientation of the links is changed, and the central pair is suddenly uniquely connected. If the central yellow saddle is now next in line, its connectivity is recomputed and the central green saddle is determined as the partner with the smallest height difference ($\omega = 1$). Since the connection between them is now unique, the path allows for an augmentation. However, the weight of the augmentation is smaller than in the prior operations.

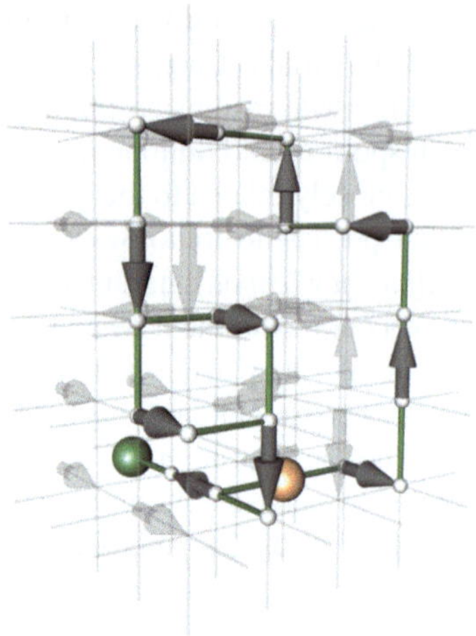

Fig. 6 Illustration of the 3D flow (*grey arrows*) along a split-merge of a 1-separation *line* (*green*)

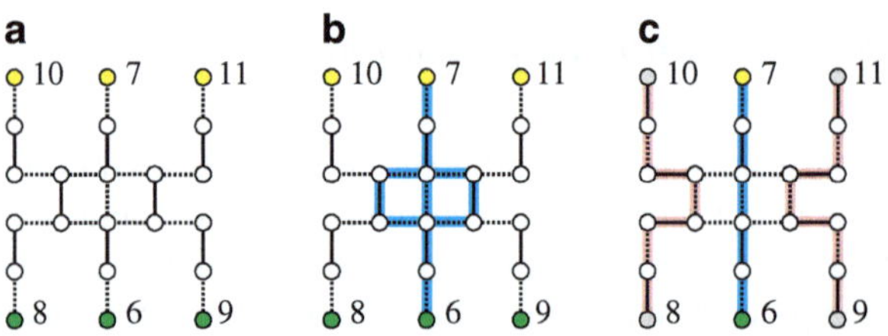

Fig. 7 Sketch of a combinatorial gradient field in a subgraph of G (3D) connecting 2-saddles (*yellow*) and 1-saddles (*green*). The critical nodes are labeled with their assigned scalar value. The links of the matching are depicted as *black solid lines*. (**a**) Subgraph of G containing only 1- and 2-nodes. (**b**) Multiple paths (*blue*) between two saddles. (**c**) Unique path (*blue*) after several augmentations (*red*)

In practice, the breaks of monotony in the weight sequence are caused by noise. The perturbations introduced by it yield to short split-merge sequences in the 1-separation lines as depicted in Fig. 7b. To analyze this behavior, we sampled the artificial function g defined in [8] on an 128^3 grid and added two levels of uniform noise to it. Figure 8 shows the results. The 1-separation lines in the pure function g are well distributed. No deferring of saddles in the above sense can be observed. The weights of the augmenting paths are monotonically increasing, see Fig. 8a. However, adding a small amount of noise in the range of $[-0.5, 0.5]$ results in 12 monotony breaks, see Fig. 8b. If we add noise in the range of $[-1, 1]$, the number of breaks increases further to 59, see Fig. 8c. This indicates that the level of noise heavily influences the number of splits in the 1-separation lines.

In contrast, the height difference is monotonically increasing in the explicit representation. A single simplification only considers the critical points. It does not consider if the connection between a pair partially overlaps with other separation

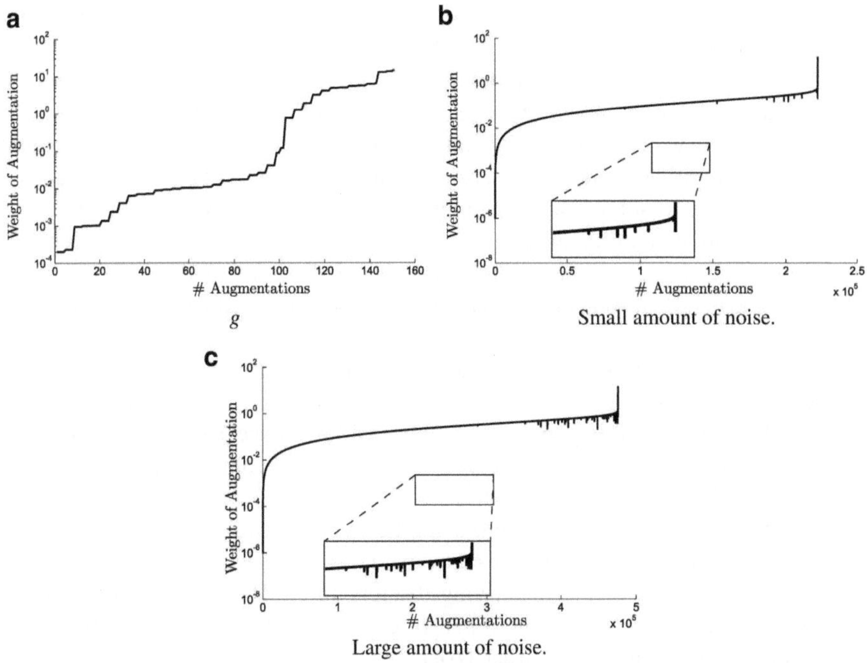

Fig. 8 Graph of the weights of the augmenting paths. (**a**) shows the weights of augmentations for the artificial function g [8] over the number of augmentations. The weights behave monotonically increasing. The monotony is broken in (**b**) where a small amount of noise is added to g. The number of monotony breaks further increases if the level of noise is increased (**c**)

lines (since all connections are treated independently). Hence, a removal of a critical point pair does not affect other connections. A saddle pair which is connected by multiple separation lines can never be removed from the MS-complex, in contrast to the implicit representation. The height difference during the simplification process is therefore monotonically increasing in the explicit representation.

4.5 Computational Complexity and Memory Consumption

In the following, we give a brief discussion about the computational complexity and the memory consumption of the explicit and implicit simplification in case of a 3D input. We denote the number of vertices by n and the number of saddles by m. Note that in the worst-case scenario there holds $m \approx n$.

We begin with the explicit representation. Let (p, q) be a critical point pair with indices ℓ and $\ell+1$, respectively. Let N_q^ℓ be the set of adjacent critical points of index ℓ in the graph representation. A removal of (p, q) deletes all incident links of p and q and creates new links between the nodes $N_q^\ell \setminus \{p\}$ and $N_p^{\ell+1} \setminus \{q\}$. Using efficient data structures with constant *insert* and *delete* operations [9], the complexity of a removal is therefore $O(|N_q^\ell| \times |N_p^{\ell+1}|)$. Since a saddle can be connected to m other saddles [22], the worst-case complexity of a single simplification step is $O(m^2)$. Hence, the overall complexity of performing all simplifications is $O(m^3)$.

The adjacency of the critical points is explicitly stored. Each critical point contains a list with its neighboring critical points. Since a saddle can be connected to m saddles, the total memory consumption is $O(m^2)$ in the explicit scheme.

In the implicit representation, the connectivity of a critical point needs to be recomputed since previous simplification steps may have affected its incident separatrices. The neighborhood of a critical point can be computed with a complexity of $O(n)$ using a restricted breadth-first search [7]. Since the simplification procedure is guided by an importance measure such as the height difference of critical points, it needs to be checked if the current pair represents the *best* pair. In the worst case, this check involves the comparison to m other saddles. Hence, a single simplification step has a worst-case complexity of $O(n\,m)$. The overall complexity of performing all simplifications is therefore $O(n\,m^2)$.

The adjacency of the critical points is implicitly represented by the combinatorial gradient field. The representation of such a field can be realized by a Boolean vector representing all cells of the underlying discretization. The size of this vector is therefore given by the size of the input. In contrast to the explicit representation, the memory consumption is of order $O(n)$.

5 Discussions and Conclusions

We investigated the properties of the explicit and implicit simplification procedures and showed that the corresponding results may be different in terms of geometry and topology. In particular, the simplification procedures generally lead to different hierarchies of MS-complexes in 3D data sets. Nevertheless, it should be stressed that both simplification procedures yield valid MS-complexes.

Theoretically, the simplification procedures would give the same results in 3D if separation lines would not partially overlap. However, this is rarely the case for real-world data. The different results, as discussed in this chapter, are caused by saddle-saddle simplifications. Saddles correspond to tunnels and cavities of the level sets and our practical observations are that qualitative differences regarding these features are often negligible. As shown in several examples [7,9], both simplification procedures are able to distill the essential features of the input.

Algorithmically, the main difference between the two simplification procedures is their memory consumption. In the worst case, the implicit method has a linear memory consumption while the explicit method has a quadratic memory consumption. Additional heuristics controlling the order of simplifications can mitigate this issue for the explicit case [10]. However, this also introduces a computational parameter, which depends on the input data and needs to be adjusted by the user. In the implicit case, contrarily, no computational parameter is needed.

In 2D and 3D, the geometric embedding of the separation lines can differ between the two schemes. In the explicit scheme, two lines can merge into a single partially overlapping line, i.e., cells in the underlying domain are visited multiple times. In the implicit representation cells can only be visited once allowing no overlap. This might be useful for noisy data since spurious lines are automatically removed.

We want to stress that both simplification schemes are based on the height difference heuristic. In 3D, the initial MS-complex cannot be perfectly simplified. As shown by Joswig and Pfetsch [12] and illustrated by Bauer [1], this is an NP-complete task. In practice, this results in the situation that some critical points – which represent spurious features – cannot be removed.

In the both schemes critical points are always removed in pairs. However, it would be interesting how clusters of critical points could be consistently canceled. Such a removal might also allow for the removal of spurious critical points in 3D. However, this needs to be done without creating closed combinatorial streamlines.

Acknowledgements This research is supported and funded by the Digiteo *unTopoVis* project, the *TOPOSYS* project FP7-ICT-318493-STREP, and MPC-VCC.

References

1. U. Bauer, Persistence in discrete Morse theory. PhD thesis, University of Göttingen, 2011
2. T. Dey, K. Li, C. Luo, P. Ranjan, I. Safa, Y. Wang, Persistent heat signature for pose-oblivious matching of incomplete models. CGF **29**(5), 1545–1554 (2010)
3. H. Edelsbrunner, J. Harer, V. Natarajan, V. Pascucci, Morse-Smale complexes for piecewise linear 3-manifolds, in *19th Annual Proceedings of SoCG*, San Diego (ACM, New York, 2003), pp. 361–370
4. H. Edelsbrunner, J. Harer, A. Zomorodian, Hierarchical Morse complexes for piecewise linear 2-manifolds. Discret. Comput. Geom. **30**, 87–107 (2003)
5. H. Edelsbrunner, D. Letscher, A. Zomorodian, Topological persistence and simplification. Discret. Comput. Geom. **28**, 511–533 (2002)
6. R. Forman, A user's guide to discrete Morse theory, in *Proceedings of the 2001 International Conference on Formal Power Series and Algebraic Combinatorics, USA*. Advances in Applied Mathematics (2001)
7. D. Günther, Topological analysis of discrete scalar data. PhD thesis, Saarland University, Saarbrücken, Germany, 2012
8. D. Günther, J. Reininghaus, H. Wagner, I. Hotz, Efficient computation of 3D Morse-Smale complexes and persistent homology using discrete Morse theory. Vis. Comput. **28**, 959–969 (2012)
9. A. Gyulassy, Combinatorial construction of Morse-Smale complexes for data analysis and visualization. PhD thesis, University of California, Davis, 2008
10. A. Gyulassy, P.-T. Bremer, V. Pascucci, B. Hamann, Practical considerations in Morse-Smale complex computation, in *Proceedings of the TopoInVis*, Zurich (Springer, 2011), pp. 67–78
11. A. Gyulassy, V. Natarajan, V. Pascucci, B. Hamann, Efficient computation of Morse-Smale complexes for three-dimensional scalar functions. TVCG **13**, 1440–1447 (2007)
12. M. Joswig, M.E. Pfetsch, Computing optimal Morse matchings. SIAM J. Discret. Math. **20**(1), 11–25 (2006)
13. H. King, K. Knudson, N. Mramor, Generating discrete Morse functions from point data. Exp. Math. **14**(4), 435–444 (2005)

14. T. Lewiner, Geometric discrete Morse complexes. PhD thesis, PUC-Rio, 2005
15. T. Lewiner, H. Lopes, G. Tavares, Optimal discrete Morse functions for 2-manifolds. Comput. Geom. **26**(3), 221–233 (2003)
16. J. Milnor, *Morse Theory* (Princeton University Press, Princeton, 1963)
17. M. Morse, *The Calculus of Variations in the Large*. Colloquium Publications, vol. 18 (AMS, New York, 1934)
18. J. Reininghaus, Computational discrete Morse theory. PhD thesis, Freie Universität, 2012
19. V. Robins, P.J. Wood, A.P. Sheppard, Theory and algorithms for constructing discrete Morse complexes from grayscale digital images. IEEE PAMI **33**(8), 1646–1658 (2011)
20. N. Shivashankar, V. Natarajan, Parallel computation of 3D Morse-Smale complexes. Comput. Graph. Forum **31**(3pt1), 965–974 (2012)
21. S. Smale, On gradient dynamical systems. Ann. Math. **74**, 199–206 (1961)
22. H. Theisel, T. Weinkauf, H.-C. Hege, H.-P. Seidel, On the applicability of topological methods for complex flow data, in *Proceedings of the TopoInVis*, Grimma (Springer, 2007), pp. 105–120
23. G. Weber, S. Dillard, H. Carr, V. Pascucci, B. Hamann, Topology-controlled volume rendering. IEEE Trans. Vis. Comput. Graph. **13**(2), 330–341 (2007)

Measuring the Distance Between Merge Trees

Kenes Beketayev, Damir Yeliussizov, Dmitriy Morozov, Gunther H. Weber, and Bernd Hamann

Abstract Merge trees represent the topology of scalar functions. To assess the topological similarity of functions, one can compare their merge trees. To do so, one needs a notion of a distance between merge trees, which we define. We provide examples of using our merge tree distance and compare this new measure to other ways used to characterize topological similarity (bottleneck distance for persistence diagrams) and numerical difference (L_∞-norm of the difference between functions).

K. Beketayev (✉)
Lawrence Berkeley National Laboratory, One Cyclotron Rd, Berkeley, CA 94720, USA

Nazarbayev University, 53 Kabanbay Batyr Ave, Astana, 010000, Kazakhstan
e-mail: kenes.b@gmail.com; KBeketayev@lbl.gov

D. Yeliussizov
Kazakh-British Technical University, 59 Tole Bi St, Almaty, 050000, Kazakhstan
e-mail: yeldamir@gmail.com

D. Morozov
Lawrence Berkeley National Laboratory, One Cyclotron Rd, Berkeley, CA 94720, USA
e-mail: DMorozov@lbl.gov

G.H. Weber
Lawrence Berkeley National Laboratory, One Cyclotron Rd, Berkeley, CA 94720, USA

Department of Computer Science, Institute for Data Analysis and Visualization (IDAV), University of California, Davis, CA 95616-8562, USA
e-mail: GHWeber@lbl.gov

B. Hamann
Department of Computer Science, Institute for Data Analysis and Visualization (IDAV), University of California, Davis, CA 95616-8562, USA
e-mail: hamann@cs.ucdavis.edu

P.-T. Bremer et al. (eds.), *Topological Methods in Data Analysis and Visualization III*, Mathematics and Visualization, DOI 10.1007/978-3-319-04099-8_10,

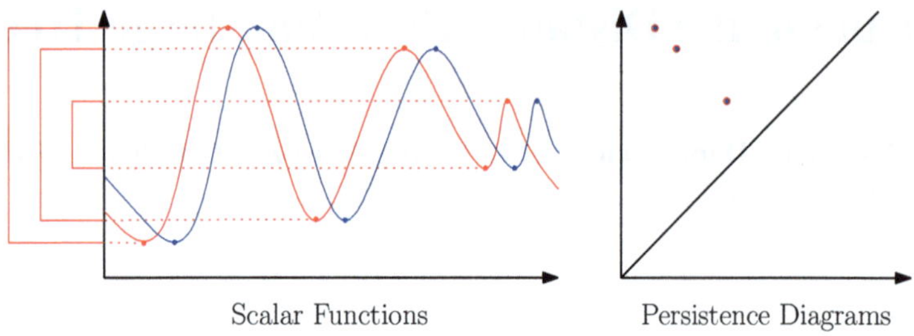

Fig. 1 Consider two scalar functions, where one is a slightly shifted version of the other. Comparing them directly, e.g., via the L_∞ norm, results in a large difference. Their persistence diagrams are the same, thus capturing the topological similarity of these functions

1 Introduction

Many aspects of physical phenomena are described and modeled by scalar functions. Computational and experimental capabilities allow us to approximate scalar functions at increasing levels of detail and resolution. This fact makes it necessary to analyze and also compare such function automatically, when possible, and to include more abstract analysis methods. Topological methods, based on the characterization of a scalar function via its critical point behavior, are gaining in importance, and we were therefore motivated to investigate the feasibility of comparing scalar functions using their topological similarity. Computational chemistry, physics and climate sciences are just a few applications where our ideas presented here should be valuable.

We address the generic problem of comparing the topology of scalar functions. Figure 1 demonstrates this problem. The figure shows slightly shifted versions of the same function, colored red and blue. Commonly used analytical distances (e.g., norms of the difference) between these functions would result in a non-zero value, failing to highlight the fact that they have the same sub-level set topology.

One well-established distance that expresses the topological similarity in the above example is the bottleneck distance between persistence diagrams, introduced by Cohen-Steiner et al. [8]. Computing the bottleneck distance for the example in Fig. 1 results in zero. Originally motivated by the shape matching problem, where the goal is to find how similar shapes are based on similarity of their topology, the bottleneck distance also has an important property – robustness to noise; see Fig. 2.

However, the bottleneck distance does not incorporate sub-level set nesting information, often necessary for analysis. Figure 3 shows two functions that differ by the nesting of the maximum m. The bottleneck distance between the corresponding persistence diagrams is again zero. Nevertheless, the corresponding merge trees cannot be matched exactly, hinting at a positive difference.

To resolve this problem, we introduce a new definition of the distance between merge trees. This distance resembles the bottleneck distance between the persistence diagrams of sub-level sets of the function, but it also respects the nesting relationship

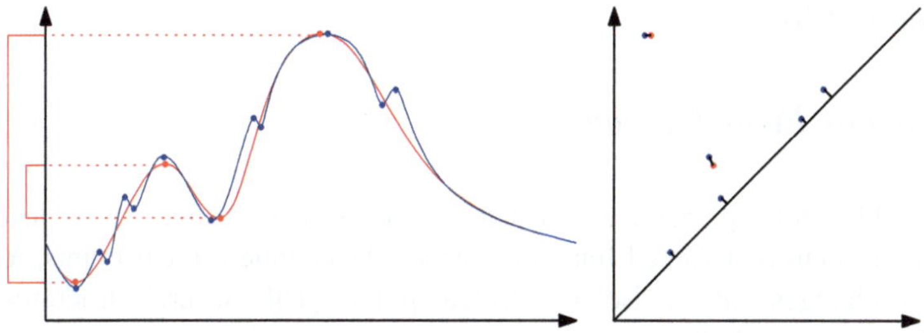

Fig. 2 Consider two close scalar functions on the *left*, where one contains additional noise. If we construct their persistence diagrams and find the bottleneck distance (which corresponds to the longest *black line* segment between paired points on the *right*), the result is small, correctly reflecting the closeness of the functions. In fact, the difference is the same as the level of the noise, which in this example is small

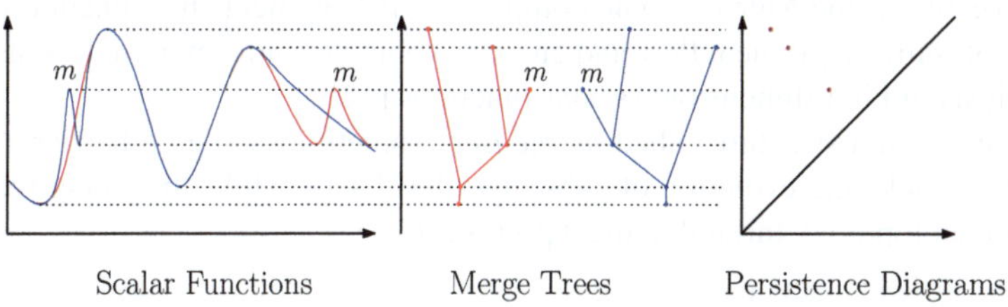

Fig. 3 Consider two scalar functions on the *left*. The bottleneck distance between persistence diagrams on the *right* equals zero, as points of two diagrams overlap. However, comparing the corresponding merge trees reveals a difference, since we cannot match them exactly. This difference highlights existence of additional nesting information in merge trees. Quantifying it is the main goal of this work

between sub-level sets. Furthermore, the proposed distance implicitly distinguishes the noise in the data, similar to the bottleneck distance, resulting in robust measurements resilient to perturbations of the input. This property is crucial when working with scientific data, where noise is a serious problem for any analysis.

The main contributions of this chapter are: a definition and an algorithm for computing the distance between merge trees; computation of the number of branch decompositions of the merge tree; an experimental comparison between the proposed distance, the bottleneck distance, and the L_∞ norm on analytical and real-world data sets.

Section 2 presents related work and background in scalar field topology, persistent homology, graph theory, and shape matching. Section 3 provides the definition and the algorithm for computing the distance between merge trees. Section 4 demonstrates several use cases and presents the results of comparing the distance between merge trees to the bottleneck distance between persistence diagrams, as well as the L_∞ norm. Finally, Sect. 5 summarizes the work and suggests ideas for future work.

2 Related Work

2.1 *Scalar Field Topology*

Scalar field topology characterizes data by topological changes of its level sets. Given a smooth, real-valued function without degenerate critical points, level set topology changes only at isolated critical points [16]. Several structures relate critical points to each other.

The *contour tree* [5, 7] and the *Reeb graph* [20, 21] track the level sets of the function by recording their births (at minima), merges or splits (at saddles), and deaths (at maxima). The contour tree is a special case of the Reeb graph, as the latter permits loops in the graph to handle holes in the domain. Both structures are used in a variety of high-dimensional scalar field visualization techniques [18, 23].

Alternatively, the Morse–Smale complex [9, 10] segments the function into the regions of uniform gradient flow and encodes geometric information. It is also used for analysis of high-dimensional scalar functions [12].

We focus on a structure called *merge tree* (sometimes called a barrier tree [11, 13]), as it tracks the evolution of sub-/super-level sets, while still being related to the level-set topology through critical points [16].

2.2 *Persistent Homology*

The concept of homology in algebraic topology offers an approach to studying the topology of the sub-level sets. We refer to Munkres [17] for the detailed introduction to homology. Informally, it describes the cycles in a topological space: the number of components, loops, voids, and so on. We are only interested in 0-dimensional cycles, i.e., the connected components.

Persistent homology tracks changes to the connected components in sub-level sets of a scalar function. We say that a component is born in the sub-level set $f^{-1}(-\infty, b]$ when its homology class does not exist in any sub-level set $f^{-1}(-\infty, b - \epsilon]$. This class dies in the sub-level set $f^{-1}(\infty, d]$ if its homology class merges with another class that exists in a sub-level set $f^{-1}(-\infty, b']$ with $b' < b$. When a component is born at b and dies at d, we record a pair (b, d) in the (0-dimensional) persistence diagram of the function f, denoted $D(f)$. For technical reasons, we add to $D(f)$ infinitely many copies of every point (a, a) on the diagonal.

Persistence diagrams reflect the importance of topological features of the function: the larger the difference $d - b$ of any point, the more we would have to change the function to eliminate the underlying feature. Thus, persistence diagrams let us distinguish between real features in the data and noise.

In Cohen-Steiner et al. [8], the authors prove the stability of persistence diagrams with respect to the bottleneck distance, $d_B(D(f), D(g))$. This distance is defined as the infimum over all bijections, $\gamma : D(f) \to D(g)$, of the largest distance between the corresponding points,

$$d_B(D(f), D(g)) = \inf_{\gamma} \sup_{u \in D(f)} ||u - \gamma(u)||_\infty.$$

Their result guarantees that the bottleneck distance is bounded by the infinity norm between functions:

$$d_B(D(f), D(g)) \leqslant \|f - g\|_\infty.$$

We use the bottleneck distance between persistence diagrams as a comparison baseline for the distance between merge trees.

2.3 Distance Between Graphs

Graph theory offers several approaches for comparing graphs and defining a notion of a distance between them.

A common approach for measuring a distance between graphs is based on an edit distance. It is computed as a number of edit operations (add, delete, and swap in the case of a labeled graph) required to match two graphs [6], or, in a special case, trees [4]. The edit distance focuses on finding an isomorphism between graphs/subgraphs, while for merge trees we can have two isomorphic trees with a positive distance (see the example in Fig. 3).

Alternatively, in a specific case of rooted trees, one can consider the generalized tree alignment distance [15], which, in addition to the edit distance, considers the minimization of the sum of distances between labeled end-points of any edge in trees. However, it is not clear how to adapt this distance definition for our purposes.

2.4 Using Topology of Real Functions for Shape Matching

The field of shape matching offers several methods related to our work. Generally, these methods focus on developing topological descriptors by treating a shape as a manifold, defining some real function on that manifold, and computing topological properties of the function. The selection of the particular function usually depends on which specific topological and shape properties of interest [2].

While the majority of the mentioned descriptors are not directly related to our work, two topological descriptors use similar approaches in defining a similarity measure. One is called a multiresolution Reeb graph, proposed by Hilaga et al. [14], which encodes nesting information into nodes of a Reeb graph for different hierarchy resolutions. Here, the hierarchy is defined by the simplification of Reeb graph.

Another descriptor is based on an extended Reeb graph (ERG), proposed by Biasotti et al. [3]. It starts by computing the ERG of the underlying shape, which is basically a Reeb graph with encoded quotient spaces in its vertices. It couples various geometric attributes with the ERG, resulting in an informative topological descriptor. In both cases, similarity of shapes is measured by applying a specialized graph matching (based on embedded/coupled information) to descriptors. However, we focus only on the sub-level set topology information, and design a matching algorithm, tailored specifically for this case.

Thomas and Natarajan [22] focus on symmetry discovery in a scalar function based on its contour tree. The authors develop a similarity measure between subtrees of the contour tree, which in some regards is similar to our proposed measure. However, they consider a single pre-processed branch decomposition, and focus on discovering symmetry in a sole function.

3 Defining a Distance Between Merge Trees

In this section, we provide a formal definition of the distance between merge trees and provide an algorithm (with optimizations) for computing it. In short, to compute the distance between two merge trees, we consider all branch decompositions of both trees and try to find a pair that minimizes the matching cost between them. Additionally, we provide the details of computing the number of branch decompositions of a merge tree, used in complexity analysis of our algorithm.

3.1 Definition

Let K be a simplicial complex; let $f : K \to \mathbb{R}$ be a continuous piecewise-linear function, defined on the vertices and interpolated in the interior of the simplices. Furthermore, assume all vertices have unique function values; in practice, we can simulate this by breaking ties lexicographically.

Let T_f be a merge tree of the function f; every vertex of K is mapped to a vertex in the merge tree. Every vertex of the merge tree has a degree of either one, two, or more, corresponding to a minimum, a regular point, or a merge saddle. Our definition works for higher-dimensional saddles (degenerate critical points) as well, and they need explicit consideration only in the complexity analysis of the algorithm (Sect. 3.4). A merge tree with purged regular vertices is called *reduced*.

A branch decomposition B [19] of a reduced merge tree T is a pairing of all minima and saddles such that for each pair there exists at least one descending path from the saddle to the minimum. We consider a rooted tree representation R of the branch decomposition B, such that the rooted tree representation R is obtained by translating each branch $b = (m, s) \in B$ into a vertex $v \in R$, where m and s

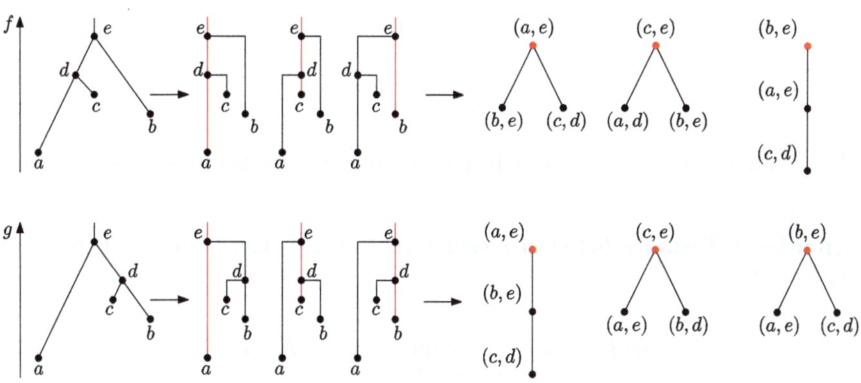

Fig. 4 Merge trees T_f (*top*) and T_g (*bottom*), all their possible branch decompositions, and corresponding rooted tree representations. Root branches are colored *red*, demonstrating the mapping of branches to vertices

are minimum and saddle that form the branch b. The edges of the rooted tree representation describe parent–child relationships between branches, see Fig. 4.

Given two merge trees, T_f and T_g, consider all their possible branch decompositions, $B_{T_f} = \{R_1^f, \ldots, R_k^f\}$ and $B_{T_g} = \{R_1^g, \ldots, R_k^g\}$, respectively; see Fig. 4. We need two auxiliary definitions to describe the matching of rooted branch decompositions.

Definition 1 (Matching cost). The cost of matching two vertices $u = (m_u, s_u) \in R_i^f$ and $v = (m_v, s_v) \in R_j^g$ is the maximum of the absolute function value difference of their corresponding elements,

$$mc(u, v) = max(|m_u - m_v|, |s_u - s_v|).$$

Definition 2 (Removal cost). The cost of removing a vertex $u = (m_u, s_u) \in R^{f,g}$ is

$$rc(u) = |m_u - s_u|/2.$$

We say that a partition (M^f, E^f) of the vertices of a rooted branch decomposition R^f is *valid*, if the subgraph induced by the vertices M^f is a tree. Here, the vertices M^f are mapped vertices, while the vertices E^f are reduced vertices. We say that an isomorphism of two rooted trees preserves order when it maps children of a vertex in one tree to the children of its image in the other tree.

Definition 3 (ϵ-Similarity). Two rooted branch decompositions R^f, R^g are ϵ-similar, if we can find two valid decompositions (M^f, E^f) and (M^g, E^g) of their vertices, together with an order-preserving isomorphism γ between the trees induced by the vertices M^f and M^g, such that the distance between each matched pair of vertices and the maximum cost for reduced vertices does not exceed ϵ:

$$\max_{u \in M^f} mc(u, \gamma(u)) \leqslant \epsilon \tag{1}$$

$$\max_{u \in E^f \cup E^g} rc(u) \leqslant \epsilon \tag{2}$$

The smallest epsilon, for which the above two inequalities hold, denoted $\epsilon_{min}(R_i^f, R_j^g)$.

Definition 4 (Distance between merge trees). The distance between two merge trees T_f, T_g is:

$$d_M(T_f, T_g) = \min_{R_i^f \in B_{T_f}, R_j^g \in B_{T_g}} (\epsilon_{min}(R_i^f, R_j^g)).$$

3.2 Distance Computation

To compute the distance d_M, we design an algorithm that is based on Definition 4. In particular, our algorithm constructs all possible pairs of branch decompositions, computes ϵ_{min} for each pair, and selects the minimum among them.

We use a recursive construction of the branch decompositions of a merge tree. The main operation is to pair a given saddle, one by one, with each minimum in its subtree. We start by pairing the highest saddle s_r with all minima in a tree. Each pair acts as a root branch (s_r, m_i) in the recursive operation. For each child saddle s_j on the root branch, we recursively repeat the pairing until all the saddle–minimum pairs are fixed, producing a unique branch decomposition b_i.

To compute $\epsilon_{min}(R_i^f, R_j^g)$, we design a function ISEPSSIMILAR(ϵ, R_i^f, R_j^g) that, for a predefined ϵ, determines whether two branch decompositions match. We start by setting ϵ to a high value – for example, the maximum of the amplitudes of the two functions – and perform a binary search to determine ϵ_{min}.

The function ISEPSSIMILAR is the core of the algorithm. It works by matching the vertices and the edges at each level of the tree. We recall that each vertex $u = (m_u, s_u) \in r_i, v = (m_v, s_v) \in r_j$ is a minimum-saddle pair. There are only two vertices at the root levels of R_i^f and R_j^g, so we determine whether their endpoints can be matched, i.e., $\max(|m_u - m_v|, |s_u - s_v|) \leqslant \epsilon$. If not, ISEPSSIMILAR returns false. Otherwise, we consider all the child vertices (see Fig. 5). Since there are several potential matches, we compute a bipartite graph between the child vertices such that the edge between a pair of children $u \in R_i^f, v \in R_j^g$ exists if and only if they can be matched within given ϵ, and ISEPSSIMILAR returns true for their subtrees. We also add ghost vertices for each vertex in the rooted branch decomposition when it can be reduced within ϵ. When there exists a perfect matching in the bipartite graph, the function returns true; otherwise, it returns false. If one or both of the current pair of children has children of their own, we recursively call ISEPSSIMILAR. The matching is perfect when there exists an edge cover such that its edges are incident to all the non-ghost vertices and do not share any of them.

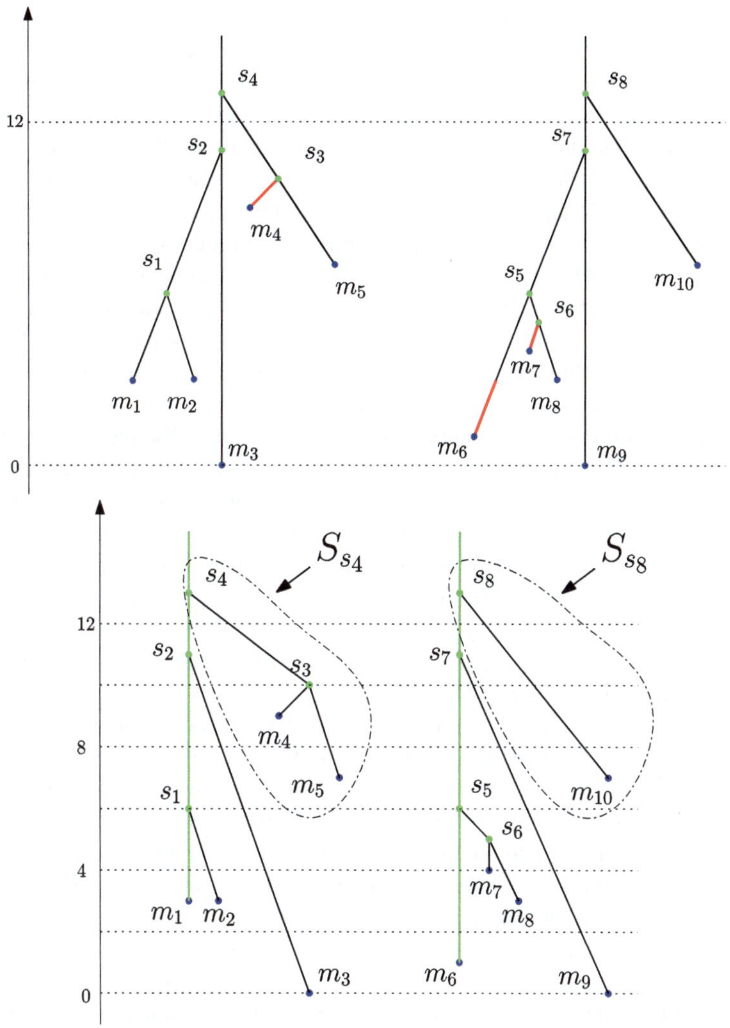

Fig. 5 *Top*: For the merge trees T_f, T_g, the smallest manually identifiable difference is shown as *red* segments. *Bottom*: The first iteration of the ISEPSSIMILAR function chooses (s_4, m_1) and (s_8, m_6) as root branches (depicted in *green*)

3.3 Optimized Algorithm with Memoization

The naive algorithm described above has exponential complexity. Indeed, there exist $O(2^{N-1})$ branch decompositions for a tree with N extrema (see Sect. 3.4 for details). Consequently, comparing all branch decompositions of two trees to each other would require a total of $O(2^{N+M-2})$ operations, where N, M are the numbers of extrema in each tree. This computational cost makes it infeasible to compare even small trees using this method. To alleviate this problem, we have designed an optimization, which reduces the number of explicitly considered branch decompositions, thus improving the complexity of the function ISEPSSIMILAR from exponential to polynomial. (Details are given at the end of this section.)

We demonstrate the optimized version of the function ISEPSSIMILAR using the example in Fig. 5. The function starts by iterating over all possible root branches $(s_4, m_i), i \in 1 \dots 5$, and $(s_8, m_j), j \in 6 \dots 10$. Once the pair of root branches is fixed as (s_4, m_1) and (s_8, m_6), the function is called recursively for every possible pairing of child subtrees in each tree. Fixing the root branches leads to two sets of child subtrees, $\{S_{s_4-s_3}, S_{s_2-m_3}, S_{s_1-m_2}\}$ and $\{S_{s_8-m_{10}}, S_{s_7-m_9}, S_{s_5-s_6}\}$. A subtree (e.g., $S_{s_4-s_3}$) needs two vertices to be uniquely identified, a child saddle (e.g., s_4), and the immediate child vertex (e.g., s_3) that can be either a saddle or minimum.

A key observation allowing us to reduce cost is that each pair of subtrees, for which the function is called recursively, also appears in subsequent iterations over other root branches. For example, the pair $(S_{s_4-s_3}, S_{s_8-s_{10}})$ that appears in the first iteration that chooses (s_4, m_1) and (s_8, m_6) as root branches, also reappears in 11 subsequent iterations, e.g., in the iteration that chooses (s_4, m_2) and (s_8, m_7) as root branches. Therefore, it is sufficient to compute the matching for subtrees S_{s_4}, S_{s_8} once, and reuse the result in subsequent iterations. More generally, for any subtree pair $S_{s_i - s_{childofi}} \in T_f$ and $S_{s_j - s_{childofj}} \in T_g$, one of the three possibilities is recorded in the array $match[S_{s_i - s_{childofi}}][S_{s_j - s_{childofj}}]$: not yet compared – (0), comparison returned false – (1), or true – (2).

Furthermore, the same pair of subtrees also reappears in other binary search iterations as well, i.e., when the function ISEPSSIMILAR is called with other values of ϵ. However, the reuse of previous results in this case is selective. If two subtrees were matched for some ϵ, they would stay matched only if the value of ϵ stayed the same or gets larger. If it gets smaller, we will have to recompute the matching result. And correspondingly, if two subtrees were unmatchable for some ϵ, they would stay unmatchable, only if the value of ϵ was the same or lower. However, if the value gets higher, we will have to recompute the matching result.

Returning to the example in Fig. 5, consider the case where $\epsilon = 5$. The first pair of root branches (s_4, m_1) and (s_8, m_6) (depicted in green in Fig. 5) do match, as $|f(s_4) - g(s_8)| = 0 < 5$ and $|f(m_1) - g(m_6)| = 2 < 5$, hence the function proceeds to their child subtrees. The first pair $S_{s_4-s_3}, S_{s_8-m_{10}}$ is matchable, thus there is an edge in the bipartite graph (similar to the naive algorithm) between the nodes that correspond to these subtrees. In fact, from nine pairs, only pairs $S_{s_2-m_3}, S_{s_8-m_{10}}$ and $S_{s_4-s_3}, S_{s_7-m_9}$ are unmatchable, thus there exists a perfect matching in the bipartite graph, and two merge trees are ϵ-similar for $\epsilon = 5$. Consequently, we continue the search with decreasing value of ϵ, until it converges to $\epsilon = 2$, in which case for the root branches (s_4, m_1) and (s_8, m_6), the only pairs of subtrees that match are (S_{s_4}, S_{s_8}), (S_{s_2}, S_{s_7}), (S_{s_1}, S_{s_5}). No lower value of ϵ would lead to the ϵ-similarity of the merge trees, making the value $\epsilon_{min} = 2$ the distance between merge trees.

Such optimization reduces the run time complexity from exponential to polynomial. Indeed, the function ISEPSSIMILAR performs $N \cdot M$ iterations over the root branches, multiplied by the sum of processing $n_{c_f} \cdot n_{c_g}$ explicit pairs of subtrees, and the complexity of a maximal matching algorithm $(n_{c_f} + n_{c_g}) \cdot n_{c_f} \cdot n_{c_g}$. The latter complexity dominates the former term. Hence, assuming that a look-up operation of previous results is done in a constant time via memoization, the resulting run time complexity of the function ISEPSSIMILAR is $O(N^2M^2(N + M))$. This complexity is multiplied by the number of iterations of the binary search algorithm, which we found to be moderate, given a reasonable selection of the search range and the precision. The worst-case memory complexity of the optimized algorithm is $O(N \cdot M)$, which is computationally less prohibitive than its run time complexity.

3.4 The Number of Branch Decompositions of a Merge Tree

We provide the details of computing the number of branch decompositions of the merge tree, used for the naive algorithm complexity analysis in the beginning of Sect. 3.3. We calculate the number of branch decompositions $P(N)$ for a merge tree with N minima in two steps. First, we compute the number $P(N)$ for the case when the merge tree is binary, in which case the tree has maximum possible number of saddles. Second, we show that for fewer saddles the number $P(N)$ decreases, leading to the worst case $P(N) = 2^{N-1}$ branch decompositions for any merge tree.

Theorem 1. *The number of branch decompositions of the binary merge tree with N minima equals $P(N) = 2^{N-1}$.*

Proof. For any saddle s, the number of branch decompositions in its subtree is $P_s = 2 \cdot P_{c_1} \cdot P_{c_2}$, where c_1 and c_2 are the children of the saddle s. Indeed, if the saddle s is paired with a minimum in a subtree of child c_1, then for each such pairing we have all the possible branch decompositions of a subtree of child c_2, resulting in $P_{c_1} \cdot P_{c_2}$ possibilities. Symmetrically, for the child c_2 we have $P_{c_2} \cdot P_{c_1}$ possibilities.

Using this fact we construct a proof by induction:

- For the base case of $N = 1$, the number of branch decompositions is one. On the other hand, $P(1) = 2^{1-1} = 1$. Hence, the formula holds.
- We assume that for all $N = 1, \dots, k$ the formula $P(N) = 2^{N-1}$ holds true.
- Now let's consider the case with $N = k + 1$, for which we have to prove that $P(k+1) = 2^k$. For the root saddle r of the tree with $k+1$ minima, we remember that $P_r = 2 \cdot P_{c_1} \cdot P_{c_2}$. If to denote the number of minima in the subtree of the child c_1 as $i \in [1, k]$, with $k-i+1$ denoting the number of minima in the subtree of the child c_2, we can expand $P(k + 1) = 2 \cdot P(i) \cdot P(k - i + 1)$. Since both i and $k - i + 1$ are not greater than k, we can substitute $P(i)$ and $P(k - i + 1)$ in accordance with assumptions for $N = 1, \dots, k$:

$$\begin{aligned} P(k+1) &= 2 \cdot P(i) \cdot P(k-i+1) \\ &= 2 \cdot 2^{i-1} \cdot 2^{k-i+1-1} \\ &= 2 \cdot 2^{i-1+k-i+1-1} = 2^k. \end{aligned}$$

□

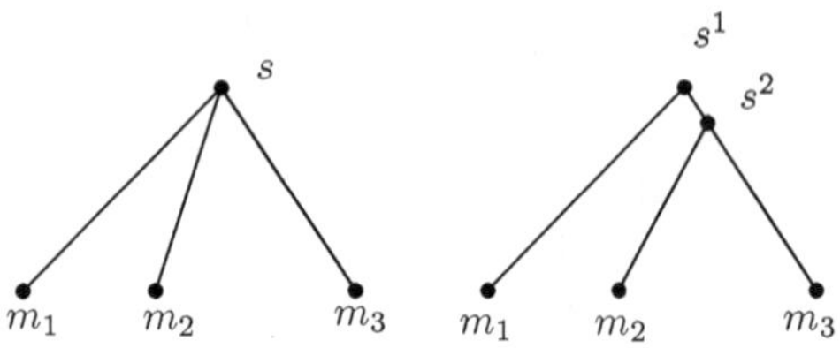

Fig. 6 Splitting the higher-degree saddle (with degree >2) always creates more branch decompositions. The saddle s has three branch decompositions, while after splitting it into two saddles s^1 and s^2, we get four branch decompositions

We consider the case with a number of saddles less than $N - 1$, i.e., the merge tree has saddles with degree higher than two. The number of minima N remains the same, while some of the saddles have more than two children. Any saddle of degree $d > 2$, can be split into $d - 1$ saddles of degree two, such that the structure of the tree changes only around the selected saddle, see Fig. 6. Such split leads to 2^{d-1} possible branch decompositions instead of the d for the selected saddle. Since $d > 2$, the inequality $2^{d-1} > d$ holds true, which means having degree-two saddles always leads to more branch decompositions. At the extreme, if all saddles become degree-two saddles, we obtain the binary merge tree, for which we already computed number of branch decompositions as 2^{N-1}.

4 Results

In this section we demonstrate the use of the proposed distance d_M. First, we apply it to simple data sets, observing its difference from the bottleneck distance and the L_∞ norm, as it captures additional information. We consider performance data sets obtained for a ray tracing program, and demonstrate how the proposed distance correctly captures the similarity of data sets.

4.1 Analytical Functions

We consider a set of simple functions that have a fixed number of maxima. Each function is constructed by creating a random set of maxima generators. For each maximum generator, function values decrease as the distance from the source grows, which results in a corresponding peak. The upper envelope of such peaks results in the required function.

We generate three bivariate (or 2D) functions f_1, f_2, f_3, and three trivariate (or 3D) functions f_4, f_5, f_6. We set the number of maxima to five, to keep them simple for visual exploration; see Fig. 7. The resulting distances, presented in Table 1, lead to two interesting observations.

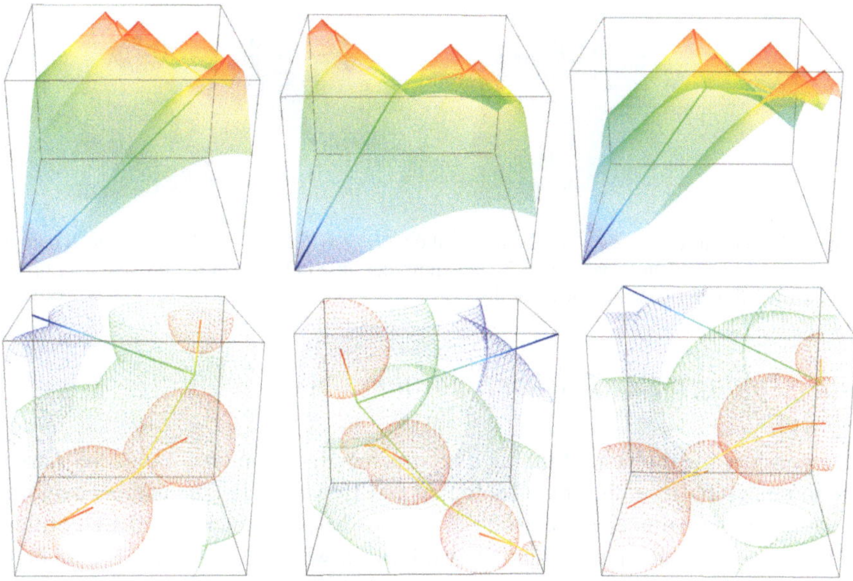

Fig. 7 Analytical functions. Rendering with embedded merge tree of three 2D functions f_1, f_2, f_3 (first row), and three 3D functions f_4, f_5, f_6 (second row)

Table 1 Resulting distances from Fig. 7

Metric	f_1^{2D}	f_2^{2D}	f_3^{2D}	f_4^{3D}	f_5^{3D}	f_6^{3D}
d_B	4.525	8.647	7	3.011	2.598	4.031
d_M	8.398	8.664	7	5.031	2.604	4.833
L_∞	67.561	43.015	65.956	29.586	20.495	22.632

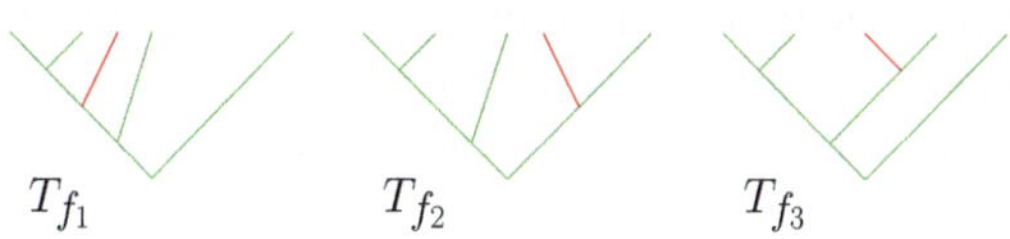

Fig. 8 Simplified view of nesting of merge trees for functions f_1, f_2, f_3. Unmatchable *red* edges cause the non-zero distance

For the 2D functions, the bottleneck distance d_B for functions f_1, f_2 is about two times lower than for functions f_1, f_3 and f_2, f_3, suggesting relative closeness of the first pair. However, the distance between merge trees d_M suggests that all three functions are equally different. Closer investigation confirms this hypothesis. In Fig. 8, we see simplified depictions of merge trees having equally different nesting.

For the 3D functions, visually the function f_5 seems different from the other two. This fact is again captured by the distance between merge trees, as the resulting distance d_M is almost two times lower for functions f_4, f_6, than from them to function f_5. The bottleneck distance again fails to capture this distinction.

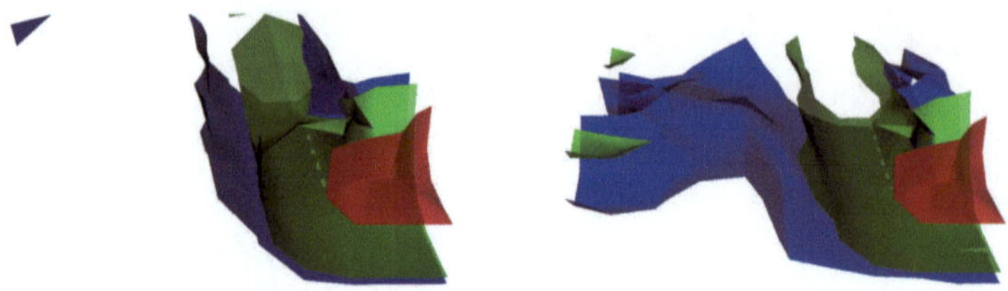

Fig. 9 Performance data. $d_B = 0.027, d_M = 0.027, L_\infty = 0.13$. The difference is small relative to the value range of functions (about 2.7 %), implying little influence of the ray sampling option on the overall performance of the algorithm

4.2 Tuning a Ray Tracing Algorithm

We consider the problem of tuning a ray tracing algorithm on a multicore shared-memory system from the study by Bethel and Howison [1]. The authors explored various tuning parameters and their effect on the performance of the algorithm, with a focus on three parameters: the work block width $\{1, 2, \ldots, 512\}$ and height $\{1, 2, \ldots, 512\}$, and the concurrency level $\{1, 2, 4, 8\}$.

We generated two data sets with the same parameter space, but slightly different algorithm, based on the selection of a ray sampling method, which is either based on nearest neighbor or trilinear approximation. For each option, the performance of the algorithm (in terms of running time) was recorded.

In this example, one is interested in studying optimal run configurations that correspond to low run times of the algorithm. Figure 9 shows two data sets using isosurfaces, such that isovalues are the same for both data sets. The similarity of data sets, implying that the selection of the chosen ray sampling method does not significantly influence the performance of the algorithm. This fact is confirmed by the resulting distance. The measured distances allow us to capture the similarity, regardless of shifted optimal configurations (minima) and the noise.

5 Conclusions

We presented a novel distance between merge trees, including the definition and the algorithm. We demonstrated the use of the proposed distance for several data sets.

We plan to perform a theoretical investigation of the proposed distance, including the concerns about its stability. We also plan to explore the use of the proposed distance for error analysis in the context of approximated scalar functions.

Acknowledgements The authors thank Aidos Abzhanov. This work was supported by the Director, Office of Advanced Scientific Computing Research, Office of Science, of the U.S. DOE under Contract No. DE-AC02-05CH11231 (Berkeley Lab), and the Program 055 of the Ministry of Edu. and Sci. of the Rep. of Kazakhstan under the contract with the CER, Nazarbayev University.

References

1. E.W. Bethel, M. Howison, Multi-core and many-core shared-memory parallel raycasting volume rendering optimization and tuning. Int. J. High Perform. Comput. Appl. **26**, 399–412 (2012)
2. S. Biasotti, L. De Floriani, B. Falcidieno, P. Frosini, D. Giorgi, C. Landi, L. Papaleo, M. Spagnuolo, Describing shapes by geometrical-topological properties of real functions. ACM Comput. Surv. **40**, 12:1–12:87 (2008)
3. S. Biasotti, M. Marini, M. Spagnuolo, B. Falcidieno, Sub-part correspondence by structural descriptors of 3D shapes. Comput. Aided Des. **38**(9), 1002–1019 (2006)
4. P. Bille, A survey on tree edit distance and related problems. J. Theor. Comput. Sci. **337**, 217–239 (2005)
5. R.L. Boyell, H. Ruston, Hybrid techniques for real-time radar simulation, in *Proceedings of the Fall Joint Computer Conference*, Las Vegas (IEEE, 1963), pp. 445–458
6. H. Bunke, K. Riesen, Graph edit distance: optimal and suboptimal algorithms with applications, in *Analysis of Complex Networks: From Biology to Linguistics*, ed. by M. Dehmer, F. Emmert-Streib (Wiley-VCH Verlag GmbH & Co. KGaA, Weinheim, 2009), pp. 113–143
7. H. Carr, J. Snoeyink, U. Axen, Computing contour trees in all dimensions. Comput. Geom. Theory Appl. **24**(2), 75–94 (2003)
8. D. Cohen-Steiner, H. Edelsbrunner, J. Harer, Stability of persistence diagrams, in *Proceedings of 21st Annual Symposium on Computational Geometry*, Pisa (ACM, 2005), pp. 263–271
9. H. Edelsbrunner, J. Harer, V. Natarajan, V. Pascucci, Morse-Smale complexes for piecewise linear 3-manifolds, in *Proceedings of the 19th Symposium on Computational Geometry*, San Diego, 2003, pp. 361–370
10. H. Edelsbrunner, J. Harer, A. Zomorodian, Hierarchical Morse-Smale complexes for piecewise linear 2-manifold. Discret. Comput. Geom. **30**, 87–107 (2003)
11. C. Flamm, I.L. Hofacker, P. Stadler, M. Wolfinger, Barrier trees of degenerate landscapes. Phys. Chem. **216**, 155–173 (2002)
12. S. Gerber, P.T. Bremer, V. Pascucci, R. Whitaker, Visual exploration of high dimensional scalar functions. IEEE Trans. Vis. Comput. Graph. **16**(6), 1271–1280 (2010)
13. C. Heine, G. Scheuermann, C. Flamm, I.L. Hofacker, P.F. Stadler, Visualization of barrier tree sequences. IEEE Trans. Vis. Comput. Graph. **12**(5), 781–788 (2006)
14. M. Hilaga, Y. Shinagawa, T. Kohmura, T.L. Kunii, Topology matching for fully automatic similarity estimation of 3D shapes, in *SIGGRAPH'01*, Los Angeles (ACM, 2001), pp. 203–212
15. T. Jiang, E. Lawler, L. Wang, Aligning sequences via an evolutionary tree: complexity and approximation, in *Symposium on Theory of Computing*, Montréal, 1994, pp. 760–769
16. J.W. Milnor, *Morse Theory* (Princeton University Press, Princeton, 1963)
17. J.R. Munkres, *Elements of Algebraic Topology* (Addison-Wesley, Redwood City, 1984)
18. P. Oesterling, C. Heine, H. Janicke, G. Scheuermann, G. Heyer, Visualization of high-dimensional point clouds using their density distribution's topology. IEEE Trans. Vis. Comput. Graph. **17**, 1547–1559 (2011)
19. V. Pascucci, K. Cole-McLaughlin, G. Scorzelli, Multi-resolution computation and presentation of contour trees. Technical report UCRL-PROC-208680, LLNL, 2005
20. V. Pascucci, G. Scorzelli, P.T. Bremer, A. Mascarenhas, Robust on-line computation of Reeb graphs: simplicity and speed. ACM Trans. Graph. **26**(3), 58.1–58.9 (2007)
21. G. Reeb, Sur les points singuliers d'une forme de pfaff complement intergrable ou d'une fonction numerique. C. R. Acad. Sci. Paris **222**, 847–849 (1946)
22. D.M. Thomas, V. Natarajan, Symmetry in scalar field topology. IEEE Trans. Vis. Comput. Graph. **17**(12), 2035–2044 (2011)
23. G.H. Weber, P.T. Bremer, V. Pascucci, Topological landscapes: a terrain metaphor for scientific data. IEEE Trans. Vis. Comput. Graph. **13**(6), 1416–1423 (2007)

Topological Integrity for Dynamic Spline Models During Visualization of Big Data

Hugh P. Cassidy, Thomas J. Peters, Horea Ilies, and Kirk E. Jordan

Abstract In computer graphics and scientific visualization, B-splines are common geometric representations. A typical display method is to render a piecewise linear (PL) approximation that lies within a prescribed tolerance of the curve. In dynamic applications it is necessary to perturb specified points on the displayed curve. The distance between the perturbed PL structure and the perturbed curve it represents can change significantly, possibly changing the underlying topology and introducing unwanted artifacts to the display. We give a strategy to perturb the curve smoothly and keep track of the error introduced by perturbations. This allows us to refine the PL curve when appropriate and avoid spurious topological changes. This work is motivated by applications to visualization of Big Data from simulations on high performance computing architectures.

1 Introduction

In geometric modeling B-splines are frequently used to model complex geometric objects [5]. The spline models are smooth structures but PL approximations are typically used to render the spline. Aeronautical, automotive and chemical simulations rely on topological algorithms to provide mathematically correct visualization. These topological algorithms typically enforce that the display curve (i.e. the PL structure) will preserve crucial topological characteristics [3, 14]. A sufficiently refined PL model will preserve topological characteristics of the initial static model. But as points on the PL model are perturbed over the course of the simulation, the

H.P. Cassidy (✉) • T.J. Peters • H. Ilies
University of Connecticut, Storrs, CT, USA
e-mail: hugh.cassidy@uconn.edu

K.E. Jordan
T.J. Watson Research Center, IBM, Cambridge, MA, USA

P.-T. Bremer et al. (eds.), *Topological Methods in Data Analysis and Visualization III*, Mathematics and Visualization, DOI 10.1007/978-3-319-04099-8_11,

PL model may diverge significantly from the smooth model that it represents. This may introduce topological artifacts to the display, resulting in a flawed image that could mislead domain scientists.

Our formal analysis is motivated by graphics experiments, which are summarized in Experiment 1 (Sect. 3). We observed that the PL approximation used for graphics could be perturbed for more time steps, *while still preserving ambient isotopic equivalence* than might be expected from previously published bounds [8]. This data-specific a posteriori analysis led us to question whether we could develop *rigorous, predictive* methods for the permissible number of time steps. A method based upon second centered differences is developed for that predictive capability to support efficient frame generation, where this new method is motivated by Experiment 1, with a formal analysis in Example 4.

Many perturbation strategies are possible, but in dynamic visualization, retaining differentiability over time is often desirable, so our predictive method is presented in the context of a representative differentiable perturbation strategy. However, the formal analysis is quite general, and other perturbation strategies could easily be integrated by a user interested in other applications. Our exposition first uses a non-differentiable strategy to introduce some central concepts within this simplified context, but the ensuing differentiable strategy is then used in the rest of the development. Our distinctive contributions are analyses of the amount of error introduced by each perturbation. This error can be monitored and the PL model can be refined as necessary to avoid unwanted topological changes. For ease of notation the investigation below is performed on Bézier curves, however the analysis is identical for general B-spline curves [5]. The motivating graphics experiments are summarized in Sect. 3 and a representative analysis is presented as Example 4 in Sect. 7.

2 Background, Motivation and Notation

In this section we introduce some fundamental definitions and notation.

2.1 Curves and Control Polygons

Definition 1. A degree d Bézier curve with control points $X = \{q_0, \cdots, q_d\}$ is given by

$$c(t) = \sum_{i=0}^{d} \binom{d}{i} (i - t)^{d-i} q_i$$

where the PL curve connecting $q_0, \cdots, q_d$ is called the control polygon of c.

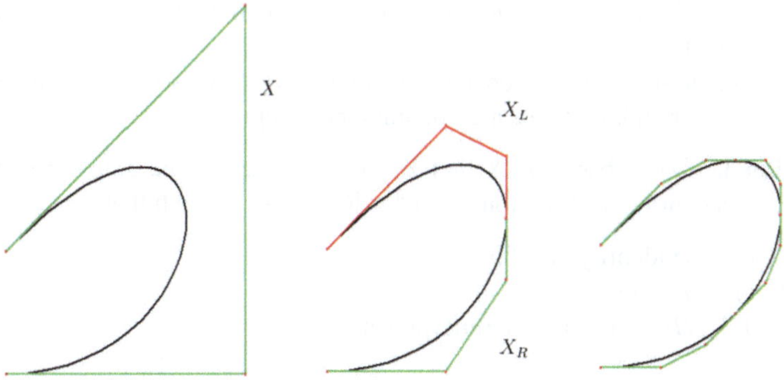

Fig. 1 Subdivision produces refined PL approximation

A subdivision algorithm operates on X to generate two PL curves, each having $d + 1$ vertices, denoted, respectively as X_L and X_R, as shown in Fig. 1. The union $X_L \cup X_R$ is also a control polygon for c but lies closer to c than the original control polygon. This process can be repeated to obtain a PL graphical approximation that is within a prescribed distance, ϵ_d, of the curve c.

Definition 2. Given the polygon generated by $X = \{q_0, \cdots, q_d\}$, the second centered difference of a given control point q_i is defined as

$$\Delta_2 q_i = q_{i-1} - 2q_i + q_{i+1}.$$

We define $\Delta_2 q_0 = \Delta_2 q_d = 0$. The *maximal second centered difference* of the polygon generated by X is given by

$$\|\Delta_2 X\|_\infty = \max_{0 \leqslant i \leqslant d} \|\Delta_2 q_i\|.$$

Here $N_\infty(d) = \dfrac{\lceil d/2 \rceil \lfloor d/2 \rfloor}{2d}$. Note that this distance is actually attained [15]. So subdividing α times guarantees that the PL structure is within the specified tolerance for display, ϵ_d, where

$$\alpha = \left\lceil -\frac{1}{2} \log_2 \left(\frac{\epsilon_d}{\|\Delta_2 X\|_\infty N_\infty(d)} \right) \right\rceil$$

2.2 Equivalence Relation

The traditional measure of topological equivalence is homeomorphism. Homeomorphic equivalence does not capture the embedding of a curve within $\mathbb{R}^3$, for example, all simple closed curves are homeomorphic even though there can be fundamentally different embeddings.

We use the stronger equivalence of *ambient isotopy* to also preserve embedding of c in $\mathbb{R}^3$. Different knot types are not ambient isotopic.

Definition 3. Two subspaces, X and Y, of $\mathbb{R}^n$ are said to be ambient isotopic if there exists a continuous function $H : \mathbb{R}^n \times [0,1] \longrightarrow \mathbb{R}^n$ such that

1. $H(\cdot, 0)$ is the identity on $\mathbb{R}^n$,
2. $H(X, 1) = Y$, and
3. $\forall t \in [0,1]$, $H(\cdot, t)$ is a homeomorphism.

2.3 Related Work

Molecular simulations are run on high performance computing (HPC) architectures, often generating petabytes of data, initiating a typical 'Big Data' problem. This data output is too voluminous for standard numerical analytic techniques and dynamic visualization has become a common zero-th order analysis. The supportive dynamic visualization techniques are well-established [12, 13] and will not be addressed further. The vitally important and novel support from this work is to provide rigorously proven numerical assurances that the frames being viewed have appropriate approximation in order to avoid topological artifacts in the images that could prove misleading to the domain scientists [4, 9]. To establish context for this work, a brief overview will be given of the three primary facets of *supportive mathematics, geometric models and molecular simulations*. The emphasis here is upon the new mathematics to meet the new Big Data challenges posed by the recent prevalence of these petabytes of simulation output, where this emerging mathematics is developing a blend of theory and experimentation.

At the highest level of viewing this work, there are so many tools available for molecular visualization, that it suffices to provide two broad summary portals [12, 13]. Often protein data is of interest, which appears publicly in an international resource [2]. The indicated resources do not directly provide geometric models of the molecules visualized – only images are produced.

The molecular simulation research [16, 18–21] closely aligns with the work presented here, with [21] being of particular interest because of its use of splines to model molecules, as also assumed here. Alternate geometric representations have been considered [10,11,17] for molecules, but the choice of splines here is offered as a very broad, fundamental representation, which could be examined for adaptation to these alternate representations. The more contemporary Big Data issues had not yet appeared when this earlier work had already been completed.

The emphasis here upon geometric representations echoes much work in computer-aided geometric design [5]. In particular, this dynamic molecular visualization has been synergistically pursued with an emerging virtual reality (VR) engineering design laboratory [7]. A fascinating common use is of 1-dimensional geometry to model the molecule writhing proteins and design features [6], where the latter application is integrated with a constraint solver.

Motivating Applications The mathematics proven here was motivated by design of dynamic visualization for molecular simulations in HPC. As a zero-th order analysis, a dynamic visualization is synchronized with the ongoing simulation. The graphics at each frame are displayed by PL approximations, raising the possibility that an image could show an intersection on a writhing molecule where none occurs on the more accurate spline model. The isotopic analysis presented is designed to integrate the necessary numerical accuracy with sufficient performance for dynamic visualization. Subdivision is chosen for the PL approximation, but the analysis presented here could easily be adapted to other PL approximation techniques, such as PL interpolation through selected points on the curve. Proteins are typical objects of interest, modeled as spline curves. Public data bases [2] provide spatial co-ordinates for interpolation to create a spline model. However, there can easily be hundreds of thousands of such co-ordinates, so that interpolation by a single segment spline would be also be on the order of hundreds of thousands – typically prohibitive for interactive graphics, where much lower degree is preferred (often as low as degree 3, but rarely higher than degree 8). Sufficiently accurate, low degree models can be created by the composite curves [5] used here. Since these geometric molecular models are not readily available in the public resources [2, 12, 13] prototype software is also being developed to provide those models, but reports on those tools will appear elsewhere.

The presented mathematical analysis has guided our algorithmic design so that we are now confident that we can use splines of sufficiently low degree, while maintaining desired topological characteristics. It remains to integrate these topology preserving techniques into the supporting dynamic visualizations discussed. That full experimental work is beyond the scope of the present paper and remains as subject for future publications.

3 Graphics Efficiency Experiment

The efficient use of PL approximations in dynamic visualizations has previously appeared [8], as a way to ensure correct graphics topology during animation, as previously presented relative to isotopic equivalence. That previous strategy [8] will now be briefly summarized, where this work adds the additional perspective of practical limits on the number of frames where this aggressive strategy can be invoked. As perspective on the extreme data and performance demands of this environment, it is instructive to note the order of 30–60 frames per second to synchronize dynamic visualization with a simulation producing peta-bytes of output.

During simulation, the molecule moves as reflected by movement of a spline. Each frame will use PL approximation. Here are two graphics display options to consider:

Option 1: At each time step, perturb the spline and create a new PL approximation for display.

Option 2: Create a PL approximation of the spline at some initial time step. Continue to perturb this PL approximation until it is no longer sufficiently accurate for graphics display.

Clearly, Option 2 can eliminate the approximation algorithm at some time steps. The previous work [8] provided existence theorems for maintaining isotopic equivalence during continued perturbation of these PL approximations. This work refines [8] by now providing specific numerical analyses to show *exactly* how many subsequent frames can invoke this aggressive strategy, *before* it becomes necessary to create a new PL approximation to ensure ongoing topological fidelity between the spline and its graphics approximation.

A representative graphics experiment will be summarized to show implications of Option 2. A sufficient[1] perturbation bound [8, Proposition 5.2] to preserve ambient isotopy is $(1/2)\nu$, with ν defined as the minimal distance between points and edges of a PL curve [1]. With the control points here, we note that $(1/2)\nu = 1/2$. We will show, later, that this upper bound, while sufficient to preserve ambient isotopy, leaves open the possibility of more aggressive perturbation strategies.

Experiment 1 *Consider the non-self intersecting C^1 composite cubic Bézier curve in $\mathbb{R}^2$, as depicted on the left hand side of Fig. 2. The following points together with their reflection through the line $y = 3$ form the control points:*

$$(0,6), (1,5), (2,4.5), (3,5.25), (4,6), (5,7), (6,8), (7,9), (9,10), (11,11), (13,12), (15,13),$$
$$(17,13.35), (19,13.7), (21,13), (22,12), (23,11), (24,10), (24.5,8), (25,6), (25,4), (25,3).$$

The control polygon is green with red control points, the underlying curve is black. Perturbing p_u and p_v over ten time steps introduces a self intersection to the PL structure that is not present in the underlying spline curve, as illustrated on the right hand side of Fig. 2. For brevity of presentation, the example of Fig. 2 presents the graphics of the original and perturbed Bézier curves to show that both are non-self-intersecting, which can be rigorously verified [1]. We return to this example for a detailed analysis in Sect. 6.3.

We note that the previous bound with of $(1/2)\nu = 1/2$ would have guaranteed that the first five time steps were permissible. When these visual experiments showed that topological fidelity could be preserved until the 10th step, we pursued a deeper analysis to explicate identification of this longer preservation of topology.

[1] There is an obvious typographical error [8, Proposition 5.2].

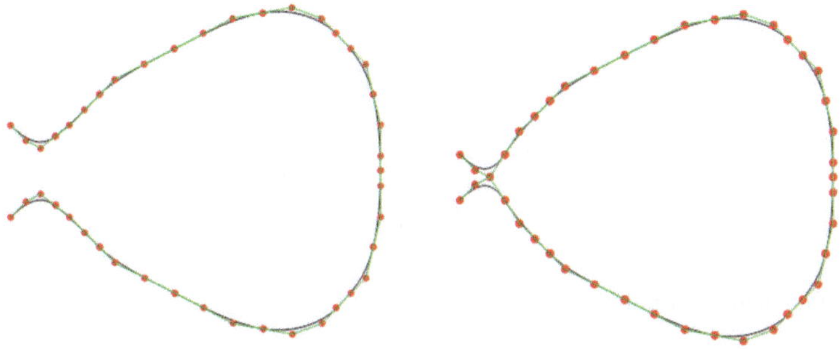

Fig. 2 Spurious self-intersection in PL structure

4 Notation for Perturbation Analysis

We now define the notation required for the perturbation analysis.

We shall examine n time steps denoted $\{t_1, \cdots, t_n\}$, t_0 denotes the time at initialization.

We assume that we are given a refined control polygon so that it is within ϵ_d of the represented curve. Note if α subdivisions are required then, from the original set of control points $\{q_0, \ldots, q_d\}$, there are generated w control points where $w = 2^\alpha d + 1$. Denote the subdivided, *but unperturbed*, control polygon by

$$X_0 = \{p_0, p_1, \cdots, p_w\}.$$

Let X_i denote the perturbed control polygon at time t_i. Assume we are supplied with a $(w + 1) \times n$ perturbation matrix, Γ, where each row contains perturbation vectors for a corresponding control point and each column contains the perturbation vectors for all control points at the corresponding time step, i.e.

$$\Gamma = \begin{array}{c} \\ p_0 \\ p_1 \\ \vdots \\ p_w \end{array} \begin{array}{c} \begin{array}{cccc} t_1 & t_2 & \cdots & t_n \end{array} \\ \begin{pmatrix} \gamma_{0,1} & \gamma_{0,2} & \cdots & \gamma_{0,n} \\ \gamma_{1,1} & \gamma_{1,2} & \cdots & \gamma_{1,n} \\ \vdots & \vdots & \ddots & \vdots \\ \gamma_{w,1} & \gamma_{w,2} & \cdots & \gamma_{w,n} \end{pmatrix} \end{array}$$

where $\gamma_{i,j}$ denotes the perturbation vector applied to p_i at time t_j (may be the zero vector). Let $\delta_j p_i$ denote the coordinates of the point that originated at p_i at t_j, i.e.

$$\delta_j p_i = p_i + \sum_{k=1}^{j} \gamma_{i,k}.$$

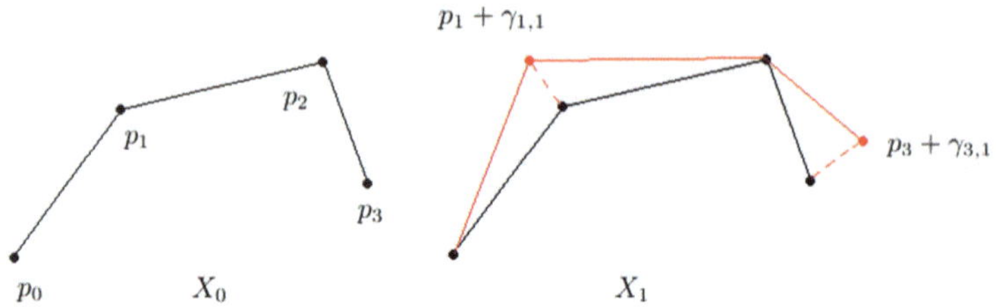

Fig. 3 Perturbation over a single time step

5 Non-differentiable Perturbations

In cases where maintaining differentiability of the curve is not required, we may simply perturb each point by the prescribed vector. At t_0 we are given X_0 and Γ as described above. At each t_i we can calculate X_i from Γ and X_{i-1}.

Example 1. Given the points $X_0 = \{p_0, p_1, p_2, p_3\}$ and the perturbation matrix,

$$\Gamma = \begin{bmatrix} 0 & 0 \\ \gamma_{1,1} & \gamma_{1,2} \\ 0 & 0 \\ \gamma_{3,1} & \gamma_{3,2} \end{bmatrix}$$

as depicted in Fig. 3. We can calculate

$$X_1 = \{\delta_1 p_0, \delta_1 p_1, \delta_1 p_2, \delta_1 p_3\} = \{p_0, p_1 + \gamma_{1,1}, p_2, p_3 + \gamma_{3,1}\}, \text{ and}$$
$$X_2 = \{\delta_2 p_0, \delta_2 p_1, \delta_2 p_2, \delta_2 p_3\} = \{p_0, p_1 + \gamma_{1,1} + \gamma_{1,2}, p_2, p_3 + \gamma_{3,1} + \gamma_{3,2}\}.$$

5.1 Perturbing a Single Point

First we consider perturbing a single point over a single time step. At initialization we have

$$X_0 = \{p_0, \cdots, p_j, \cdots, p_w\}.$$

Note that

$$\|\Delta_2 X_0\|_\infty N_\infty(d) \leqslant \epsilon_d$$

Let p_j be the point being perturbed. At time t_1 the point p_j is perturbed to $p_j + \gamma_{j,1}$ and all other points remain in their original positions.

$$X_1 = \{p_0, \cdots, \delta_1 p_j, \cdots, p_w\} = \{p_0, \cdots, p_j + \gamma_{j,1}, \cdots, p_w\}$$

The only second differences affected are $\Delta_2(p_{j-1})$, $\Delta_2(p_j)$ and $\Delta_2(p_{j+1})$.

$$\|\Delta_2 X_1\|_\infty = \max\{\|\Delta_2 X_0\|_\infty, \|\Delta_2(\delta_1 p_{j-1})\|, \|\Delta_2(\delta_1 p_j)\|, \|\Delta_2(\delta_1 p_{j+1})\|\}$$

where

$$\begin{aligned}
\Delta_2(\delta_1 p_{j-1}) &= p_{j-2} - p_{j-1} + p_j + \gamma_{j,1} = \Delta_2(p_{j-1}) + \gamma_{j,1}, \\
\Delta_2(\delta_1 p_j) &= \Delta_2(p_j) - 2\gamma_{j,1}, \text{ and} \\
\Delta_2(\delta_1 p_{j+1}) &= \Delta_2(p_{j+1}) + \gamma_{j,1}.
\end{aligned}$$

This approach extends easily to n time steps

$$\|\Delta_2 X_n\|_\infty = \max\{\|\Delta_2 X_0\|_\infty, \|\Delta_2(\delta_n p_{j-1})\|, \|\Delta_2(\delta_n p_j)\|, \|\Delta_2(\delta_n p_{j+1})\|\}$$

where $\Delta_2(\delta_n p_{j-1}) = \Delta_2(p_{j-1}) + \sum_{i=1}^{n} \gamma_{j,i}$, $\Delta_2(\delta_n p_j) = \Delta_2(p_j) - 2\sum_{i=1}^{n} \gamma_{j,i}$ and $\Delta_2(\delta_n p_{j+1}) = \Delta_2(p_{j+1}) + \sum_{i=1}^{n} \gamma_{j,i}$.

5.2 Perturbing Multiple Points

To perturb multiple points over multiple time steps, using the information supplied by Γ, sort the points being perturbed into adjacency chains, i.e. sets of adjacent control points denoted $Q_0, \cdots, Q_s$ where each Q_i contains either a single point or a list of adjacent points to be perturbed. This is necessary as chains of different length have different effects on the second differences that involve points in that chain. Let $|Q_i| = u$. If $u = 1$ then this is treated as in the single point case above. If $u = 2$ then we write $Q_i = \{p_k, p_{k+1}\}$, and we compute the affected centered differences as follows:

$$\Delta_2(\delta_n p_{k-1}) = \Delta_2(p_{k-1}) + \sum_{j=1}^{n} \gamma_{k,j}$$

$$\Delta_2(\delta_n p_k) = \Delta_2(p_k) + \sum_{j=1}^{n} (\gamma_{k+1,j} - 2\gamma_{k,j})$$

$$\Delta_2(\delta_n p_{k+1}) = \Delta_2(p_{k+1}) + \sum_{j=1}^{n} (\gamma_{k,j} - 2\gamma_{k+1,j})$$

$$\Delta_2(\delta_n p_{k+2}) = \Delta_2(p_{k+2}) + \sum_{j=1}^{n} \gamma_{k+1,j}$$

If $u \geqslant 3$ then $Q_i = \{p_k, \cdots, p_{k+v}\}$ for some $v \geqslant 2$. The affected centered differences are computed:

$$\Delta_2(\delta_n p_{k-1}) = \Delta_2(p_{k-1}) + \sum_{j=1}^{n} \gamma_{k,j}$$

$$\Delta_2(\delta_n p_k) = \Delta_2(p_k) + \sum_{j=1}^{n} (\gamma_{k+1,j} - 2\gamma_{k,j})$$

$$\vdots$$

$$\Delta_2(\delta_n p_s) = \Delta_2(p_s) + \sum_{j=1}^{n} (\gamma_{s-1,j} - 2\gamma_{s,j} + \gamma_{s+1,j})$$

$$\vdots$$

$$\Delta_2(\delta_n p_{k+v}) = \Delta_2(p_{k+v}) + \sum_{j=1}^{n} (\gamma_{k+v-1,j} - 2\gamma_{k+v,j})$$

$$\Delta_2(\delta_n p_{k+v+1}) = \Delta_2(p_{k+v+1}) + \sum_{j=1}^{n} \gamma_{k+v,j}$$

6 Differentiable Perturbations

It may be desirable to maintain a degree of differentiability either for appearances, analysis or both. We define a perturbation strategy that guarantees C^1 continuity (assuming the original curve is at least C^1).

6.1 Perturbation Strategy

We are given a composite Bézier curve to perturb. Recall that a *junction point* is a point where curve segments meet. We identify three types of point:

- *Type 1:* A point adjacent to a junction point.
- *Type 2:* A junction point.
- *Type 3:* A point that is neither a junction point nor adjacent to a junction point.

To maintain C^1 continuity we must require that the tangent edges with a shared junction point be collinear and have the same length [5].

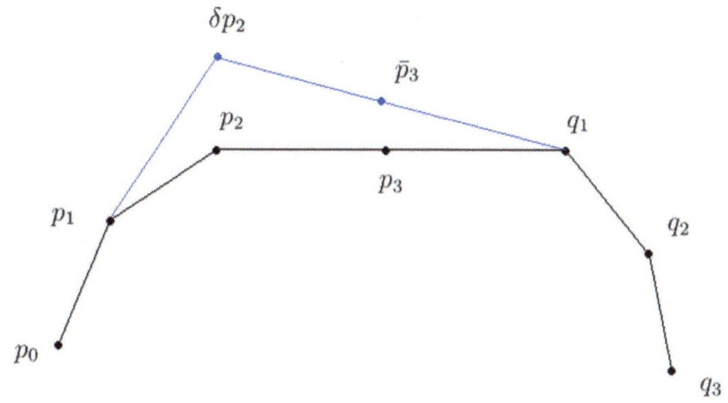

Fig. 4 Type 1

Type 1 If we perturb a Type 1 point in order to satisfy the C^1 criteria we perturb the junction point to the midpoint of the line segment joining its adjacent points. This approach is illustrated in the following example, where p_2 is being perturbed, relative to the junction point of p_3.

Example 2. Given a composite cubic control polygon with sub polygons $\{p_0, p_1, p_2, p_3\}$ and $\{p_3, q_1, q_2, q_3\}$ as depicted in Fig. 4. If we perturb p_2 by a vector γ:

$$p_2 \to \delta p_2 = p_2 + \gamma,$$

then to maintain C^1 continuity we perturb p_3 as follows:

$$p_3 \to \bar{p}_3 = \frac{\delta p_2 + q_1}{2}.$$

Type 2 To maintain C^1 differentiability when perturbing a Type 2 point we must also perturb its adjacent points by the same vector so the tangent edges are collinear and have the same length and are collinear.

Example 3. Here we have a composite cubic control polygon with sub polygons $\{p_0, p_1, p_2, p_3\}$ and $\{p_3, q_1, q_2, q_3\}$ as shown in Fig. 5. Perturbing p_3 by γ has the following effect:

$$p_2 \to \delta p_2 = p_2 + \gamma,$$
$$p_3 \to \delta p_3 = p_3 + \gamma,$$
$$q_1 \to \delta q_1 = q_1 + \gamma.$$

Type 3 Since Type 3 points do not affect tangent edges we can just perturb them as normal without perturbing neighboring points.

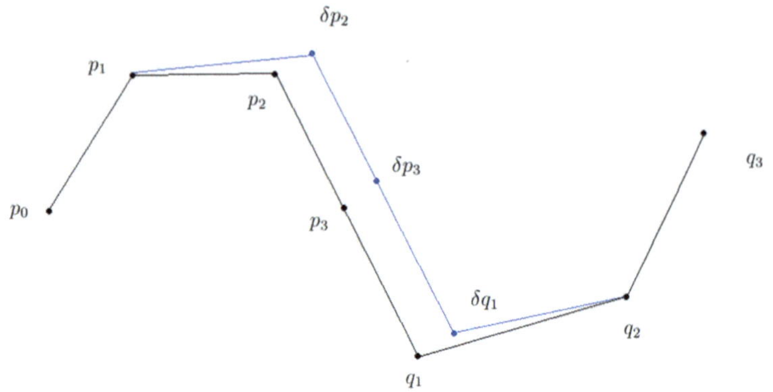

Fig. 5 Type 2

6.2 *Perturbing a Single Point*

We can now examine the effect of perturbing a single point using the strategy outlined above. Given

$$X_0 = \{p_0, \cdots, p_j, \cdots, p_w\}.$$

Note that

$$\|\Delta_2 X_0\|_\infty N_\infty(d) \leqslant \epsilon_d$$

Let $Y = \{p_j\}$ for some $j \in \{0, 1, \cdots, w\}$. At time t_1 the point p_j is perturbed to $p_j + \gamma_{j,1}$, note that adjacent points may be perturbed depending on the type.

Type 1. After all time steps are completed we have

$$X_n = \left\{ p_0, \cdots, p_j + \sum_{k=1}^{n} \gamma_{j,k}, \frac{p_j + \sum_{k=1}^{n} \gamma_{j,k} + p_{j+2}}{2}, \cdots, p_w \right\}.$$

The second centered differences are affected as follows:

$$\Delta_2(\delta_n p_{j-1}) = \Delta_2 p_{j-1} + \sum_{k=1}^{n} \gamma_{j,k}$$

$$\Delta_2(\delta_n p_j) = \left(p_{j-1} - \frac{3}{2} p_j + \frac{1}{2} p_{j+2} \right) - \frac{3}{2} \sum_{k=1}^{n} \gamma_{j,k}$$

$$\Delta_2(\delta_n p_{j+1}) = 0$$

$$\Delta_2(\delta_n p_{j+2}) = \left(\frac{1}{2}p_j - \frac{3}{2}p_{j+2} + p_{j+3}\right) + \frac{1}{2}\sum_{k=1}^{n} \gamma_{j,k}.$$

Type 2. After the first time step

$$X_1 = \{p_0, \cdots, p_{j-1} + \gamma_{j,1}, p_j + \gamma_{j,1}, p_{j+1} + \gamma_{j,1} \cdots, p_w\},$$

p_{j-1}, p_j and p_{j+1} are each perturbed by $\gamma_{j,i}$ at time t_i, $i \in \{1, \cdots, n\}$. At time t_n we have

$$\|\Delta_2 X_n\|_\infty = \max\{\|\Delta_2 X_0\|_\infty, \max\{\|\Delta_2(\delta_n p_k)\|\}_{k=j-2}^{j+2}\}.$$

The changes to the second centered differences are as follows:

$$\Delta_2(\delta_n p_{j-2}) = \Delta_2(p_{j-2}) + \sum_{k=1}^{n} \gamma_{j,k}$$

$$\Delta_2(\delta_n p_{j-1}) = \Delta_2(p_{j-1}) - \sum_{k=1}^{n} \gamma_{j,k}$$

$$\Delta_2(\delta_n p_j) = 0$$

$$\Delta_2(\delta_n p_{j+1}) = \Delta_2(p_{j-1}) - \sum_{k=1}^{n} \gamma_{j,k}$$

$$\Delta_2(\delta_n p_{j+2}) = \Delta_2(p_{j+2}) + \sum_{k=1}^{n} \gamma_{j,k}$$

Type 3. If we are perturbing a Type 3 point then at time t_n we have

$$\|\Delta_2 X_n\|_\infty = \max\{\|\Delta_2 X_0\|_\infty, \|\Delta_2(\delta_n p_{j-1})\|, \|\Delta_2(\delta_n p_j)\|, \|\Delta_2(\delta_n p_{j+1})\|\},$$

with the changes in second centered differences:

$$\|\Delta_2(\delta_n p_{j-1})\|_\infty = \|\Delta_2(p_{j-1}) + \sum_{i=1}^{n} \gamma_{j,i}\|_\infty$$

$$\|\Delta_2(\delta_n p_j)\|_\infty = \|\Delta_2(p_j) - 2\sum_{i=1}^{n} \gamma_{j,i}\|_\infty$$

$$\|\Delta_2(\delta_n p_{j+1})\|_\infty = \|\Delta_2(p_{j+1}) + \sum_{i=1}^{n} \gamma_{j,i}\|_\infty$$

6.3 Perturbing Multiple Points

Let p_j and p_{j+2} be Type 1 point, so p_j is a Type 2. We consider the illustrative case where p_j, P_{j+1} and p_{j+2} are each being perturbed over n time steps:

$$p_j \to p_j + \sum_{k=1}^{n} (\gamma_{j,k} + \gamma_{j+1,k})$$

$$p_{j+1} \to \frac{1}{2}\left(p_j + p_{j+2} + \sum_{k=1}^{n}(\gamma_{j,k} + 2\gamma_{j+1,k} + \gamma_{j+2,k})\right)$$

$$p_{j+2} \to p_{j+2} + \sum_{k=1}^{n}(\gamma_{j+1,k} + \gamma_{j+2,k})$$

The effect on the second differences is as follows:

$$\Delta_2(\delta_n p_{j-1}) = \Delta_2 p_{j-1} + \sum_{k=1}^{n}(\gamma_{j,k} + \gamma_{j+1,k})$$

$$\Delta_2(\delta_n p_j) = p_{j-1} - \frac{3}{2}p_j + \frac{1}{2}p_{j+2} + \sum_{k=0}^{n}\left(-\frac{3}{2}\gamma_{j,k} - \gamma_{j+1,k} + \frac{1}{2}\gamma_{j+2,k}\right)$$

$$\Delta_2(\delta_n p_{j+1}) = 0$$

$$\Delta_2(\delta_n p_{j+2}) = \frac{1}{2}p_j - \frac{3}{2}p_{j+2} + p_{j+3} + \sum_{k=1}^{n}\left(\frac{1}{2}\gamma_{j,k} - \gamma_{j+1,k} - \frac{3}{2}\gamma_{j+2,k}\right)$$

$$\Delta_2(\delta_n p_{j+3}) = \Delta_2 p_{j+3} + \sum_{k=1}^{n}(\gamma_{j+1,k} + \gamma_{j+2,k})$$

7 An Example Predictive Analysis

Our predictive method is now applied to formalize the empirical observations of Experiment 1, explicating extensions beyond previous bounds [8].

Example 4. The cubic Bézier curve of Example 1 was specifically synthesized to permit more aggressive PL graphics perturbations than previously known [8]. Given control points $X_0 = \{(0,6),(1,5),(2,4.5),\cdots,(2,1.5),(1,1),(0,0)\}$. Denote $\{(0,6),(1,5),(2,4.5)\}$ by $U = \{p_{u-2}, p_{u-1}, p_u\}$ and $\{(2,1.5),(1,1),(0,0)\}$ by $V = \{p_v, p_{v+1}, p_{v+2}\}$. Let the display tolerance, $\epsilon_d = 1.9167$. The maximal distance between the curve and the control polygon is $5/12$ which we trivially note is less than the given ϵ_d. Say we wish to perturb the points in U and V over 10 time steps with perturbation vectors $\{\gamma_{u-2,k}\}_{k=1}^{10} = \{\gamma_{u-1,k}\}_{k=1}^{10} = \{\gamma_{u,k}\}_{k=1}^{10}$ and $\{\gamma_{v,k}\}_{k=1}^{10} = \{\gamma_{v+1,k}\}_{k=1}^{10} = \{\gamma_{v+2,k}\}_{k=1}^{10}$ where

$$\begin{aligned}\{\gamma_{u,k}\}_{k=1}^{10} = \{ & (0, 5/20), (0, 4/20), (0, 4/20), (0, 4/20), (0, 3/20), \\ & (0, 2/20), (0, 1/20), (0, 1/20), (0, 1/20), (0, 5/20)\}, \text{ and}\end{aligned}$$

$$\begin{aligned}\{\gamma_{v,k}\}_{k=1}^{10} = \{ & (0, -5/20), (0, -4/20), (0, -4/20), (0, -4/20), (0, -3/20), \\ & (0, -2/20), (0, -1/20), (0, -1/20), (0, -1/20), (0, -5/20)\}\end{aligned}$$

For this curve, $1/2\nu = 1/2$, a value which is clearly exceeded after five steps of this strategy. Since previous criteria [8] were only sufficient, the rest of this example demonstrates that greater perturbation is possible to support efficiency in Strategy 2. Since the analysis for points in U and V is identical we shall focus on V. Notice that

$$\sum_{k=1}^{10} \gamma_{v,k} = \left(0, \frac{3}{2}\right).$$

Since p_v is a Type 2 point, the junction point $p_j = (3, 3/4)$ will also be perturbed as described above. Denote the control point following p_j by p_{j+1}.

$$p_v = (2, 3/2) \to \delta_{10} p_v = (2, 3)$$

$$p_j = (3, 3/4) \to \delta_{10} p_j = (3, 3/2)$$

We require that for each i,

$$\|\Delta_2 X_i\|_\infty + \| \sum_{j=1}^{i} \gamma_{X,j} \| < \epsilon_d.$$

Here $\sum \gamma_{X,j}$ is the sum of the perturbation vectors applied to the control point that yields $\|\Delta_2 X_i\|_\infty$. These quantities are easily calculated using the analysis above. It is easy to see that $\|\Delta_2 X_i\|_\infty + \| \sum_{j=1}^{i} \gamma_{X,j} \| < \epsilon_d$ for $i = 1, \cdots, 9$. At the ninth time step we have At the tenth time step

$$\|\Delta_2 X_{10}\|_\infty + \| \sum_{j=1}^{10} \gamma_{X,j} \| = 0.667 + 1.5 = 2.167 > \epsilon_d$$

Observing this we are now aware of the need to refine the control polygon by subdivision. Note that [3] and [14] allow us to determine the amount of subdivision required so that an ambient isotopic approximation is guaranteed

8 Conclusions and Future Work

For dynamic visualization of molecular simulations it is important to ensure that the rendered curve and the underlying spline are ambient isotopic at each time step. That global bounds on these perturbations can be exceeded if only local perturbations are executed is obvious, the performance imperatives for dynamic visualization make such data-specific refinements relevant, as is explored here. This can be achieved by keeping track of changes to the second centered differences and applying further subdivision as required.

The above analysis was performed for B-spline curves, the surface case was not pursued but we expect that the results can be extended to B-spline surfaces easily.

The molecules modeled certainly have 3-dimensional structure that is not captured by the 1-dimensional spline models. The reduction in dimension was chosen to support the performance demands of dynamic visualization of an ongoing simulation producing peta-bytes of output, while still being able to capture essential topological characteristics needed for zero-th order analyses. A similar reduction of dimension was undertaken to simplify engineering design studies [6]. The user identifies boundaries, that are modeled as 1-dimensional curves, as abstractions to convey design intent. This low-order geometry affords interactive manipulation and constraint satisfaction. Emerging VR techniques rely upon hand and finger gestures to express design variations. It would be desirable to adapt such gestures to interactive steering of these molecular simulations, providing further opportunities to share these research perspectives. Indeed, some of the required emphasis on graphics manipulation is being pursued, concurrently, under associate technology transfer projects [7] for gesture based editing during production of computer animations in the film making industry. An ideal outcome would be the effective merging from these three fronts of molecular simulation, engineering design and film making – as remains the subject of planned activities.

Acknowledgements The authors thank the referees, both for the conference presentation, as well as for this final book, for their helpful and constructive comments, which led to many improvements. The three UConn authors were partially supported by NSF grants CMMI 1053077 and CNS 0923158. T. J. Peters was also partially supported by an IBM Faculty Award and IBM Doctoral Fellowships. All statements here are the responsibility of the author, not of the National Science Foundation nor of IBM.

References

1. L.-E. Andersson, T.J. Peters, N.F. Stewart, S.M. Doney, Polyhedral perturbations that preserve topological form. Comput. Aided Geom. Des. **12**(8), 785–799 (1995)
2. Anonymous, The protein data bank (2013), http://www.rcsb.org/pdb/home/home.do
3. H. Cassidy, T. Peters, K. Jordan, Dynamic computational topology for piecewise linear curves, in *Proceedings of the Canadian Conference on Computational Geometry 2012*, Charlottetown, 8–10 Aug 2012, pp. 279–284

4. T. Etiene, L.G. Nonato, C.E. Scheidegger, J. Tierny, T.J. Peters, V. Pascucci, R.M. Kirby, C.T. Silva, Topology verification for isosurface extraction. IEEE Trans. Vis. Comput. Graph. **18**(6), 952–965 (2012)
5. G. Farin, *Curves and Surfaces for Computer Aided Geometric Design: A Practicle Guide*, 2nd edn. (Academic, San Diego, 1990)
6. R. Gal, O. Sorkine, N.J. Mitra, D. Cohen-Or, iwires: an analyze-and-edit approach to shape manipulation. ACM Trans. Graph. **28**(3), 33:1–33:10 (2009)
7. H. Ilies, MRI: development of a gesture based virtual reality system for research in virtual worlds. NSF award 0923158 (2009), http://www.nsf.gov
8. K.E. Jordan, L.E. Miller, E.L.F. Moore, T.J. Peters, A.C. Russell, Modeling time and topology for animation and visualization with examples on parametric geometry. Theor. Comput. Sci. **405**, 41–49 (2008)
9. R.M. Kirby, C.T. Silva, The need for verifiable visualization. IEEE Comput. Graph. Appl. **28**(5), 78–83 (2008)
10. R. Kolodny, L. Guibas, M. Levitt, P. Koehl, Inverse kinematics in biology: the protein loop closure problem. J. Robot. Res. **24**, 151–162 (2005)
11. R. Kolodny, P. Koehl, L. Guibas, M. Levitt, Small libraries of protein fragments model native protein structures accurately. J. Mol. Biol. **323**, 297–307 (2002)
12. E. Martz, T.D. Kramer, World index of molecular visualization resources (2010), http://web.archive.org/web/20101029215032/http://molvis.sdsc.edu/visres/#srvrs
13. E. Martz, T.D. Kramer, Molviz (2013), http://molvis.sdsc.edu/
14. L. Miller, E. Moore, T. Peters, A. Russell, Topological neighborhoods for spline curves: practice and theory, in *Reliable Implementation of Real Number Algorithms: Theory and Practice*, ed. by P. Hertling et al. Volume 5045 of LNCS (Springer, New York, 2008), pp. 149–161
15. D. Nairn, J. Peters, D. Lutterkort, Sharp, quantitative bounds on the distance between a polynomial piece and its Bézier control polygon. Comput. Aided Geom. Des. **16**, 613–631 (1999)
16. G. Ramachandran, T. Schlick, Solvent effects on supercoiled DNA dynamics explored by Langevin dynamics simulations. Phys. Rev. E **51**(6), 6188–6203 (1995)
17. D. Russel, L. Guibas, Exploring protein folding conformations using spanners, in *Pacific Symposium on Biocomputing*, Hawaii, 2005, pp. 40–51
18. T. Schlick, Dynamic simulations of biomolecules, http://www.searlescholars.net/people/1991/schlick.html
19. T. Schlick, Modeling superhelical DNA: recent analytical and dynamic approaches. Curr. Opin. Struct. Biol. **5**, 245–262 (1995)
20. T. Schlick, *Molecular Modeling and Simulation: An Interdisciplinary Guide* (Springer, New York, 2002)
21. T. Schlick, W.K. Olson, Trefoil knotting revealed by molecular dynamics simulations of supercoiled DNA. Science **21**, 1110–1115 (1992)

Part IV
Time-Dependent Analysis

A Comparison of Finite-Time and Finite-Size Lyapunov Exponents

Ronald Peikert, Armin Pobitzer, Filip Sadlo, and Benjamin Schindler

Abstract Finite-time and finite-size Lyapunov exponents are related concepts that have been used for the purpose of identifying transport structures in time-dependent flow. The preference for one or the other concept seems to be based more on a tradition within a scientific community than on proven advantages. In this study, we demonstrate that with the two concepts highly similar visualizations can be produced, by maximizing a simple similarity measure. Furthermore, we show that results depend crucially on the numerical implementation of the two concepts.

1 Introduction

The finite-time Lyapunov exponent (FTLE) has been proposed by Haller [13] as an indicator of Lagrangian coherent structures (LCSs). The finite-size Lyapunov exponent (FSLE) is an alternative and especially popular in oceanography [9, 19]. It seems that the preference for FTLE or FSLE is based more on the tradition within a particular scientific community than on an evaluation of the two approaches. Direct comparisons between FTLE and FSLE have been made by Boffetta et al. [5] and by Sadlo and Peikert [21]. Boffetta argued that the FTLE is not capable of recognizing the relevant structures, namely the boundaries between chaos and large-scale mixing

R. Peikert (✉) • B. Schindler
Department of Computer Science, ETH Zurich, Zurich, Switzerland
e-mail: peikert@inf.ethz.ch; bschindler@inf.ethz.ch

A. Pobitzer
University of Bergen, Bergen, Norway
e-mail: armin.pobitzer@uib.no

F. Sadlo
University of Stuttgart, Stuttgart, Germany
e-mail: Filip.Sadlo@visus.uni-stuttgart.de

P.-T. Bremer et al. (eds.), *Topological Methods in Data Analysis and Visualization III*, Mathematics and Visualization, DOI 10.1007/978-3-319-04099-8_12,

regime. A recent paper by Karrasch and Haller [16] lists a number of theoretical limitations of FSLE and proposes the use of infinitesimal-time Lyapunov exponents.

We show, based on two examples, that FTLE and FSLE, if appropriately calibrated, produce comparable results which can be interchangeably used for most purposes in numerical flow visualization. However, in terms of computational effort FTLE is slightly more efficient, because computation can be based on a single precomputed flow map. In oceanographic research, FSLE and FTLE are typically computed from particle pairs consisting of horizontal or vertical pairs of grid points. This means a quantization of the range of directions into a set of only four directions. As we show, the effect of this quantization is that FTLE and FSLE values are underestimated by a factor which can be large, especially for short integration times. We give an example where this factor is unbounded.

Another important algorithmic aspect is the correct choice of a scale, given a grid resolution. There are two approaches to FSLE and FTLE computation, both of which require a scale parameter. In methods based on particle pairs, this is the initial separation of the particles. In methods based on the Cauchy-Green tensor, a scale is needed in gradient estimation. We demonstrate the effect of the scale parameter.

We define a similarity measure for pairs of FTLE and FSLE images, based on which image pairs can be computed using an optimization process. For a given FTLE or FSLE image, a corresponding image of the other field can be produced. We discuss results of this process in Sects. 5 and 6.

2 Related Work

Particle transport is a central feature of fluid flows. In this context, the investigation of mixing properties and the existence of transport barriers is of great interest, e.g., for the distribution of chemicals in the atmosphere [18] or in a bay area [8].

Transport barriers and mixing can be studied by means of the local attracting or repelling properties of trajectories. These properties are connected to the concept of *stable* and *unstable manifolds* in dynamical systems [12, 26]. Along these lines, Haller proposed FTLE as a quantification method of these properties [12, 13]. FTLE is a finite-time version of the well-known *Lyapunov exponent* [11]. An alternative is the *finite-size Lyapunov exponent*, as described by Boffetta et al. [5].

In the same paper, Boffetta et al. [5] compare FSLE to FTLE and two Eulerian-based techniques for the identification of transport barriers in two-dimensional incompressible fluids. The comparison is angled towards geophysical flows, like ocean currents and atmospheric flows. The authors conclude that both FSLE and FTLE are superior to the Eulerian-based methods in uncovering the transport barriers. Furthermore, they find FSLE to be more expressive for large-scale transport, while FTLE captures mainly small-scale properties of the particle dispersion. It is worthwhile noticing that the comparison conducted by Boffetta et al. is based on particle pairs, and may hence be affected by sampling issues. We comment on this issue in Sect. 3.3.1. Joseph and Legras [15] cite Haller [12] stating that the distribution of pair separation is typically a "fuzzy" view of the hyperbolic manifolds. They claim that FSLE results in a less fuzzy distribution and allows the manifold structures to emerge more naturally.

While FSLE is usually preferred in the analysis of mixing and transport in oceans [9,15] and atmosphere [18], FTLE has been applied to these domains as well. Coulliette et al. [8] make use of FTLE to investigate the existing transport barriers in a bay area in order to achieve optimal timing for the relief of contaminants into the bay. Beron-Vera et al. [4] identify so-called *invariant tori*, which act as transport barriers, in geophysical flows by means of FTLE. Üffinger et al. [25] compare the FTLE computation based on linearization and dense direction sampling. They conclude that a dense sampling of directions produces more accurate results than the classical approach based on linearizion, while preserving its advantages in the detection of topological features.

3 Finite-Time and Finite-Size Lyapunov Exponents

Given a velocity field $\mathbf{u}(\mathbf{x}, t)$, the *flow map* $\Phi_t^{t+\tau}(\mathbf{x})$ describes the advection of a "particle" released at time t and location $\mathbf{x}$ over a time span τ. Both t and the integration time τ are considered to be parameters, hence the flow map is mapping from the spatial domain to itself, mapping initial locations to final locations.

3.1 *Intuitive Definitions*

Given a particle seeded at location $\mathbf{x}$ and time t, and advected over time τ, the *dispersion* of a neighbor particle seeded at $\mathbf{y}$ is observed. The location $\mathbf{y}$ is chosen on a sphere of radius d around $\mathbf{x}$, such that the dispersion is maximized. This leads to the definition of the *finite-time Lyapunov exponent* at location $\mathbf{x}$ and time t, for a given integration time τ, which is

$$FTLE_\tau(\mathbf{x}, t) = \lim_{d\to 0} \max_{\|\mathbf{y}-\mathbf{x}\|=d} \frac{1}{|\tau|} \ln \frac{\left\|\Phi_t^{t+\tau}(\mathbf{y}) - \Phi_t^{t+\tau}(\mathbf{x})\right\|}{d}. \quad (1)$$

For a given *dispersion factor* r, the *finite-size Lyapunov exponent* [1] at location $\mathbf{x}$ and time t is:

$$FSLE_r(\mathbf{x}, t) = \lim_{d\to 0} \max_{\|\mathbf{y}-\mathbf{x}\|=d} \frac{1}{|\tau_r|} \ln r \quad (2)$$

where τ_r is the minimum time for which

$$\left\|\Phi_t^{t+\tau_r}(\mathbf{y}) - \Phi_t^{t+\tau_r}(\mathbf{x})\right\| \Big/ d = r. \quad (3)$$

3.2 Equivalent Definitions

The above equations contain a maximization over directions and a limit, which both have to be approximated in computation. This can be avoided by using the equivalent definitions [7]

$$FTLE_\tau(\mathbf{x}, t) = \frac{1}{|\tau|} \ln \sqrt{\lambda_{\max}(\mathbf{C}_t^{t+\tau}(\mathbf{x}))} \tag{4}$$

where $\mathbf{C}_t^{t+\tau}(\mathbf{x}) = \nabla \Phi_t^{t+\tau}(\mathbf{x})^* \nabla \Phi_t^{t+\tau}(\mathbf{x})$ is the (right) Cauchy-Green tensor, and

$$FSLE_r(\mathbf{x}, t) = \frac{1}{|\tau_r|} \ln r \tag{5}$$

where $|\tau_r|$ is the minimum time for which

$$\sqrt{\lambda_{\max}(\mathbf{C}_t^{t+\tau_r}(\mathbf{x}))} = r. \tag{6}$$

3.3 Computation Based on Particle Pairs

One approach to FTLE and FSLE computation is the explicit use of particle pairs, according to the definitions from Sect. 3.1. The limits occurring in Eqs. (1) and (2) are approximated in practice by choosing a small d, while maximization is approximated by testing a finite set of directions. While for a good approximation quality sufficiently many directions are needed, it is established practice to use only the four grid neighbors [14].

The approach based on Eqs. (1) and (2) is popular especially in the oceanography community for FSLE computation. D'Ovidio et al. [9] describe the investigation of mixing structures in the Mediterranean sea by FSLE, using a four neighbor-based approximation of the maximal growth rate of particle separation. FSLE is used to identify hyperbolic points, which indicate areas of strong mixing. The authors point out the natural compatibility of FSLE with data collected by *Lagrangian drifters*, and the possibility to suppress small-scale mixing by the dispersion factor r.

Hernández-Carrasco et al. [14] discuss the reliability of FSLE with respect to sampling density for particles seeding and the grid resolution of the underlying velocity field. While noting the theoretical need to investigate "all" possible separation directions, also their computations consider two directions only. They find that FSLE are consistent under grid refinement for both seed point and velocity field, while the amount of details captured increases with finer resolution.

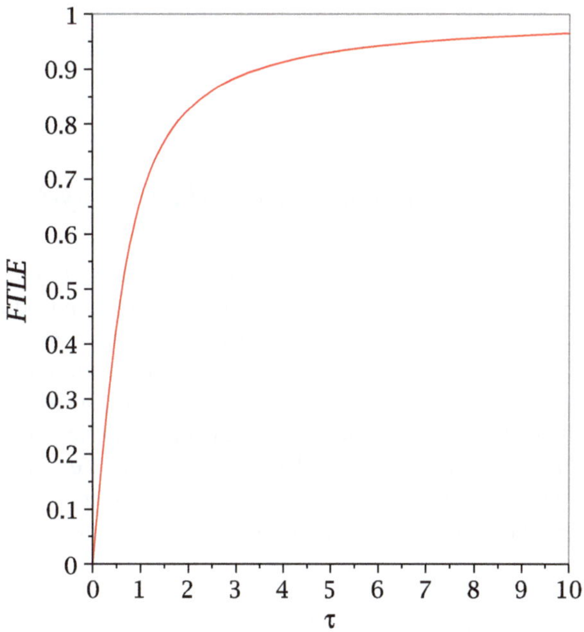

Fig. 1 FTLE of the velocity field $u = y, v = x$, computed with particle pairs on uniform grid. For larger τ they converge to the correct value of 1.0

Fig. 2 Effects of quantization of directions: Same as Fig. 6a, but with computation based on particle pairs (four neighbors only)

3.3.1 Shortcoming of the Computation Based on Four Neighbors

Using the simple linear velocity field $u = x, v = -y$, we demonstrate that computation using only four neighbors can lead to large errors in FTLE (and FSLE) values. The flow map of the field after integration time τ is $\Phi_0^\tau(x, y) = (xe^\tau, ye^{-\tau})$, its right Cauchy-Green tensor is $diag(e^{2\tau}, e^{-2\tau})$, and finally, the FTLE is the constant 1. Two particles seeded at $(0, 0)$ and $(h, 0)$ are mapped to $(0, 0)$ and $(he^\tau, 0)$, therefore $r = e^\tau$, and the particle-based method gives the correct result. A wrong result is obtained, however, if a 45° rotation is applied to the velocity field. Then, the seed points $(\pm h, 0)$ and $(0, \pm h)$ are mapped to $(\pm h \cosh \tau, \pm h \sinh \tau)$ and $(\pm h \sinh \tau, \pm h \cosh \tau)$, respectively, therefore $r = \sqrt{\cosh(2\tau)}$, and the FTLE value is $\ln \cosh(2\tau)/2\tau$, which deviates from 1 especially for small τ, as is shown in Fig. 1. Such quantization artifacts in an FSLE computation are shown in Fig. 2.

In order to reduce this error, one would have to use a finer quantization of directions. This could be done by using auxiliary particles on a circle around the reference particle as in [25].

3.4 Computation Based on the Cauchy-Green Tensor

Errors due to quantization of directions are eliminated if the definitions from Sect. 3.2 are used. This is at the cost of errors introduced by gradient estimation, which are typically much smaller. Therefore, computation based on the Cauchy-Green tensor is the preferred method both in the fluid dynamics and the scientific visualization community [7, 20, 24]. The first stage of the method is to sample the flow map, which requires a high-quality numeric integrator. The sampling grid can be chosen independent of the discretization of the velocity data, therefore, a uniform regular grid is typically used. In the second stage, the flow map gradient is estimated at each sample position, and Eqs. (4) or (5) is used to generate the final values.

3.5 Scale Considerations

Both of the above approaches contain a "scale" parameter which affects the computed scalar field. In Sect. 3.3 this is the distance d between particle pairs, while in Sect. 3.4 a distance d is needed for gradient estimation using, e.g., finite differences. For setting the parameter d, in either one of its meanings, there are two opposite approaches:

1. By choosing d very small, a good approximation of the FTLE or FSLE at the point $\mathbf{x}$ is obtained. Accuracy can be further improved with Benettin's *renormalization* algorithm [2], where during integration particles are reset to a fixed distance from the reference particle. Renormalization was originally used for computing (non-finite) Lyapunov exponents. A variant of renormalization is the L-FTLE proposed by Kasten et al. [17], where numerical integration and differentiation are swapped. While pointwise exact, these methods may miss thin ridges of high values passing between samples. Even without renormalization, using a d smaller than the sampling distance creates artifacts, since it is equivalent to first computing the FTLE or FSLE field at sampling distance d and then doing a subsampling, which obviously causes aliasing in the presence of high frequencies. This type of artifacts can be noticed in Fig. 5 of [5].
2. By choosing d in the order of the sampling distance, the obtained FTLE or FSLE values are less accurate, but they are representative for the space between samples. This results in unbroken ridges, which is a requirement for subsequent ridge extraction. Figure 6b has been produced using gradient estimation from a 5-point finite-difference stencil. For the fine structures, resulting from the

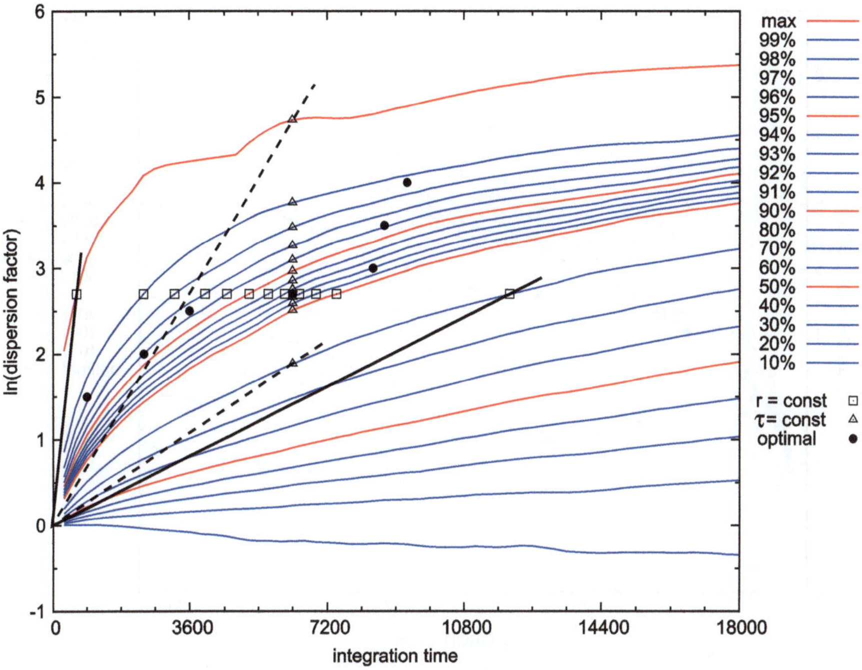

Fig. 3 Percentiles of $\ln(r)$ as functions of integration time of the Vatlestraumen data (with forward integration). FTLE (FSLE) values are obtained by taking a vertical (horizontal) section. The actual values are the slopes of the *dashed* (*solid*) *lines* passing through the origin and the respective point. The *filled circles* mark a few pairs (τ, r), where for a given integration time τ the dispersion factor r has been computed which maximizes the similarity measure

long integration time, the chosen d is barely sufficient to avoid aliasing. For even longer integration a smaller d would be needed [10]. The same holds if subsequent ridge extraction is needed, where it is advantageous to compute gradients by convolution with derivatives of a Gaussian [23].

4 Comparing FTLE and FSLE Images

Computation of FSLE samples on a given grid can be done by simultaneous integration of trajectories and computing the dispersion factors r after each integration step. This way, a function $r(\tau)$ is obtained, from which both FTLE and FLSE can be derived, by computing $\ln(r(\tau))/|\tau|$ for some fixed τ or fixed r, respectively. Figure 3 shows percentiles of $\ln(r)$, computed for the data set of Sect. 6 on a 868 by 868 grid. At each point (x, y) in this plot, the actual FTLE or FSLE value is given by the slope y/x. Taking a vertical section (marked with triangles) yields percentiles of FTLE values for this integration time, while a horizontal section (marked with boxes) yields FLSE values for the corresponding r. In the example FTLE at $\tau = 6{,}300$ s and FSLE at $\ln(r) = 2.7$ are shown.

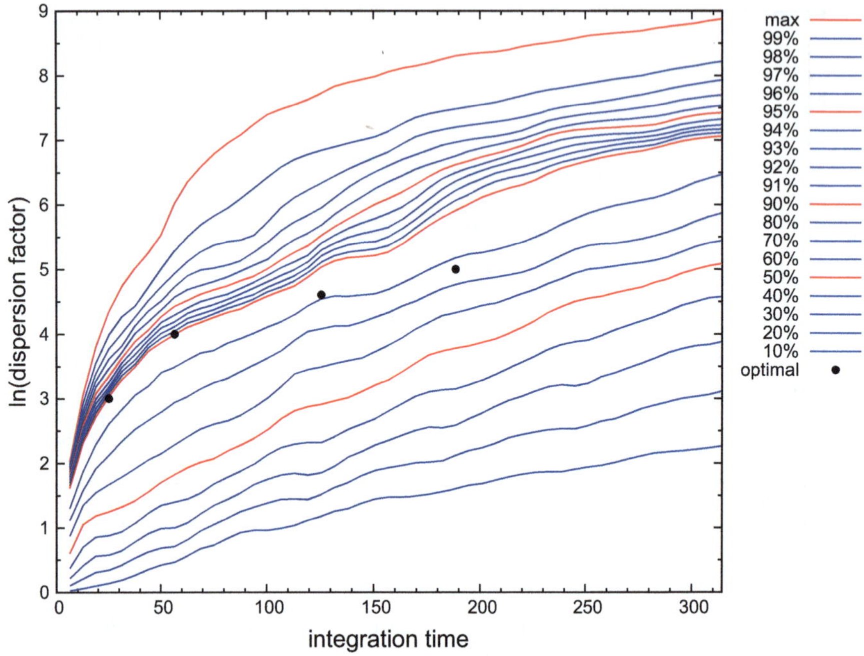

Fig. 4 Percentiles and optimal parameter pairs for meandering jet model. $(\omega, \epsilon) = (0.1, 0.3)$

The fact that the horizontal line does not intersect some of the curves indicates that for some of the trajectories the dispersion factor r has not been reached. At these points, an FSLE sample cannot be computed within the given maximum integration time (here $\tau_{max} = 18{,}000$), but it is known to be less than $\ln r/\tau_{max}$. For visualization purposes, a value of either 0 or $\ln r/\tau_{max}$ can be assigned to such points.

While FSLE and FTLE values are the same at the point where the vertical and horizontal lines intersect, their range is in general different. In our example in Fig. 3 the range between the 80- and 100-percentiles is wider for FSLE (marked with solid lines) than for FTLE values (marked with dashed lines).

We observed this difference in all velocity fields that we tested. A similarity measure for comparing FTLE and FSLE fields has to account for this difference. Therefore, the straightforward choice is to define the *similarity* of a pair of fields $f(\mathbf{x}) = FTLE_\tau(\mathbf{x})$ and $g(\mathbf{x}) = FSLE_r(\mathbf{x})$ as their correlation

$$\mathrm{corr}(f, g) = \mathrm{cov}(f, g)/\sqrt{\mathrm{var}(f)\,\mathrm{var}(g)}. \tag{7}$$

Given one FTLE or FLSE field, a matching field of the other type can now be found by maximizing the similarity. A few such pairs are represented in Figs. 3 and 4 by filled circles. Such optimal pairs, resulting from maximization of similarity, are shown in Figs. 6 and 7.

In principle, it is possible to maximize similarity over both parameters simultaneously. However, this tends to result in extreme parameter values, which are not practically usable. By manually choosing one of the two parameters, the user can balance the visualization with respect to the length and "sharpness" of structures and the amount of folding.

Typically, only the higher FTLE or FSLE values are of interest, therefore the transfer function should clamp data at a lower threshold. Furthermore, to account for the different ranges of FTLE and FSLE data, the transfer function should either be based on normalized data or on percentiles. We chose the second approach, and we clamped the analytic data in Sect. 5 at the 50-percentile and the numerical data in Sect. 6 at the 90-percentile, meaning that we focused on the highest 10 % of FTLE and FSLE values. Which percentile range is of interest depends on the data, or more precisely, on the size and frequency of transport barriers. The percentile plots (Figs. 3 and 4) can then be used to map the chosen percentile range to a range in the parameters τ or r. Finally, a fast heuristic way to avoid the brute-force optimization would be to define the midpoints of these two ranges as a matching parameter pair (τ, r). However we experienced that the similarity values obtained this way were in some cases not near the optimum.

5 Analytic Test Case: The Meandering Jet Model

Bower [6] introduced an analytic flow field for studying the mixing behavior of oceanic jets. Its stream function is defined as

$$\psi = -\tanh\left(\frac{y - B\cos kx}{\sqrt{1 + (kB\sin kx)^2}}\right) + cy. \tag{8}$$

The velocity components are then $u = -\frac{\partial \psi}{\partial y}, v = \frac{\partial \psi}{\partial x}$. This flow is steady in the moving frame where it is formulated. Samelson [22] derived an unsteady version of it by making the meander amplitude B depend on time:

$$B = B_0 + \epsilon\cos(\omega t). \tag{9}$$

The stream function of the flow, at times $t = 0$ and $t = 10\pi$ (where the amplitude takes its maximum and minimum) is shown in Fig. 5. Parameter values have been chosen as $B_0 = 1.2, L = 10.0, k = 2\pi/L, c = 0.1, \omega = 0.1$ and $\epsilon = 0.3$. In these images, contour lines are the streamlines of the velocity field.

5.1 Comparison of FTLE and FSLE

The meandering jet flow has been used by Boffetta et al. [5] in their comparison of FTLE and FSLE. They used Samelson's model with parameters $B_0 = 1.2, L = 10.0$, $k = 2\pi/L$ and $c = 0.1$. For the parameter pair (ω, ϵ) the two settings

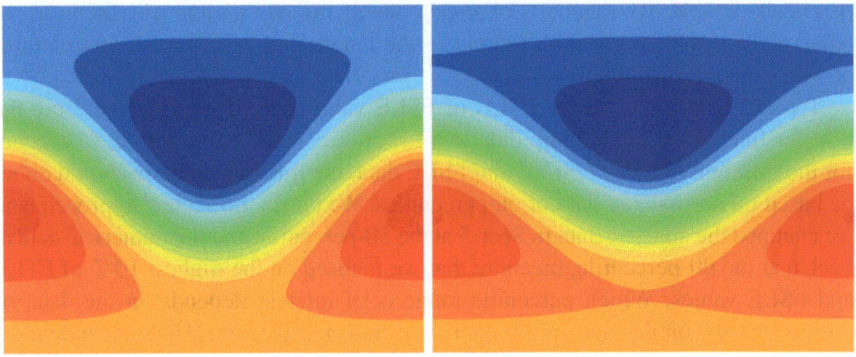

Fig. 5 Stream function of the meandering jet ($0 \leq x \leq 10, -4 \leq y \leq 4$) at $t = 0$ (*left*) and $t = 10\pi$ (*right*)

(0.1, 0.3) and (0.4, 0.3) were investigated. The resulting images have led the authors to the conclusion that FSLE is preferable over FTLE, because the FTLE image (Fig. 4 in [5]) fails to discern the structures seen in the FSLE image (Fig. 5 in [5]).

We reproduced FTLE and FSLE images for the same parameter settings, but using computation based on the Cauchy-Green tensor. In order to obtain comparable structures, it is necessary to adjust the parameters of each technique, i.e., the integration time τ for FTLE and the dispersion factor r for FSLE. For this purpose we used the similarity measure from Sect. 4. Starting from the FSLE parameter $r = 100$, used in [5] (see their Fig. 2a), we obtained first the FSLE image Fig. 6a. Then, a maximally similar FTLE image 6b was obtained for the parameter value $\tau = 40\pi$ (having tested only multiples of 2π). Boffetta et al. were using an integration time of only $\tau = 4\pi$ for their FTLE image (see their Fig. 4a in [5]). The reason for the unsharp FTLE structures is clearly the short integration time, which does not match well the parameter chosen for the FSLE image. Our FTLE result for a short integration time ($\tau = 8\pi$) is shown in Fig. 6d, and we obtained a similar FSLE image for $\ln(r) = 3$ in Fig. 6c.

Aliasing artifacts, which are present in Boffetta's FSLE images, can be avoided by using a sufficient mesh resolution. We used 1,000 by 800, as compared to the 10,000 points mentioned in [5]. Another difference to be noted is that Fig. 4a in [5] is missing the glide reflection symmetry, which is present in the underlying velocity field. The explanation given for this is the dependency on the phase of the flow. However, we obtained an FTLE field having the correct symmetry for all tested phase shifts.

Finally, FSLE computation requires as a secondary parameter a maximum integration time $|\tau_{max}|$. If this is exceeded, this means that the FSLE value can be at most $\ln(r)/|\tau_{max}|$, and it is set to zero (black color in Fig. 6a, c). Choosing $|\tau_{max}|$ too small results in missing structures at lower FSLE values.

By computing FTLE and FSLE from the Cauchy-Green tensor and choosing pairs of parameters based on our proposed similarity measure, comparable structures can be found in this test data. This holds for long (Fig. 6a, b) and short (Fig. 6c, d) integration times.

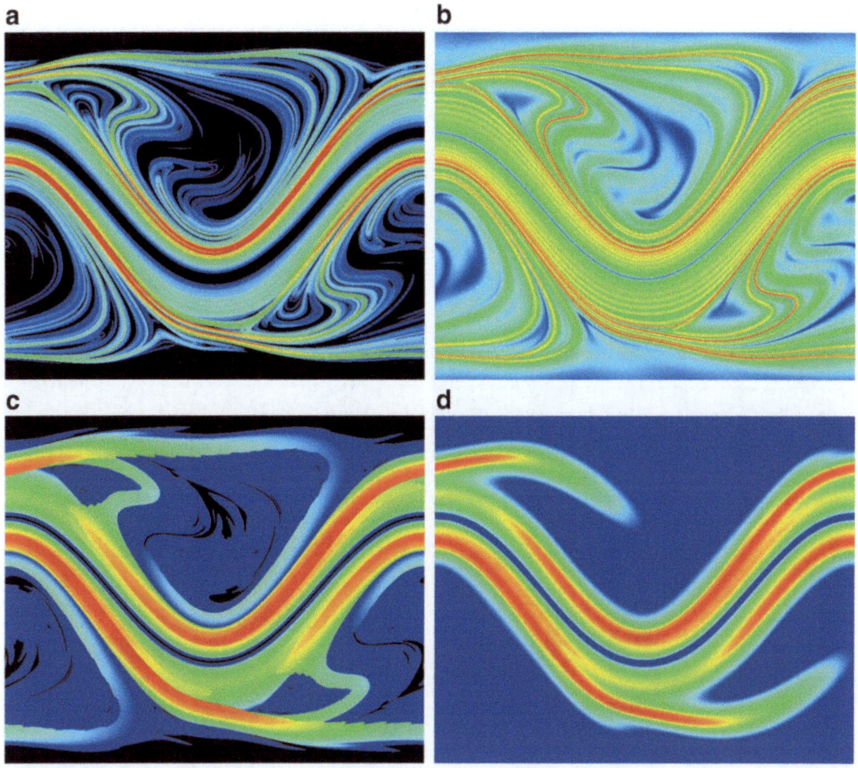

Fig. 6 Meandering jet with $(\omega, \epsilon) = (0.1, 0.3)$, FSLE and FTLE computed with two methods. (**a**) FSLE computed with Cauchy-Green tensor. Time $t = 0$, dispersion factor $r = 100$, maximum time 100π (five flow periods). Color map: 0.01 *(blue)* to 0.1 *(red)*. (**b**) FTLE computed with Cauchy-Green tensor. Time $t = 0$, integration time $\tau = 40\pi$ (two flow periods). Color map: 0.01 *(blue)* to 0.055 *(red)*. (**c**) Same as Fig. 6a, but with $r = e^3 \approx 20.1$. (**d**) Same as Fig. 6b, but with shorter integration time $\tau = 8\pi$

6 Numerical Test Case: Tidal Flow in a Narrow Passage

Our second test case is a CFD simulation of the tidal flow in the *Vatlestraumen* passage, which is a part of the main shipping lane to the harbor of Bergen, Norway. After a tragic naval accident in 2004, the tidal flow in the passage has been modeled with the *Bergen Ocean Model* (BOM) [3]. The data set used in this comparison study is a simulation of approximately one full tidal cycle (12 h). The simulation itself is 3D, but only the surface layer was used for the study. The surface layer flow is the essential one for, e.g., pollution by fuel leaking from the vessel. The time resolution of the data set is 30 s, internally the simulation uses a time step of 1.5 s.

While in this kind of application FSLE is typically used, highly similar structures are obtained with FTLE, as Fig. 7 proves. At least the strong ridges (colored red) can be recognized in the corresponding image, even though colors can differ.

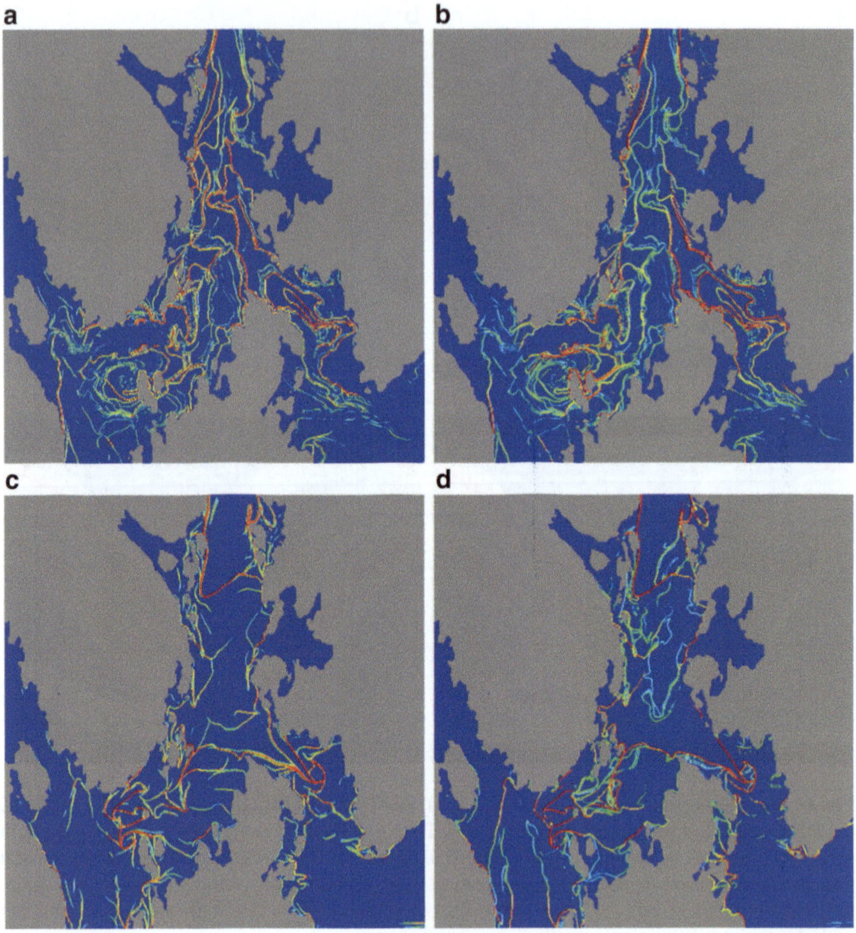

Fig. 7 Comparison of FTLE and FSLE in tidal flow. (**a**) Forward (repelling) FTLE, integration time $\tau = 2\,\mathrm{h}\,20'$. (**b**) Forward (repelling) FSLE, dispersion factor $r = \exp(3.00)$. (**c**) Backward (attracting) FTLE, integration time $\tau = -30'$. (**d**) Backward (attracting) FSLE, dispersion factor $r = \exp(3.00)$

7 Conclusion

Calibration of parameters can produce similar FTLE and FSLE fields and thus similar visualizations of the LCSs in a flow dataset. Differences still exist, and it would be a worthwhile goal to further analyze them. However, we showed that differences are less fundamental than was suggested in literature. Observed differences can be due to poor choice of parameters as well as sampling artifacts. Using our proposed calibration method, it is now possible to compare adequate pairs of fields and focus on the inherent differences of FTLE and FSLE. For this, pairs of parameters τ and r have to be found such that the corresponding FTLE and FSLE fields are similar, i.e., highly correlated. An obvious, brute-force approach is to use an optimization strategy in order to maximize correlation. However, this requires repeated computation of such fields. For practical purposes, a faster method is needed. An interesting problem for future work would be to find a heuristic that is able to find near-optimal pairs of parameters.

While the actual results that can be achieved by FTLE and FSLE are rather similar, it is worthwhile noting the subtle conceptual difference between the two methods: The FSLE allows us, by the choice of the parameter r, to focus on local separation above a certain threshold. Additional information about the typical time-scale of the observed mixing/transport is also provided, which can be a valuable additional information [9, 14]. The FTLE, on the other hand, operate with a fixed time-scale and detect separation at all spatial scales. This allows us also to assess the interplay of the detected structures [26]. Depending on prior knowledge about the time- and spatial scales of interest for a specific application, as well as whether or not the interplay of structures is of interest, FTLE may be a more appropriate choice than FSLE, and vice versa.

Acknowledgements We wish to thank Tomas Torsvik, Uni Research, Uni Computing (Bergen, Norway), for the tidal flow data. This work was funded in part by the Seventh Framework Programme for Research of the European Commission, under FET-Open grant number 226042 (project SemSeg) and the Swiss National Science Foundation, under grant number 200020_140556.

References

1. E. Aurell, G. Boffetta, A. Crisanti, G. Paladin, A. Vulpiani, Growth of noninfinitesimal perturbations in turbulence. Phys. Rev. Lett. **77**, 1262–1265 (1996)
2. G. Benettin, L. Galgani, A. Giorgilli, J.M. Strelcyn, Lyapunov characteristic exponent for smooth dynamical systems and hamiltonian systems: a method for computing all of them. Mechanica **15**, 9–20 (1980)
3. J. Berntsen, User guide for a modesplit σ-coordinate ocean model. Version 4.1. Technical report, Department of Mathematics, University of Bergen, Norway, 2004
4. F.J. Beron-Vera, M.J. Olascoaga, M.G. Brown, H. Kocak, I.I. Rypina, Invariant-tori-like lagrangian coherent structures in geophysical flows. Chaos **20**(1), 1–13 (2010)
5. G. Boffetta, G. Lacorata, G. Redaelli, A. Vulpiani, Detecting barriers to transport: a review of different techniques. Physica D **159**, 58–70 (2001)
6. A. Bower, A simple kinematic mechanism for mixing fluid parcels across a meandering jet. J. Phys. Oceanogr. **21**, 173–171 (1991)
7. S. Brunton, C. Rowley, Fast computation of finite-time Lyapunov exponent fields for unsteady flows. Chaos **20**(017503), 017503-1–017503-12 (2010)
8. C. Coulliette, F. Lekien, J.D. Paduan, G. Haller, J.E. Marsden, Optimal pollution mitigation in monterey bay based on coastal radar data and nonlinear dynamics. Environ. Sci. Technol. **41**(18), 6562–6572 (2007)
9. F. d'Ovidio, V. Fernández, E. Hernández-García, C. López, Mixing structures in the Mediterranean Sea from finite-size Lyapunov exponents. Geophys. Res. Lett. **31**(17), L17203-1–L17203-4 (2004). doi:10.1029/2004GL020328

10. R. Fuchs, B. Schindler, R. Peikert, Scale-space approaches to FTLE ridges, in *Topological Methods in Data Analysis and Visualization II*, ed. by R. Peikert, H. Hauser, H. Carr, R. Fuchs (Springer, New York, 2012), pp. 283–296
11. I. Goldhirsch, P.L. Sulem, S.A. Orszag, Stability and Lyapunov stability of dynamical systems: a differential approach and a numerical method. Physica D **27**(3), 311–337 (1987)
12. G. Haller, Finding finite-time invariant manifolds in two-dimensional velocity fields. Chaos **10**(1), 99–108 (2000)
13. G. Haller, Distinguished material surfaces and coherent structures in three-dimensional fluid flows. Physica D **149**, 248–277 (2001)
14. I. Hernández-Carracos, C. López, E. Hernández-García, A. Turiel, How reliable are finite-size Lyapunov exponents for the assessment of ocean dynamics? Ocean Model. **36**(3–4), 208–218 (2011)
15. B. Joseph, B. Legras, Relation between kinematic boundaries, stirring, and barriers for the antarctic polar vortex. J. Atmos. Sci. **59**, 1198–1212 (2002)
16. D. Karrasch, G. Haller, *Do Finite-Size Lyapunov Exponents Detect Coherent Structures?* (2013). http://arxiv.org/abs/1307.7888
17. J. Kasten, C. Petz, I. Hotz, B. Noack, H.C. Hege, Localized finite-time lyapunov exponent for unsteady flow analysis, in *Vision Modeling and Visualization*, vol. 1, ed. by M. Magnor, B. Rosenhahn, H. Theisel (Universität Magdeburg, Inst. f. Simulation u. Graph., 2009), pp. 265–274
18. T.Y. Koh, B. Legras, Hyperbolic lines and the stratospheric polar vortex. Chaos **12**(2), 382–394 (2002)
19. A.J. Mariano, A. Griffa, T.M. Özgökmen, E. Zambianchi, Lagrangian analysis and predictability of coastal and ocean dynamics 2000. J. Atmos. Ocean. Technol. **19**(7), 1114–1126 (2002)
20. A. Pobitzer, R. Peikert, R. Fuchs, B. Schindler, A. Kuhn, H. Theisel, K. Matković, H. Hauser, The state of the art in topology-based visualization of unsteady flow. Comput. Graph. Forum **30**(6), 1789–1811 (2011)
21. F. Sadlo, R. Peikert, Efficient visualization of lagrangian coherent structures by filtered AMR ridge extraction. IEEE Trans. Vis. Comput. Graph. **13**(5), 1456–1463 (2007)
22. R. Samelson, Fluid exchange across a meandering jet. J. Phys. Oceanogr. **22**, 431–440 (1992)
23. B. Schindler, R. Peikert, R. Fuchs, H. Theisel, Ridge concepts for the visualization of lagrangian coherent structures, in *Topological Methods in Data Analysis and Visualization II*, ed. by R. Peikert, H. Hauser, H. Carr, R. Fuchs (Springer, New York, 2012), pp. 221–236
24. S.C. Shadden, F. Lekien, J.E. Marsden, Definition and properties of Lagrangian coherent structures from finite-time Lyapunov exponents in two-dimensional aperiodic flows. Physica D: Nonlinear Phenom. **212**(3–4), 271–304 (2005)
25. M. Üffinger, F. Sadlo, M. Kirby, C.D. Hansen, T. Ertl, FTLE computation beyond first order approximation, in Eurographics Short Papers, ed. by C. Andujar, E. Puppo, Eurographics Association, Cagliari, pp. 61–64, 2012
26. S. Wiggins, The dynamical systems approach to lagrangian transport in oceanic flows. Annu. Rev. Fluid Mech. **37**, 295–328 (2005)

Development of an Efficient and Flexible Pipeline for Lagrangian Coherent Structure Computation

Siavash Ameli, Yogin Desai, and Shawn C. Shadden

Abstract The computation of Lagrangian coherent structures (LCS) has become a standard tool for the analysis of advective transport in unsteady flow applications. LCS identification is primarily accomplished by evaluating measures based on the finite-time Cauchy Green (CG) strain tensor over the fluid domain. Sampling the CG tensor requires the advection of large numbers of fluid tracers, which can be computationally intensive, but presents a large degree of data parallelism. Processing can be specialized to parallel computing architectures, but on the other hand, there is compelling need for robust and flexible implementations for end users. Specifically, code that can accommodate analysis of wide-ranging fluid mechanics applications, while using a modular structure that is easily extended or modified, and facilitates visualization is desirable. We discuss the use of Visualization Toolkit (VTK) libraries as a foundation for object-oriented LCS computation, and how this framework can facilitate integration of LCS computation into flow visualization software such as ParaView. We also discuss the development of CUDA GPU kernels for efficient parallel spatial sampling of the flow map, including optimizing these kernels for better utilization.

1 Introduction

The computation of Lagrangian coherent structures (LCS) originated as a means to compute stable and unstable manifold type structures in vector fields with aperiodic time dependence. This was motivated by knowledge that the interaction

S. Ameli • S.C. Shadden (✉)
University of California, Berkeley, CA, USA
e-mail: sameli@berkeley.edu; shadden@berkeley.edu

Y. Desai
Georgia Tech, Atlanta, GA, USA
e-mail: yogindesai02@gmail.com

P.-T. Bremer et al. (eds.), *Topological Methods in Data Analysis and Visualization III*, Mathematics and Visualization, DOI 10.1007/978-3-319-04099-8_13,

of such manifolds gives rise to chaotic dynamics, and hence understanding these interactions helped bring an ordered understanding to chaotic advection in fluid flow. This approach was originally applied to time periodic systems, especially using Poincaré mappings that make the dynamics autonomous. The applicability of traditional invariant manifold concepts breaks down for time aperiodic vector fields for practical and conceptual reasons. Notably, asymptotic notions associated with invariant manifold theory are neither applicable nor desirable for understanding inherently transient phenomena associated with unsteady computational or experimental fluid flow data.

LCS computational techniques do not typically solve for the stable and unstable manifolds of explicit trajectories. A global approach was developed from observations that material points straddling stable and unstable manifolds typically separate faster in forward and backward time than pairs of points not straddling such manifolds. That is, the manifold geometry may be inferred by considering the stretching associated with the hyperbolicity of these structures [14]. This led to an alternative characterization of organizing structures in fluid flows as invariant manifolds (material surfaces) that satisfy certain locally attracting or repelling properties, which became the basis for formalizing the concepts of LCS. LCS computations have been applied to diverse applications, as reviewed in [26], demonstrating wide-ranging utility in analysis of unsteady fluid advection.

Scalar measures are often used to identify LCS, such as the (largest) finite time Lyapunov exponent FTLE field [29]. This is accomplished by plotting the field and visually identifying such structures, or using an algorithmic approach to extract features such as ridges in the field [19,21,24]. The FTLE is derived from the largest eigenvalue of the Cauchy Green (CG) strain tensor

$$\mathbf{C}(\mathbf{x}_0, t_0, t_f) = \nabla \mathbf{F}_{t_0}^{t_f}(\mathbf{x}_0)^{\mathsf{T}} \cdot \nabla \mathbf{F}_{t_0}^{t_f}(\mathbf{x}_0) \, , \tag{1}$$

where $\mathbf{F}_{t_0}^{t_f} : \mathbf{x}(t_0) \mapsto \mathbf{x}(t_f)$ denotes the flow map, and $\mathbf{x}(t)$ is a fluid element trajectory with $\mathbf{x}_0 = \mathbf{x}(t_0)$. The full CG tensor encodes direction-dependent stretching information that can be leveraged, for example in defining normally hyperbolic LCS, i.e. material surfaces that are locally the most normally repelling over a chosen time interval [9]. Therefore, the computation of the CG tensor over the flow domain can be thought of as a common, or at least representative, target in LCS identification strategies.

The global approach to LCS identification requires a highly resolved sampling of the CG tensor over the fluid domain to locate potential LCS. One typically starts with fluid velocity field data $\mathbf{u}(\mathbf{x}, t)$, obtained from computation or measurement, and the flow map is computed by seeding the fluid domain with tracers and integrating the advection equation

$$\dot{\mathbf{x}}(\mathbf{x}_0, t) = \mathbf{u}(\mathbf{x}, t) \, , \tag{2}$$

over a finite time interval (t_0, t_f) for a grid of seed points $\mathbf{x}_0$. This is typically the most computationally intensive aspect of LCS identification since the seed grid may be composed of millions of material points especially in 3D flows. Furthermore, the integration of Eq. (2) for each seed point typically requires thousands of integration steps or more, and each integration step can require several space-time interpolations of $\mathbf{u}(\mathbf{x}, t)$. Furthermore, this process may be repeated for a time series of seed grids at different t_0 to obtain the time evolution of LCS. We note that the high resolution of seed points is needed in part for finite differencing to compute $\nabla \mathbf{F}_{t_0}^{t_f}(\mathbf{x}_0)$. Alternatively, integration of the variational equations, as performed in [15], may also be used to obtain the linearized flow map directly. However, sufficient resolution is still needed for LCS detection, especially when ridge extraction is performed [24].

It has been our experience that two major deterrents for wider adoption of LCS computations for flow post-processing are (1) computational time and (2) ease of use. With regards to (1), we will discuss the use of general purpose computing on graphics processing units (GPGPU) for accelerating flow map computations. This work builds on previous works of [7, 8, 13]. With regards to (2), we will discuss the use of the Visualization Toolkit (VTK) for creating an LCS computational pipeline that on the "front-end" is capable of handling various input data, and on the "back-end" facilitates visualization of LCS with standard flow analysis tools. In Sect. 2 we describe the main component of the LCS computational pipeline. In Sect. 3 we describe the implementation of the flow map computation on the GPU and results for various applications, ranging from 2D Cartesian to 3D unstructured velocity data processing. In Sect. 4 we will discuss some results and relationship with previous work.

2 LCS Pipeline

Here we describe a pipeline developed to process velocity field data for the purpose of computing LCS. This pipeline was written in C++ and will be made openly available on GitHub (www.github.com/FlowPhysics). This pipeline was developed using an object-oriented approach to create filters (classes that modify data) that can be extended or modified with minimal effort. Specifically, this pipeline makes extensive use of the popular Visualization ToolKit (www.vtk.org). VTK is a broad collection of open-source libraries for 3D computer graphics, image processing and visualization. The filters that we developed adhere to VTK coding standards and conventions for modularity, portability and debugging. We also chose to develop our pipeline using VTK for two additional reasons. First, VTK has a number of classes for reading standard file formats and data grid types commonly used by the fluid mechanics community. We could therefore leverage existing functionality to better support input from a variety of applications. Second, open-source programs for visualizing scientific data such as ParaView and VisIt are built on VTK.

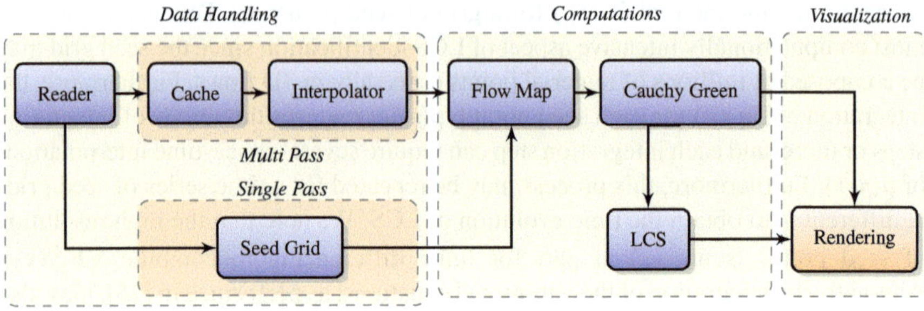

Fig. 1 Pipeline architecture

Developing an LCS pipeline in VTK facilitates integration with these programs, and alternatively development of custom rendering for a stand-alone software.

VTK is highly object-oriented and supports a pipeline architecture that generalizes the connection and execution of algorithms. The benefits of this methodology are (1) a universal interface between different filters in the pipeline, (2) greater control over information flow to better manage cache. The pipeline enables streaming of data, which is important for the application here because unsteady velocity field data are typically stored by a series of files that are often altogether too large to be loaded in memory. Streaming requires the use of *time-aware* filters and for this we developed time-related request keys, similar to the approach described by Biddiscombe et al. [4]. This pipeline does not support spatial streaming of velocity data for applications where individual time frames are too large for memory.

Our pipeline architecture includes the following custom filters, which are connected as shown in Fig. 1. We note that the pipeline does not represent data flow, but rather connectivity of filters and the management of requests.

Reader. A reader filter was developed to accept a variety of input data. VTK contains a number of classes to load various types of data (e.g. structured and unstructured grid data) and file formats. There are few readers in VTK that are time-aware, and those may only support one or a few file formats. Our filter implements an reader class that casts to the appropriate readers of both legacy and XML file types to support 11 file formats commonly used in VTK. This filter provides additional functionality by providing two different types of output based on requests of the pipeline. Output can be either a single data object or a data object that encapsulates multiple data objects depending on if the request comes from the `Seed` or the `Cache` filter. This filter also adds metadata on output objects such as times and indices that are needed throughout the pipeline to facilitate temporal streaming and interpolation. Lastly, this filter provides a framework for implementing necessary pre-processing of velocity data, such as tetrahedralization as motivated below.

Cache. This filter is connected to the output of `Reader`. The `Cache` filter is essentially a wrapper that manages VTK's `vtkTemporalDataSetCache` class.

It stores a window of the last requested data time frames. The amount of data cached depends on the upstream request. This filter is used to avoid repetitive streaming processes in the pipeline and deletes unnecessary data on memory.

Interpolator. This filter is used to interpolate the velocity field data in space and time. Data in the pipeline before this filter are provided discretely and all filters after `Interpolator` treat velocity data as continuous. Tracers are updated altogether each time step. This ensures that tracers are requesting the same flow field information each update step to best handle data streaming and memory utilization. This is also consistent with our GPU implementation whereby tracers are updated altogether to keep threads short-lived. We note that both space and time interpolation is performed on the GPU. Specifically, a window of velocity data frames (nominally two velocity files for linear interpolation) are loaded on the GPU's memory, and interpolation is performed on the GPU as necessary as integration proceeds between these time frames.

Seed. This filter is used to initialize the seed points where the CG tensor is to be sampled. Currently these locations are defined as structured grid data. This filter defines their initial conditions, other attributes, and for unstructured velocity data computes which cell in the velocity field mesh each seed point is located in. While the seed grid are nominally the locations where the CG tensor is evaluated, highly localized auxiliary points may also be defined about each seed point for improving flow map gradient computation as discussed below.

Flow Map. This filter maps the seed points forward in time. Therefore it is connected to both `Seed` and `Interpolator`. It is capable of implementing various types of integration routines, e.g. single step (Runge Kutta) and multi step (e.g. Adams Bashforth) methods. The same points passed from the input (`Seed`) are passed to the output with the final seed point locations added as a vector attribute to the point data. If a trajectory leaves the spatial domain of the velocity data, two options are possible. The CG tensor evaluation can be performed early at all locations relying on this trajectory. Alternatively, velocity extrapolation can be performed inside the interpolation filter, however one must take care is this regard [26, 30].

Cauchy-Green. This filter calculates the finite time `Cauchy-Green` strain tensor over the seed points using the initial and final point locations output from `Flow Map`. Currently, this is performed using standard central difference formula for the flow map gradient matrix entries. This differencing can be applied directly to the evolution of the seed point grid, or by differencing auxiliary points attributed to each seed location as described above. The auxiliary point method is more computationally expensive, but can provide more accurate CG tensor computation at less computational expense than an equivalent increase of seed grid resolution when seed spacing is reduce by more than 1/2. However, an equivalent increase in seed grid resolution also improves resolution of the CG tensor field, which is not accomplished through the auxiliary points method. Additionally, the auxiliary point method can result in missed structures, when LCS are not properly straddled by points used in finite differencing [29, 31]. This filter also computes the eigenvalues and eigenvectors of the CG tensor at each seed point, which are commonly used for LCS detection. The seed points are passed to the output with the eigenvalue/vector fields added as point data attributes.

LCS. This filter processes the output from the `Cauchy Green` filter to apply criterion for LCS detection, for example, further processing of FTLE field data, such as Gaussian smoothing to remove computational noise as well as C-Ridge computation [23, 24]. While not currently implemented, strainline and shearline computation as described in [10] could be implemented in this filter.

Visualization. This is not a separate filter in our pipeline, but we have separated this conceptually. One important benefit of using the new pipeline functionality of VTK is that we can compile each filer to a shared library that can be loaded by the ParaView plugin manager, so an end user can simply import the filters as plugins in ParaView and then use these to create a pipeline for computing and visualization of fields defined from the CG filter (e.g. FTLE), or visualize features from the LCS filter. We note that support for FTLE computation was recently added to VisIt [2] as used by Özgökmen et al. [18].

3 Flow Map Computation on the GPU

The computation of the CG tensor field requires the advection of a large set of tracer trajectories. Since each tracer moves independently, these calculations present a parallel workload. GPGPU was used to perform acceleration/parallelization of this computation. On the hardware side, GPUs are widely available, cheap and scalable and on the software side GPU programming has become widely accessible, portable and supported by various compilers.

We used NVIDIA's compute unified device architecture (CUDA) platform [1]. Optimization of a CUDA program depends somewhat on the GPU architecture used. We report results run on a consumer-level graphics card with the NVIDIA GeForce GTX 670 GPU and the higher-end NVIDIA Tesla K20 GPU. Both GPUs are based on the Kepler architecture. GeForce GTX 670 has peak theoretical double floating point performance of 0.19 TFlops whereas Tesla K20 has peak theoretical double floating point performance of 1.17 TFlops. While GPU threads are plentiful and can be launched and terminated with minimal overhead, performance of parallelized flow map (trajectory) computation is primarily limited by memory bandwidth.

For the GPU implementation only registers, global memory and constant memory were utilized. Registers were used by individual threads to store local variables in specific interpolation and integration kernels discussed below, and were reused in each kernel to the greatest extent possible. Since constant memory is read only and limited to 64 kb but can be accessed by all threads, we used this for mesh parameters and other constants. Tracer and velocity field arrays were always stored on global memory.

Each streaming multiprocessor (SMX) can run up to 16 resident blocks or 2,048 resident threads concurrently for the architecture we used. A common target in optimizing the code was to make sure every streaming processor had roughly 2,048 resident threads running at any given time. The number of registers available per SMX limits the number of threads that can reside on a SMX. Since the number of registers were limited to 64 K per multiprocessor, this implied that every thread needed to use only 31 registers for 100 % occupancy. We developed our kernels, as discussed below, to keep register usage nominally within this bound.

3.1 Implementation

For the purpose of explaining the implementation and performance results we focus discussion on the 4th order explicit Runge-Kutta integration method (RK4). To decrease register usage and hence improve GPU utilization, we broke up the integration kernel into smaller kernels. For example, RK4 requires the vector field to be evaluated at four different locations in space-time for a single update step. These interpolations are done as separate kernels. Furthermore, for unstructured grid data each interpolation requires a cell-search, i.e. determination of which velocity grid cell the interpolation point is located. Therefore, for each integration step, eight kernels are run; four kernels to evaluate the velocity field, and each of these requires a preceding cell-search kernel call. This strategy involves the use of a large number of threads that are short-lived, which are what GPUs are designed to handle. Using this strategy also enables the CG tensor to be evaluated at any time step during the integration interval (t_0, t_f) with minimal effort. This can be advantageous if the CG tensor is to be evaluated early for trajectories that leave the fluid domain before the full integration interval is reached. This is common with velocity data coming from modeling or measurement over a truncated domain. But moreover, some LCS detection strategies require certain criteria to be evaluated on the fly as integration proceeds. This need also motivated updating tracers synchronously for each integration step, which is consistent with maintaining efficient data flow in our pipeline, Sect. 2. That is, the pipeline is most efficient if all upstream requests require the same velocity data loaded in memory. We note however that velocity data is loaded into GPU memory only every time a new velocity data file is needed, not every integration time step. Figure 2 shows the runtime comparison of the single-kernel code and split-kernel code. It can be seen that split-kernel strategy executes in roughly 30 % less time for all by very small seed grid sizes.

We use an efficient local cell search strategy for unstructured grids [27]. We have demonstrated that this method outperforms other cell search methods we have tested such as the structured auxiliary mesh approach, and VTK's Kd-tree and Oct-tree methods. This method is designed for tetrahedral grids (or triangular grids in 2D). Therefore, other grid topologies of velocity data that might need to be

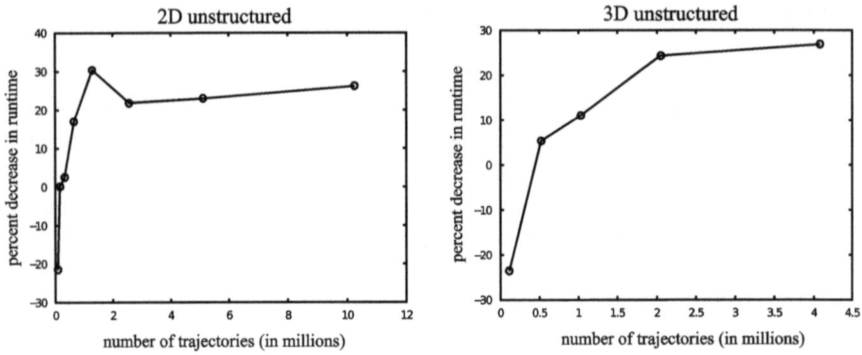

Fig. 2 Runtime comparison between a single-kernel RK4 time step implementation and a split-kernel RK4 time step implementation. Graphs plot the percent decrease in runtime of the split kernel implementation from that of the single kernel implementation. Runs were performed on both the 2D and 3D unstructured data examples described below

processed are tetrahedralized using built in VTK functionality as a preprocessing step in the `Reader` filter. This enables a single, efficient cell search algorithm to be used, which alleviates the need to develop and optimize different interpolation and integration kernels for different velocity grid topologies. However the tradeoff is that piece-wise linear representation using a tetrahedral grid may degrade interpolation accuracy when the native CFD grid has higher order elements.

3.2 *Performance*

Most of the processing time for the CG tensor field computation is spent inside the velocity field interpolation function. This function gets called one or more times each integration step for each trajectory update. Efficiency of this function is a main determinant of computational time. The interpolation function is mainly dependent on the grid topology of the velocity data, e.g. interpolation on Cartesian grids differs in strategy and performance than interpolation on unstructured grids. Different kernels were developed to handle interpolation on different grid types. The efficient cell search algorithm for tetrahedralized unstructured grid data that is used readily facilitates spatial interpolation as described in [27]. We present results for four different grid types considered: 2D Cartesian, 3D Cartesian, 2D unstructured, 3D unstructured. Specifically, the example applications include the double-gyre flow [29], which has become a standard test case for LCS computations (2D Cartesian); the 3D Rayleigh-Bénard convection cell [17] (3D Cartesian); a coronary stenosis model [28] (2D unstructured); and an abdominal aortic aneurysm (AAA) model [3] (3D unstructured). These examples are shown in Fig. 3.

Figure 4 plots the performance results from the GeForce and Tesla GPUs. The values plotted are the runtimes on the respective GPU normalized by the serial CPU

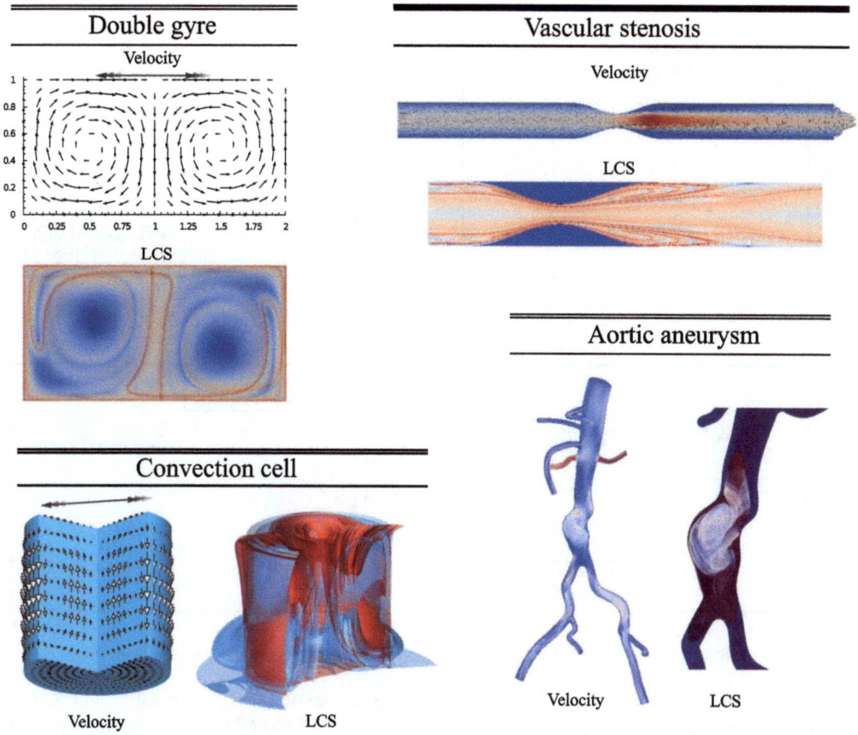

Fig. 3 Snapshots of the flow fields and LCS from the four applications used for performance testing, clockwise from *upper-left*: double gyre [29]; vascular stenosis [28]; abdominal aortic aneurysm [3]; Rayleigh-Bénard convection [17]

runtime, which gives the speedup over serial CPU processing. The CPU used was an Intel Core i7-3770 Ivy Bridge 3.5 GHz processor and the implementation was FlowVC [25], which was written in C. We observed up to roughly 70× speedup for computations on the 2D Cartesian data down to roughly 12× speed up for 3D unstructured grid data. As expected, the Tesla K20 yields higher speedup due to a higher number of streaming multiprocessors. The AAA velocity data was specified on an unstructured tetrahedral grid of 1.01 million elements. Unstructured velocity grids up to several million elements were tested and yielded similar speedup results as the AAA model.

We note that we achieved significant performance improvement by properly coalescing tracer grid arrays, which is common practice, whereby nearby threads processed tracer data coalesced in memory. We did not notice any improvement of performance by sorting velocity field data arrays. This is because as the trajectories of the seed grid evolve, trajectories with nearby indices in memory (and hence locally processed on the GPU) do not necessarily require velocity data that is localized in space. A recent solution to this problem was proposed by Chen and Hart

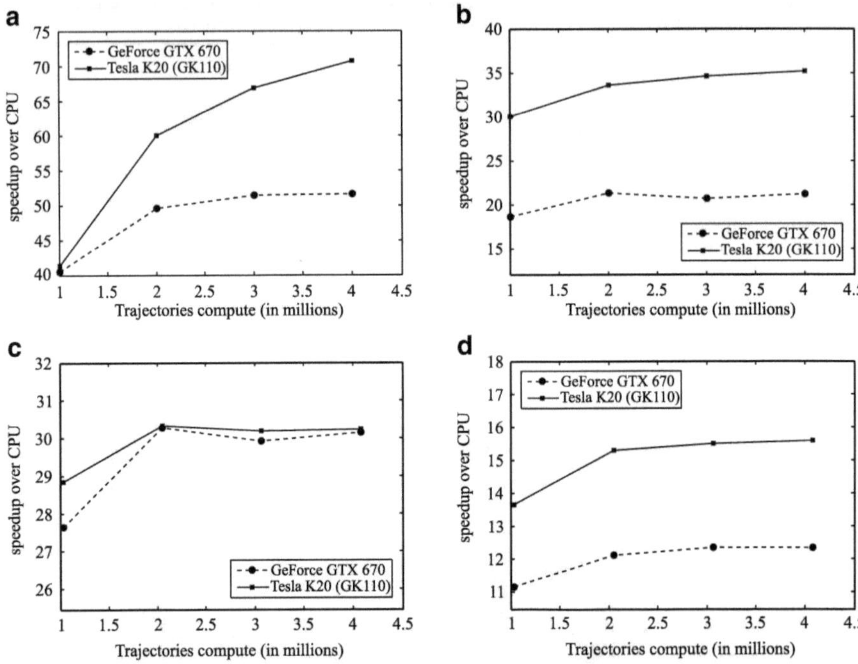

Fig. 4 Speedup of particle trajectory computation on a GeForce GTX 670 GPU and a Tesla K20 GPU compared to serial processing on an Intel Core i7-3770 CPU. All computations were performed using double precision for floating point variables. (**a**) 2D Cartesian. (**b**) 3D Cartesian. (**c**) 2D unstructured. (**d**) 3D unstructured

[6], which reorganizes particles into spatially coherent bundles as they are advected to improve memory coherence and shared-memory GPU performance.

3.3 *Verification*

Computations performed on the GPU were verified against existing CPU code that has been extensively used and tested, FlowVC. We present verification data from the 2D Cartesian and 3D unstructured grid applications, as these cases represent the extrema in results from the four examples considered. For each application we released several thousand tracers in the domain and integrated their trajectories for the nominal minimum time needed for LCS computation. Specifically, we chose this time based on the AAA model, and scaled the other integration times by the nominal edge size x_l divided by the mean velocity magnitude $\langle \mathbf{u}(\mathbf{x}, t) \rangle$ at peak flow. This defines a characteristic time scale similar to use of the CFL number. This ensured that all cases were integrated over roughly the same number of elements in their respective domains. We performed these computations using the GPU and CPU codes on the GeForce GTX 670 and Intel Core i7-3770. The average error was defined by computing of the L_2 norm of the differences in tracer locations over time between the two runs, and averaging over all tracers as follows

$$e_D(t) = \frac{1}{N}\sum_{i=1}^{N} \left\| \mathbf{x}_i^{DG}(t) - \mathbf{x}_i^{DC}(t) \right\| ,$$

where N is the number of trajectories, $\mathbf{x}_i^{DG}(t)$ is the trajectory of tracer i computed on the GPU using double precision floating point numbers, $\mathbf{x}_i^{DC}(t)$ is the trajectory of tracer i computed on the CPU using double precision floating point numbers, and $\mathbf{x}_i^{DG}(0) = \mathbf{x}_i^{DC}(0)$. In all applications the error stayed below 1×10^{-10}, indicating that both GPU and CPU implementations were performing equivalent tracer trajectory computation.

Single precision calculation can be several times faster than the double precision calculation, however accuracy may be unacceptable. The results generated using a single precision floating point GPU implementation were compared against the double precision CPU results. As described above, the average L_2 norm of the error in trajectories of several thousand tracers was computed from a single precision GPU run and a double precision CPU run as

$$e_S(t) = \frac{1}{N}\sum_{i=1}^{N} \left\| \mathbf{x}_i^{SG}(t) - \mathbf{x}_i^{DC}(t) \right\| ,$$

where $\mathbf{x}_i^{SG}(t)$ is the trajectory of tracer i computed on the GPU using single precision for floating point variables. For the 2D Cartesian data, we noticed fairly acceptable errors in tracer trajectories and subsequent LCS computation. However, for the more complex 3D unstructured grid data, we notice unacceptable degradation in accuracy, e.g. to a point where noticeable degradation of the FTLE field occurred. The errors $e_d(t)$ and $e_s(t)$ are plotted against integration time for these two cases in Fig. 5 using the NVIDIA GeForce GTX 670 GPU.

Because the plots in Fig. 5 represent averages over many tracers, some tracer trajectories may deviate to far greater extent than the mean values shown. Indeed, in the AAA flow, some tracers were advected to different arterial branches based on these errors, which has important consequences for that application. Not surprisingly, errors can be worse near LCS due to inherent sensitivity to initial conditions at these locations, which can be problematic for accurate LCS detection. In addition, while we consider the double precision CPU results as a baseline for comparison, this does not imply that these results represent the "true" trajectories. The double precision CPU computations are subject to normal truncation and round-off errors. However, since double precision computation represents the de facto method for minimizing numerical error, and since no application considered nor of practical importance has a closed-form analytic solution, this was deemed an appropriate baseline for comparison.

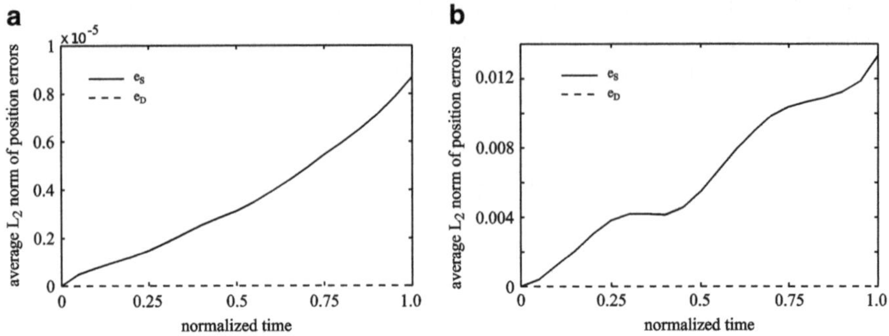

Fig. 5 Error between single precision (*solid line*) and double precision (*dashed line*) GPU computations compared to a double precision CPU computations versus integration time for the double gyre flow (*left*) and AAA flow (*right*). Integration times have been normalized and the value of 1.0 approximately represents the nominal minimum integration time needed to compute LCS for each application. (**a**) 2D Cartesian. (**b**) 3D unstructured

4 Discussion

We have developed a modular pipeline for LCS computation that is capable of loading a wide variety of fluid mechanics data, and that can be easily interfaced with ParaView for visualization of LCS computational results alongside other flow visualization tools. This pipeline was designed to be modular and flexible so that modifications and additions can be made with minimal effort.

As the LCS method has gained considerable popularity in the past few year, there have been several developments to improve computation and visualization of LCS. Strategies have been devised so that tracer seeding can be adapted to cluster sampling of the flow map near LCS for improved detection [8,16,21,22]. Also, since vector field interpolation is computationally intensive, strategies have also been developed to locally approximate the flow map [5], which can be beneficial when time series of tracer grids are considered. As well, trajectory computations obtained in a hierarchical manner have been considered for efficient FTLE computation [11]. We note that sampling the flow map (gradient) needs to be highly resolved in space but not necessarily time for LCS computations. That is, sampling in space is driven by LCS identification methods; mainly computation of the flow map gradient or subsequently the CG tensor field. Sampling time is done primarily for visualizing the evolution of the structures. Regardless of the strategy of sampling or interpolating the flow map or its derivative over the fluid domain, this process can be parallelized since advection of tracers can be performed for the most part independently. Because of this high degree of parallelism, and the fact the LCS are typically used for desktop postprocessing and flow visualization, many-core implementation on a single workstation is desirable, as opposed to a visualization cluster, which are less accessible.

Jimenez and Vankerschaver [13] discussed the computation and visualization of FTLE by GPGPU using CUDA. They also made their source code publicly available [12]. Their implementation was, by their own admission, naive since it only could handle FTLE computation for analytically defined vector fields. This greatly reduced memory accesses that, as discussed above, are the bottleneck in computing FTLE fields using GPU processing in practical applications. Garth et al. [8] and Hlawatsch et al. [11] leveraged GPU processing for FTLE computation as well, though these papers did not discuss details of their implementation or performance results as this was not a main focus. The recent paper by Conti et al. [7] described FTLE computation for bluff body flows using OpenCL, which allows implementation on mixed architectures, e.g. AMD's accelerated processing units. Their implementation was specialized to remeshed vortex methods for bluff body flows that involve mesh-particle interpolations previously tailored for GPU implementation [20]. They were able to achieve around one order magnitude speedup compared to serial CPU implementation, similar to what we observed here. However, their application was specialized to a particular flow problem. Most LCS computation is performed as a post processing step on fluid velocity field data. Since this is the most common and general scenario, it was the one which we developed our framework around, consistent with the design specification of our LCS pipeline.

We considered application to 3D flow and flow on unstructured grids, as [7, 13] reported performance results for 2D flows on structured grids. As shown and discussed above, we have run this implementation to integrate millions of tracers on velocity grids with several million element. This represents a reasonable limit for most LCS applications. For significantly larger velocity field grids, or tracer grids, one will overflow the global memory for the GPU. We generally noticed peak performance when kernels were kept within register memory bounds. This required each trajectory update to be divided into a series of kernels. Therefore, kernels were very short. Because of the way data is processed through our pipeline, we perform synchronous integration between the tracers. We believe this has advantages for LCS detection as well. For example, in truncated domains, which represent the vast majority of fluid mechanics data, tracers leave the domain before the finite time interval being considered for flow map computation. But perhaps more importantly, one may want to have access to the CG tensor at various times. Most LCS criteria are a one parameter family dependent on the chosen integration window. Using one integration window may identify a manifold as an LCS, but another integration window may not. Similarly, a manifold may satisfy one LCS criterion but not the other for a particular integration time. Therefore, having access to this information enables implementation of methods that may depend on how strain rate or direction changes over the integration time parameter. Or alternatively, schemes that adaptively sample the flow map based on CG tensor information may need sequential access to this information.

While this pipeline can function as a stand alone program, it can also be compiled as libraries to be used with ParaView. Such integration provides a natural platform for visualizing not only LCS computations, but also integrating these results with other existing flow visualization tools provided by this programs, and also GPU-based visualization techniques available. Indeed, LCS as a collection of codimension one surfaces are rarely useful in understanding the flow. Knowledge of these manifolds must be integrated with other knowledge of the flow to gain fundamental insight into the flow topology, and these platforms provide significant capability in this regard.

Acknowledgements This work was supported by the National Science Foundation, award number 1047963.

References

1. NVIDIA CUDA C Programming Guide, Version 4.2 (2012)
2. VISIT – Software that delivers Parallel, Interactive Visualization. http://visit.llnl.gov/
3. A. Arzani, S.C. Shadden, Characterization of the transport topology in patient-specific abdominal aortic aneurysm models. Phys. Fluids **24**(8), 081,901-1-16 (2012)
4. J. Biddiscombe, B. Geveci, K. Martin, K. Moreland, D. Thompson, Time dependent processing in a parallel pipeline architecture. IEEE Trans. Vis. Comput. Graph. **13**(6), 1376–1383 (2007)
5. S.L. Brunton, C.W. Rowley, Fast computation of finite-time Lyapunov exponent fields for unsteady flows. Chaos **20**(1) (2010)
6. M. Chen, J.C. Hart, Fast coherent particle advection through time-varying unstructured flow datasets (2013, Preprint). http://graphics.cs.illinois.edu/papers/fastvection
7. C. Conti, D. Rossinelli, P. Rossinelli, GPU and APU computations of finite time Lyapunov exponent fields. J. Comput. Phys. **231**, 2229–2244 (2012)
8. C. Garth, F. Gerhardt, X. Tricoche, H. Hagen, Efficient computation and visualization of coherent structures in fluid flow applications. IEEE Trans. Vis. Comput. Graph. **13**(6), 1464–1471 (2007)
9. G. Haller, A variational theory of hyperbolic Lagrangian coherent structures. Physica D **240**(7), 574–598 (2011)
10. G. Haller, F.J. Beron-Vera, Geodesic theory of transport barriers in two-dimensional flows. Physica D **241**, 1680–1702 (2012)
11. M. Hlawatsch, F. Sadlo, D. Weiskopf, Hierarchical line integration. IEEE Trans. Vis. Comput. Graph. **17**(8), 1148–1163 (2011)
12. R. Jimenez, J. Vankerschaver, CUDA_FTLE. http://www.its.caltech.edu/~raymondj/LCS/cuda_ftle-0.9.tar.bz2
13. R. Jimenez, J. Vankerschaver, Optimization of FTLE calculations using nVidia's CUDA. Technical report, California Institute of Technology, 2009. http://www.its.caltech.edu/~raymondj/LCS/FTLE_on_GPU.pdf
14. C.K.R.T. Jones, S. Winkler, Invariant manifolds and Lagrangian dynamics in the ocean and atmosphere, in *Handbook of Dynamical Systems*, vol. 2, ed. by B. Fiedler (Elsevier, Amsterdam/Boston, 2002), pp. 55–92
15. J. Kasten, C. Petz, I. Hotz, B. Noack, H.C. Hege, Localized finite-time lyapunov exponent for unsteady flow analysis, in *Vision, Modeling and Visualization*, ed. by M. Magnor, B. Rosenhahn, H. Theisel (2009), pp. 265–274
16. F. Lekien, S.D. Ross, The computation of finite-time Lyapunov exponents on unstructured meshes and for non-Euclidean manifolds. Chaos **20**(1) (2010)
17. F. Lekien, S.C. Shadden, J.E. Marsden, Lagrangian coherent structures in n-dimensional systems. J. Math. Phys. **48**, 065,404-1-19 (2007)
18. T.M. Özgökmen, A.C. Poje, P.F. Fischer, H. Childs, H. Krishnan, C. Garth, A.C. Haza, E. Ryan, On multi-scale dispersion under the influence of surface mixed layer instabilities and deep flows. Ocean Model. **56**, 16–30 (2012)

19. R. Peikert, F. Sadlo, Height ridge computation and filtering for visualization, in *Pacific Visualization Symposium*, Kyoto, ed. by I. Fujishiro, H. Li, K.L. Ma (2008), pp. 119–126
20. D. Rossinelli, C. Conti, P. Koumoutsakos, Mesh-particle interpolations on graphics processing units and multicore central processing units. Philos. Trans. R. Soc. A – Math. Phys. Eng. Sci. **369**(1944), 2164–2175 (2011)
21. F. Sadlo, R. Peikert, Efficient visualization of Lagrangian coherent structures by filtered AMR ridge extraction. IEEE Trans. Vis. Comput. Graph. **13**(5), 1456–1463 (2007)
22. F. Sadlo, A. Rigazzi, R. Peikert, Time-dependent visualization of lagrangian coherent structures by grid advection, in *Topological Methods in Data Analysis and Visualization*, ed. by V. Pascucci, X. Tricoche, H. Hagen, J. Tierny (Springer, Dordrecht, 2010), pp. 151–165
23. B. Schindler, R. Fuchs, S. Barp, J. Waser, A. Pobitzer, R. Carnecky, K. Matkovic, R. Peikert, Lagrangian coherent structures for design analysis of revolving doors. IEEE Trans. Vis. Comput. Graph. **18**(12), 2159–2168 (2012)
24. B. Schindler, R. Peikert, R. Fuchs, H. Theisel, Ridge concepts for the visualization of lagrangian coherent structures, in *Topological Methods in Data Analysis and Visualization II*, ed. by R. Peikert, H. Hauser, H. Carr, R. Fuchs (Springer, Heidelberg/New York, 2012), pp. 221–236
25. S.C. Shadden, FlowVC (Version 1) [Computer software]. Retrieved from http://shaddenlab.berkeley.edu/software/
26. S.C. Shadden, Lagrangian coherent structures, in *Transport and Mixing in Laminar Flows: From Microfluidics to Oceanic Currents*, chap. 3, ed. by R. Grigoriev (Wiley, Weinheim, 2012)
27. S.C. Shadden, M. Astorino, J.F. Gerbeau, Computational analysis of an aortic valve jet with Lagrangian coherent structures. Chaos **20**, 017,512-1-10 (2010)
28. S.C. Shadden, S. Hendabadi, Potential fluid mechanic pathways of platelet activation. Biomech. Model. Mechanobiol. **12**(3), 467–474 (2013)
29. S.C. Shadden, F. Lekien, J.E. Marsden, Definition and properties of Lagrangian coherent structures from finite-time Lyapunov exponents in two-dimensional aperiodic flows. Phys. D: Nonlinear Phenom. **212**(3–4), 271–304 (2005)
30. W. Tang, P.W. Chan, G. Haller, Accurate extraction of lcs over finite domains, with applications to flight data analyses over Hong Kong International Airport. Chaos **20**(1), 017,502-1-8 (2010)
31. M. Üffinger, F. Sadlo, M. Kirby, C. Hansen, T. Ertl, FTLE computation beyond first-order approximation, in *Short Paper Proceedings of Eurographics* 2012, Cagliari, pp. 61–64

Topological Features in Time-Dependent Advection-Diffusion Flow

Filip Sadlo, Grzegorz K. Karch, and Thomas Ertl

Abstract Concepts from vector field topology have been successfully applied to a wide range of phenomena so far—typically to problems involving the transport of a quantity, such as in flow fields, or to problems concerning the instantaneous structure, such as in the case of electric fields. However, transport of quantities in time-dependent flows has so far been topologically analyzed in terms of advection only, restricting the approach to quantities that are solely governed by advection. Nevertheless, the majority of quantities transported in flows undergoes simultaneous diffusion, leading to advection-diffusion problems. By extending topology-based concepts with diffusion, we provide an approach for visualizing the mechanisms in advection-diffusion flow. This helps answering many typical questions in science and engineering that have so far not been amenable to adequate visualization. We exemplify the utility of our technique by applying it to simulation data of advection-diffusion problems from different fields.

1 Introduction

Vector field topology is a powerful tool for the analysis of vector fields, since it reveals their overall structure and provides insights into their intrinsic dynamics. In the visualization community, the problem of extracting topological features has been extensively researched for more than two decades. Visualization by traditional (stationary) 2D vector field topology builds on the concept of critical points which represent isolated zeros of the vector field, i.e., isolated points where velocity magnitude vanishes. By additionally extracting separatrices, i.e., sets of stream lines that converge to these points in forward or reverse time, one obtains

F. Sadlo (✉) • G.K. Karch • T. Ertl
University of Stuttgart, Stuttgart, Germany
e-mail: Filip.Sadlo@visus.uni-stuttgart.de; karchgz@visus.uni-stuttgart.de; ertl@visus.uni-stuttgart.de

P.-T. Bremer et al. (eds.), *Topological Methods in Data Analysis and Visualization III*, Mathematics and Visualization, DOI 10.1007/978-3-319-04099-8_14,

the overall structure of a vector field, i.e., these sets of lines divide the domain into regions of qualitatively different behavior. Nevertheless, there arise some issues regarding the extraction of topological structures from time-dependent vector fields. The separatrices no longer provide information on the true transport of massless particles, since stream lines (or stream surfaces in 3D) only capture the instantaneous structure of the field. Thus, an alternative approach has recently gained importance that obtains a time-dependent counterpart to separatrices called Lagrangian coherent structures (LCS). The LCS represent transport barriers in time-dependent flow and can be obtained as ridges of the finite-time Lyapunov exponent (FTLE) [7], which measures the separation of neighboring particles as they are advected by the flow. This approach allows for qualitative analysis of time-dependent flows, as the LCS separate regions of different behavior over time.

So far, however, only advective transport has been taken into account in topology-based flow visualization, neglecting additional mechanisms that can cause transport, such as diffusion. Diffusion is present in a wide range of physical and mathematical processes including flows, where it leads to advection-diffusion flow problems. Traditional LCS computed from the velocity field by means of the FTLE obviously cannot depict the true transport of species of interest, such as temperature or solubles, in time-dependent advection-diffusion flow. These LCS exhibit substantial cross-flux of the species due to the involved diffusion and hence cannot separate regions of qualitatively different advection-diffusion. By including diffusion flux of the quantity under examination into the flow map computation, we obtain FTLE ridges that are consistent with the true transport of the respective quantity in advection-diffusion flow. Beyond the resulting LCS, we also show the utility of combining the velocity field with the involved diffusion fluxes. The resulting advection-diffusion field represents the instantaneous transport of the quantity of interest—and because it represents a vector field, the entire body of literature on flow visualization can be applied to it. We demonstrate this for the example of temperature diffusion (thermal conduction) in flows by extracting, additional to LCS, vortex core lines from this field and by providing respective interpretations.

This chapter is organized as follows: In Sect. 2 we discuss related work. In Sect. 3 we introduce the advection-diffusion field that provides a basis for generic visualization of advection-diffusion processes. Two existing feature extraction techniques are briefly described in Sect. 4, while in Sect. 5 these features are extracted from three advection-diffusion CFD datasets and discussed. Section 6 concludes this work.

2 Related Work

Flow visualization by topological features has proven successful in a wide range of applications. The field was founded by the works of Perry and Chong [25], Helman and Hesselink [9, 10], and Globus et al. [5]. About a decade later an increase in research in this field has taken place, in the works of Löffelmann et al. [18] regarding dynamical systems, and Weinkauf et al. with respect to new topological features—starting with saddle connectors [40], and followed by connectors of boundary switch curves [45]. They also proposed a first approach [41] to topological features that are able to reflect the true transport behavior in time-dependent vector fields—motivated by the incapability of traditional vector field topology in this respect due to its definition by (instantaneous) stream lines. By introducing the concept of the local Lyapunov exponent [19], later called direct Lyapunov exponent by Haller [7], and finally called finite-time Lyapunov exponent (FTLE), into the field of visualization, Garth et al. [4] and Sadlo and Peikert [29] provided a nowadays popular basis for scientific visualization by means of topological features in time-dependent vector fields. As proposed by Haller [7] and subsequently defined by Shadden et al. [36], the local maximizing curves in the FTLE field, called Lagrangian coherent structures (LCS), represent a time-dependent counterpart to separatrices. Several (and even contradicting) definitions for these curves have been proposed by Haller [7], Shadden et al. [36], and others. In the field of visualization, Sadlo and Peikert [28, 29] proposed the extraction of LCS by means of height ridges [2].

It has to be noted that FTLE ridges can fail to represent LCS due to two reasons: insufficient advection time [30, 36] or shear. More recently, Sadlo and Weiskopf [31] have generalized 2D vector field topology to time-dependent vector fields by replacing the role of stream lines by streak lines in the concept, and Üffinger et al. [44] have extended this approach to 3D. It is worth noticing that these approaches (and the recent work by Haller [8]) obtain only LCS caused by hyperbolic mechanisms—avoiding those caused by shear flow which do not necessarily represent transport barriers even if long advection times are used for the FTLE computation (resulting in sharp ridges). We refer the reader to the state of the art report by Pobitzer et al. [26] for further details on topology-based visualization in time-dependent vector fields. The FTLE has further been extended to tensor fields by Tricoche et al. [43] and Hlawatsch et al. [11]. Furthermore, it has been applied to a wide range of problems including video analysis by Kuhn et al. [16]. More recently the FTLE has been extended to uncertain vector fields by Schneider et al. [35], while Otto et al. [20] extended the traditional instantaneous vector field topology to uncertain vector fields.

Further works beyond hyperbolic topological structures include those focusing on critical point analogues by Kasten et al. [15] and Fuchs et al. [3], and those by Peikert and Sadlo [22–24] to reveal the structure in recirculating flow, in particular vortex breakdown bubbles. These have been followed up by techniques due to Tricoche et al. [42] and Sanderson et al. [32] for visualizing the structure of the recurrent magnetic field in Tokamak fusion reactors. More recently, [34] applied FTLE ridges for analyzing the air flow in revolving doors. In contrast to our results, [34] reports only a very loose relation between FTLE ridges and temperature, possibly because thermal conduction (numerical diffusion in the solver) or sub-scale mixing due to turbulence was not taken into account.

Only few works have so far been presented related to advection-diffusion flow. The most closely related work is by Karch et al. [14], where we present an interactive dye advection technique that can take into account diffusion fluxes, i.e., can move the virtual dye according to the true advection-diffusion transport of quantities. Other works have made use of diffusion-related concepts for visualizing advection-only flow, such as the reaction-diffusion technique by Sanderson et al. [33], or the anisotropic diffusion approach by Bürkle et al. [1].

3 Advection-Diffusion Field

In [14] we achieved a visualization of the transport of quantities in advection-diffusion flow by moving virtual dye according to the combination of advective fluxes (i.e., the velocity field) and diffusion fluxes. While we formulated the dye advection by means of the finite volume method, i.e., in terms of fluxes at the boundaries of each sampling cell, we derive here the so-called *advection-diffusion field*, a vector field describing the true transport of the respective quantity, such as temperature or solubles, due to advection-diffusion.

The advection-diffusion vector field is derived from the dye advection formulation [14] as follows. Pure advection of the virtual dye concentration $\phi = \phi(\mathbf{x}, t)$ with respect to the simulated velocity field $\mathbf{u} = \mathbf{u}(\mathbf{x}, t)$ is modeled by

$$\frac{\partial \phi}{\partial t} + (\nabla \phi)\mathbf{u} = 0.$$

Including *passive diffusion*, i.e., the motion of the dye with respect to the diffusion fluxes of the simulated quantity $\psi = \psi(\mathbf{x}, t)$, such as temperature or a soluble, with its constant of diffusivity D_ψ, according to Fick's law of diffusion, reads

$$\frac{\partial \phi}{\partial t} + (\nabla \phi)\mathbf{u} = (\nabla \phi) D_\psi \nabla \psi$$

and leads to Eq. 7 from [14]

$$\frac{\partial \phi}{\partial t} + (\nabla \phi)(\mathbf{u} - D_\psi \nabla \psi) = 0.$$

This represents again a pure advection problem, however, now with respect to the advection-diffusion field

$$\mathbf{u}_\psi = \mathbf{u} - D_\psi \nabla \psi. \tag{1}$$

Hence, $\mathbf{u}_\psi$ describes the transport of the quantity ψ by means of advection-diffusion. Since $\mathbf{u}_\psi$ represents a vector field, all techniques from flow visualization can be applied to it, however, with the difference that they then reveal the transport of ψ with respect to advection-diffusion, instead of the transport of massless particles due to $\mathbf{u}$, as in the majority of traditional flow visualization.

In this work we present the extraction of topological features from $\mathbf{u}_\psi$ and show how they can be interpreted. The contribution of this chapter is not in terms of techniques for the extraction of topological features per se, but the introduction of the advection-diffusion field and in particular the application of topological feature extraction to this field. To the best of our knowledge, there has not been any work on the visualization of features in advection-diffusion flow yet, and due to the wide presence of advection-diffusion problems and the importance of the visualization of the involved transport of quantities, we see this as an important contribution to the field of scientific visualization. The extraction of LCS, for example, enables answering questions such as where heat is transported from by advection-diffusion—since the LCS represent transport barriers with respect to advection-diffusion. While the potential of visualization by LCS from $\mathbf{u}_\psi$ is evident, we also provide examples how the topology-related concept of vortex core lines in $\mathbf{u}_\psi$ can provide insight into the transport behavior of the quantity under examination.

The advection-diffusion field $\mathbf{u}_\psi(\mathbf{x})$ can either be computed on the fly during visualization, i.e., $\mathbf{u}(\mathbf{x})$ and $\nabla\psi(\mathbf{x})$ can be interpolated/evaluated at required positions $\mathbf{x}$ and then combined according to Eq. 1, or it can be precomputed at the nodes of the simulation grid and directly fed into existing visualization algorithms. In this work we follow the latter approach since interchanging interpolation and, e.g., multiplication is a common approach in visualization, in particular in feature extraction (see, e.g., [39]). However, it has to be noted that it introduces (typically negligible) error, which did not represent a problem in our experiments. We estimate the gradient $\nabla\psi(\mathbf{x})$ using least squares fitting according to [28]. Note that the constant of diffusivity D_ψ that was used in the CFD simulation has to be available for visualization, however, it is typically uniform and hence only a single number.

4 Feature Extraction

We exemplify feature extraction from the advection-diffusion field $\mathbf{u}_\psi$ using LCS by means of FTLE ridges (Sect. 4.1) and vortex core lines (Sect. 4.2). In Sect. 5 we present the results obtained from applying the feature extraction techniques to three CFD results of advection-diffusion flow.

4.1 Lagrangian Coherent Structures

Lagrangian coherent structures serve as a replacement for separatrices from traditional vector field topology, with the important difference that they are able to correctly depict transport in time-dependent vector fields. Several—and even contradicting—definitions for coherent structures exist. Hussain [12] defined them in terms of the curl of the velocity field, while Robinson [27] provided a more general definition based on correlation of flow variables—both defining coherent structures as volumes in 3D vector fields. In contrast, in the definitions of Haller [6] and Ide et al. [13], Lagrangian coherent structures represent the counterpart to invariant manifolds, i.e., they separate regions of qualitatively different behavior. As discussed in Sect. 2 it was proposed by Haller [7] and subsequently defined by Shadden et al. [36] that ridges in the finite-time Lyapunov exponent (FTLE) field represent LCS. In this work we follow the approach by Sadlo and Peikert [28] and extract LCS from the FTLE field by means of height ridges [2]. To assure sufficiently sharp ridges—according to Shadden et al. [36] a prerequisite for FTLE ridges to represent LCS—we reject ridge regions where the modulus of the respective eigenvalue of the FTLE Hessian is too small or where the FTLE value itself does not reach a user-defined threshold. Due to space limitations, we refer the reader to [28] for all details of our LCS extraction.

The FTLE is a scalar field that describes the separation of neighboring trajectories after a finite advection time interval T. The FTLE at point $\mathbf{x}$ and time t_0 with respect to the time interval T reads

$$\sigma(\mathbf{x}, t_0, T) = \frac{1}{|T|}\sqrt{\lambda_{\max}\big((\nabla\phi_{t_0}^{t_0+T}(\mathbf{x}))^\top \nabla\phi_{t_0}^{t_0+T}(\mathbf{x})\big)}$$

where $\sqrt{\lambda_{\max}(\cdot)}$ represents the spectral norm $\|\cdot\|$ of a matrix with $\lambda_{\max}(\mathbf{A})$ being the major eigenvalue of matrix $\mathbf{A}$. $\nabla\phi_{t_0}^{t_0+T}(\mathbf{x})$ is the gradient of flow map $\phi_{t_0}^{t_0+T}(\mathbf{x})$ which maps starting points $\mathbf{x}(t_0)$ of trajectories to their end points at time $t_0 + T$:

$$\phi_{t_0}^{t_0+T}(\mathbf{x}) = \mathbf{x}(t_0) + \int_{t_0}^{t_0+T} \mathbf{u}(\mathbf{x}(\tau), \tau)\, \mathrm{d}\tau.$$

Depending on the sign of T, i.e., if forward or reverse trajectories are used for the computation of the FTLE, the FTLE ridges represent either repelling ($T > 0$) or attracting ($T < 0$) LCS. We color repelling LCS red and attracting ones blue. It has to be noted that the FTLE field can be sampled independently from the simulation grid on which $\mathbf{u}$, or $\mathbf{u}_\psi$, is given. In this work we use uniform sampling grids, although some of the used advection-diffusion CFD simulations are given on unstructured grids. The integration of the trajectories is accomplished by the 4th-order Runge-Kutta scheme.

4.2 Vortex Core Lines

Vortex core lines—sometimes also denoted as vortex axes—are in close relation to topological features for several reasons. First of all, they are typically identical to critical points of type center and focus in 2D flow. While in 3D the so-called topological vortex cores due to Globus et al. [5] directly represent 1D manifolds of spiral saddle critical points, the relation is less obvious for other definitions of vortex core lines. Since not all vortex core lines represent stream lines, it is nowadays more common to extract them according to the definitions by Levy et al. [17] or Sujudi and Haimes [38]. In this work we use the definition by Sujudi and Haimes, which identifies those points as part of a core line where velocity is parallel or antiparallel to a real eigenvector of the velocity gradient and the other eigenvectors are complex. If the flow is projected on the eigenplane spanned by the complex eigenvectors, this definition corresponds to a critical point of type center or focus in the projected field—further supporting the proximity of vortex core lines and vector field topology. Due to the same reasons, spiral saddle and spiral source critical points typically reside on vortex core lines according to Sujudi and Haimes. As with the other feature extractions, we apply the concept to $\mathbf{u}_\psi$ instead of $\mathbf{u}$ for analyzing the transport of the quantity that is governed by advection-diffusion.

We extract the vortex core lines using the parallel vectors operator due to Peikert and Roth [21], according to their algorithm that triangulates the faces of each cell and solves for the points where the core lines intersect the faces. These points are then connected by piecewise linear segments to obtain the final polyline representation of the core lines.

5 Results

We demonstrate our approach using three advection-diffusion CFD simulations—a static mixer (Sect. 5.1), a flow around a heating coil (Sect. 5.2), and a buoyant flow in a room that is heated at its bottom and cooled at the top (Sect. 5.3). The FTLE field (and its ridges) is computed on a uniform grid in the regions marked by black boxes, while the $\mathbf{u}$ and $\mathbf{u}_\psi$ fields are evaluated at the original grid without resampling.

5.1 Static Mixer

The first simulation is steady-state and was conducted on a unstructured grid consisting of 2,266,894 tetrahedra. It represents a mixing device that mixes a hot and a cold fluid by means of a strong vortex that is maintained by the tangential inflows. The mixed fluid is provided at an axial outlet at the top, see Fig. 1a. The upper half of the solid boundary is not visualized to reveal the interior flow—the removed part is illustrated by its edges and a half in wireframe representation. Hot air at 700 K enters at the left lower inlet while cold air at 300 K enters at the lower right inlet.

While in Fig. 1a the path lines follow the velocity field $\mathbf{u}$, they follow the advection-diffusion field $\mathbf{u}_\psi$ in Fig. 1b. It is apparent that the lines of the hot fluid are attracted toward the center (which exhibits average temperature due to mixing) in (b), starting right after they have left the inlet tube. This directly visualizes the transport of heat by advection-diffusion from the left inlet, in contrast to (a) where

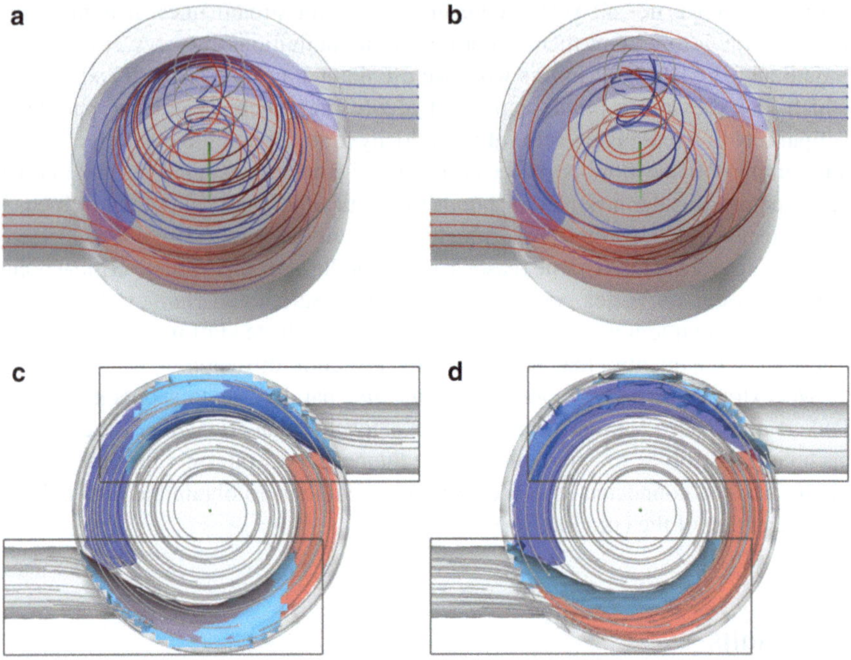

Fig. 1 Static Mixer flow example. (**a**) Visualization of **u** (advection only) by path lines (*red*: hot; *blue*: cold; seeds by spheres), a vortex core line (*green*), and two transparent isosurfaces (*red*: 550 K; *blue*: 450 K, same in (**a**)–(**d**)). (**b**) Same as (**a**), but path lines and vortex core line visualizing advection-diffusion field $\mathbf{u}_\psi$. Hot path lines are initially forced toward the center by heat diffusion while *blue path lines* are initially forced toward the wall due to heat diffusion from the hot flow. (**c**) Attracting LCS of **u** (*light blue*) are close to the isosurfaces (a subset of respective reverse FTLE trajectories in *gray*). (**d**) Attracting LCS of $\mathbf{u}_\psi$ (*light blue*) depict regions where heat is transported to by advection-diffusion (with a subset of respective reverse FTLE trajectories in *gray*)

the lines only show the advection part—using (a) for the interpretation of heat transport would give the wrong picture. Surprisingly, there is an asymmetry with respect to the cold flow in (b). Since there is no negative heat, the blue lines of the cold fluid are not attracted toward the center as they enter the mixer from the right inlet—instead, they are pushed against the outer wall (see that these lines approach the center vortex slower than the blue lines in (a)) because heat from the mixed fluid at the center of the mixer is diffusing into the cold fluid. Hence, the blue lines in (b) depict how heat contained in the cold fluid is transported by advection-diffusion—it is 'avoiding' the heat from the warmer fluid. Because the center of the vortex consists of mixed fluid, there is no detectable deviation between the vortex core line in (a) and (b), i.e., heat is transported there almost purely by advection and therefore the vortex core lines are basically identical.

To investigate the transport of heat from the inlets, we have computed LCS using reverse-time FTLE on a regular sampling grid of $17 \times 47 \times 15$ nodes from both the **u**

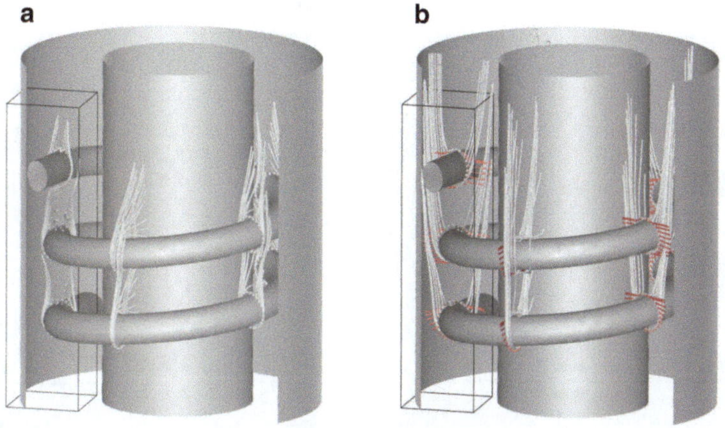

Fig. 2 Heating Coil flow example. (**a**) Path lines of $\mathbf{u}$ (*white*, advection only) seeded at circles around coil (*white spheres*). (**b**) Path lines of $\mathbf{u}_\psi$ (*white*, advection-diffusion) and of $-D_\psi \nabla \psi$ (*red*, diffusion only). Compared to (**a**), the *white lines* in (**b**) are attracted by the inner and outer wall. This is due to thermal conduction (diffusion of heat) from the heated coil to the cooled walls

(Fig. 1c) and $\mathbf{u}_\psi$ (Fig. 1d) fields in the regions of interest marked by the black boxes. Consistent with the observations and interpretation so far, the LCS of the hot flow deviates toward the center in (d) while the LCS of the cold flow is attracted toward the wall in (d). The LCS of the hot flow in (d) separates the region (bottom of the image) where heat is transported from the left inlet by advection-diffusion. The LCS of the cold flow in (d) separates the region (top of the image) that does not obtain any heat on its way from the inlet. In contrast, the LCS in (c) are very similar to the isosurfaces. Please note that the isosurface of average temperature (500 K) extends to the center of the mixer in a helical manner.

5.2 *Heating Coil*

This dataset is quasi-stationary and consists of an unstructured grid of 93,227 cells, including tetrahedra, pyramids, and prisms. Hence, we used a single time step for a steady-type analysis. The simulation represents a heat exchanger—a heated coil is immersed into an air flow with the inlet at its bottom and the outlet at the top while the inner and outer boundaries (the two vertical tubes) are cooled, see Fig. 2. Buoyant forces have not been employed, resulting in a rather simple flow.

In Fig. 3c a subset of the reverse-time trajectories of $\mathbf{u}$ is shown that were used for FTLE computation on a regular grid of $41 \times 41 \times 161$ nodes within the region of interest (ROI, black box). Figure 3a shows some vortex core lines (green) extracted from $\mathbf{u}$ and some stream lines that have been seeded in their vicinity (white tubes). In Fig. 3b, d, the same has been conducted for the advection-diffusion

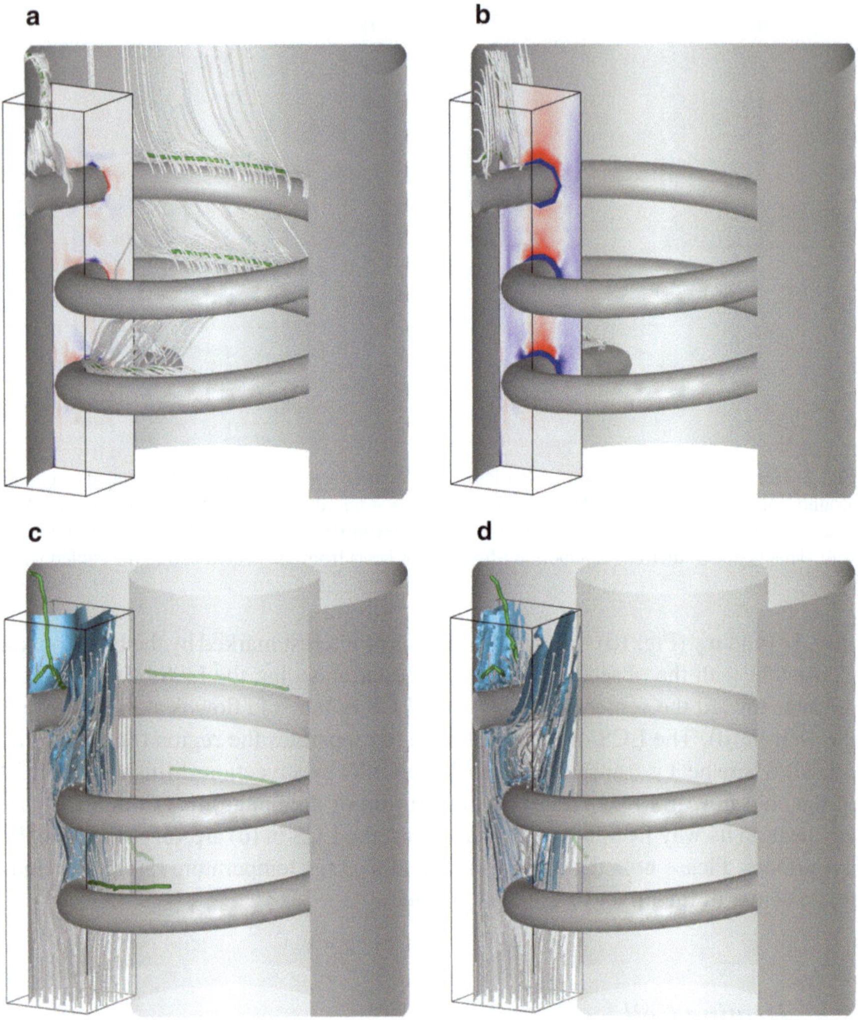

Fig. 3 Heating Coil flow example, region of interest (*black box*) from Fig. 2. (**a**) Vortex core lines (*green*), some forward and reverse stream lines (*white*) seeded therefrom, and cross section visualizing divergence (*red*: −0.5, *blue*: 0.5), all computed from $\mathbf{u}$ (advection only). (**b**) Same as (**a**) but computed from advection-diffusion field $\mathbf{u}_\psi$. The strong heat flux from the coil dominates the recirculations at the downstream edge of the coil, hence these vortices disappear. (**c**) LCS (*cyan*) from reverse FTLE of $\mathbf{u}$ for comparison with a subset of used trajectories (*gray*). (**d**) LCS (*cyan*) from reverse FTLE of $\mathbf{u}_\psi$ with a subset of used trajectories (*gray*). It is apparent that in (**d**) LCS are repelled from the coil due to thermal conduction from the coil to the cooled walls. The LCS separate the region that obtains heat by advection-diffusion from the respective part of the coil

field $\mathbf{u}_\psi$. It is apparent that the vortices along the coil are not present in (b). This is because the transport of heat away from the coil by diffusion (see also stream lines of $-D_\psi \nabla \psi$) is stronger than the advective recirculation at the downstream edge of

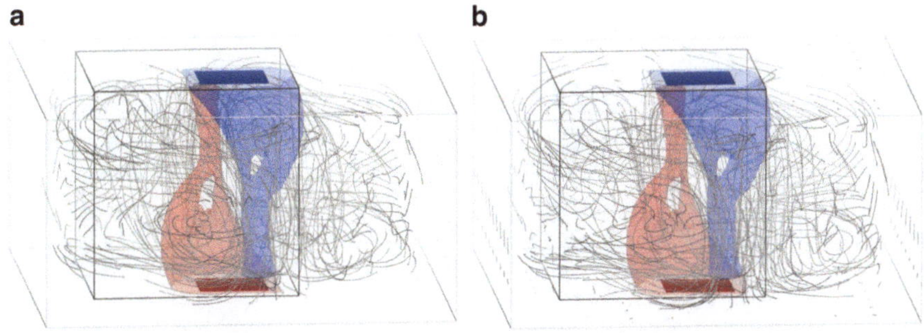

Fig. 4 Buoyant Flow example at time 24.474 s with transparent isosurfaces (*red* at 318.5 K; *blue* at 308.5 K; same as in Fig. 5), region of interest (*black*) used in Fig. 5, and hot and cold plate. (**a**) Path lines of $\mathbf{u}$ show advection. (**b**) Path lines of $\mathbf{u}_\psi$ show transport of heat due to advection-diffusion

the coil, resulting in heat transport that has no upstream component. The same can be observed at the top left vortex. It is still present because it is a longitudinal vortex oriented in direction of heat diffusion—however, one can see that the longitudinal component is much stronger with respect to heat transport. Nevertheless, it is an example where heat transport by advection-diffusion exhibits a vortex. The cross sections in Fig. 3a, b visualize $\nabla \cdot \mathbf{u}$ and $\nabla \cdot \mathbf{u}_\psi$, respectively. The 25th and 75th percentiles of $\nabla \cdot \mathbf{u}_\psi$ are -0.276 and 0.0191, respectively (i.e., rather low compared to 0.05 m/s average speed and 0.5 m ROI width). The direct impact of $\nabla \cdot \mathbf{u}_\psi$ on LCS (see below) is rather low since $\nabla \cdot \mathbf{u}_\psi$ is substantially smaller at the LCS.

We investigated the transport of heat from the heating coil by computing LCS from reverse-time FTLE using $\mathbf{u}$ (Fig. 3c) and $\mathbf{u}_\psi$ (Fig. 3d). It is apparent that in (d) the LCS are repelled from the heating coil toward the inner and outer walls. The LCS region at the lowest turn of the coil in (d) separates the region that is reached by the heat from the coil by means of advection-diffusion. The levels where the LCS reach the inner and outer walls depict the points where heat is transported from the heated coil to the cooled walls.

5.3 Buoyant Flow

The final example is a time-dependent simulation of buoyant flow consisting of 2,000 time steps simulated on a uniform grid of $60 \times 30 \times 60$ cells. There is a region heated at 348.15 K at its bottom and a cooled region at 278.15 K at its top, and gravity is pointing downward. The rest of the walls are adiabatic. The buoyant flow exhibits transient aperiodic convection, see Fig. 4.

Figure 5a shows some of the reverse-time trajectories used for FTLE computation on a regular sampling grid of $41 \times 41 \times 41$ nodes (black box) from $\mathbf{u}$, while (b) shows those computed from the advection-diffusion field $\mathbf{u}_\psi$. It is apparent that

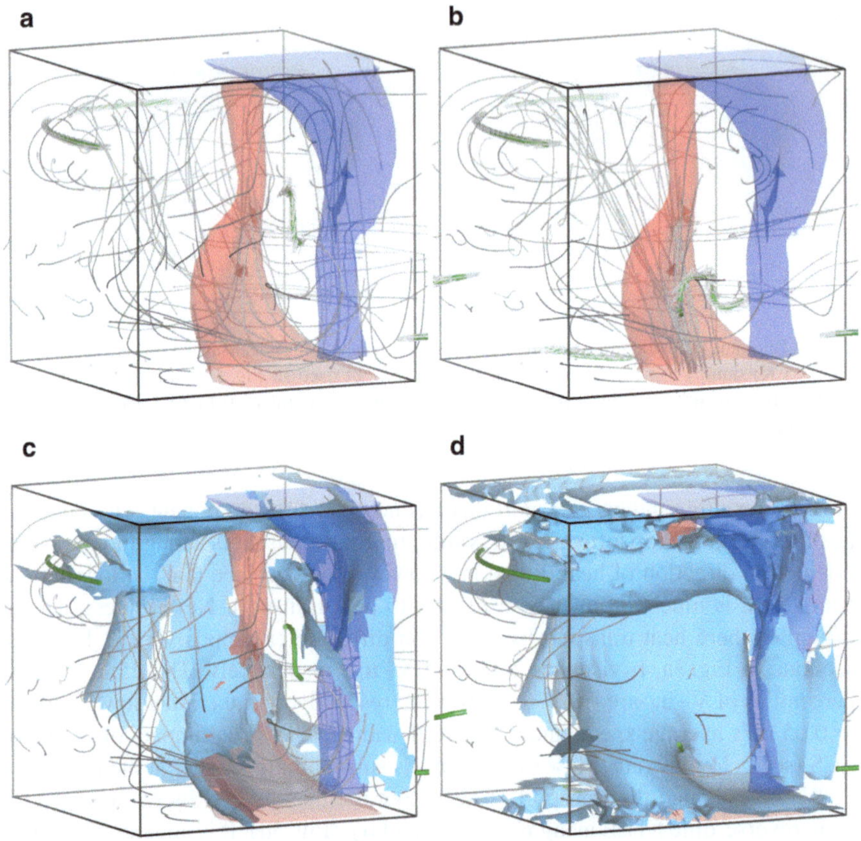

Fig. 5 Buoyant flow example. (**a**) Trajectories in **u**. (**b**) Trajectories in $\mathbf{u}_\psi$. (**c**) Reverse-time LCS in **u**. (**d**) Reverse-time LCS in $\mathbf{u}_\psi$ depict region that obtains heat from hot bottom plate

many trajectories that show the transport of heat in (b) leave the domain at the hot plate. The reason for this is that heat enters the domain in this region. In (a) there is a vortex of **u** (green) located about the center between the cold and hot plumes (visualized by the transparent red and blue isosurfaces). In (b) the vortex of $\mathbf{u}_\psi$ is shifted toward the bottom of the domain and it is intensified (see white stream lines of $\mathbf{u}_\psi$ seeded in its vicinity). Here, the heat flux due to diffusion in the advection-diffusion flow originating at the bottom hot plate rises and then diffuses partially into the cold plume, causing the vortex to shift downward and accelerating the vortex. Such vortices in heat transport can represent undesired configurations in advection-diffusion flow because they can hinder the overall transport of heat.

In Fig. 5c, d we have computed reverse-time LCS from **u** and $\mathbf{u}_\psi$, respectively. In (c) the LCS attaches to the center of the hot plate at the bottom—hence it partially separates the domain in two regions: the region that obtains its flow from the bottom right front corner of the domain and the region that obtains the flow from the bottom left back corner region. In contrast, in (d) the LCS wraps the complete hot flow that is generated by the bottom plate. In other words, the LCS in (c) shows where the flow comes from, while in (d) the LCS shows where the heat comes from—it separates the region that obtains heat from the bottom hot plate by advection-diffusion.

6 Conclusion

We proposed visualization of the advection-diffusion field, a field that describes the transport of quantities due to advection-diffusion, and presented the extraction of topological features therefrom. These features give valuable insights into transport processes due to advection-diffusion. We applied our approach to three CFD datasets and provided interpretations that gave insight into the underlying phenomena. We believe that visualization by means of topological features in the advection-diffusion field has a high potential in many fields of science and engineering, e.g., in heat exchanger design. As future work, we plan to apply our approach to real-world problems and to extend it to advection-diffusion of vector quantities. Future work could also follow [37] and compare path lines of $\mathbf{u}$ with stream lines of $-D_{\psi}\nabla\psi$.

Acknowledgements This work was supported by the Cluster of Excellence in Simulation Technology (EXC 310/1) and the Collaborative Research Center SFB-TRR 75 at University of Stuttgart.

References

1. D. Bürkle, T. Preußer, M. Rumpf, Transport and anisotropic diffusion in time-dependent flow visualization, in *Proceedings of the IEEE Visualization*, San Diego, 2001, pp. 61–67
2. D. Eberly, *Ridges in Image and Data Analysis*. Computational Imaging and Vision (Kluwer, Boston, 1996)
3. R. Fuchs, J. Kemmler, B. Schindler, F. Sadlo, H. Hauser, R. Peikert, Toward a Lagrangian vector field topology. Comput. Graph. Forum **29**(3), 1163–1172 (2010)
4. C. Garth, G.-S. Li, X. Tricoche, C.D. Hansen, H. Hagen, Visualization of coherent structures in transient 2d flows, in *Topology-Based Methods in Visualization II*, ed. by H.-C. Hege, K. Polthier, G. Scheuermann (Springer, Berlin, 2009), pp. 1–13
5. A. Globus, C. Levit, T. Lasinski, A tool for visualizing the topology of three-dimensional vector fields, in *Proceedings of the IEEE Visualization*, San Diego, 1991, pp. 33–40, 408
6. G. Haller, Finding finite-time invariant manifolds in two-dimensional velocity fields. Chaos **10**(1), 99–108 (2000)
7. G. Haller, Distinguished material surfaces and coherent structures in three-dimensional fluid flows. Physica D **149**(4), 248–277 (2001)
8. G. Haller, A variational theory of hyperbolic Lagrangian coherent structures. Phys. D: Nonlinear Phenom. **240**(7), 574–598 (2011)
9. J. Helman, L. Hesselink, Representation and display of vector field topology in fluid flow data sets. IEEE Comput. **22**(8), 27–36 (1989)

10. J. Helman, L. Hesselink, Visualizing vector field topology in fluid flows. IEEE Comput. Graph. Appl. **11**(3), 36–46 (1991)
11. M. Hlawatsch, J. Vollrath, F. Sadlo, D. Weiskopf, Coherent structures of characteristic curves in symmetric second order tensor fields. IEEE Trans. Vis. Comput. Graph. **17**(6), 781–794 (2011)
12. F. Hussain, Coherent structures and turbulence. J. Fluid Mech. **173**, 303–356 (1986)
13. K. Ide, D. Small, S. Wiggins, Distinguished hyperbolic trajectories in time-dependent fluid flows: analytical and computational approach for velocity fields defined as data sets. Nonlinear Process. Geophys. **9**(3/4), 237–263 (2002)
14. G.K. Karch, F. Sadlo, D. Weiskopf, C.-D. Munz, T. Ertl, Visualization of advection-diffusion in unsteady fluid flow. Comput. Graph. Forum **31**(3), 1105–1114 (2012)
15. J. Kasten, I. Hotz, B. Noack, H.-C. Hege, On the extraction of long-living features in unsteady fluid flows, in *Topological Methods in Data Analysis and Visualization. Theory, Algorithms, and Applications*, ed. by V. Pascucci, X. Tricoche, H. Hagen, J. Tierny (Springer, Berlin/Heidelberg, 2010), pp. 115–126
16. A. Kuhn, T. Senst, I. Keller, T. Sikora, H. Theisel, A Lagrangian framework for video analytics, in *Proceedings of the IEEE Workshop on Multimedia Signal Processing*, Banff, 2012
17. Y. Levy, D. Degani, A. Seginer, Graphical visualization of vortical flows by means of helicity. AIAA **28**(8), 1347–1352 (1990)
18. H. Löffelmann, H. Doleisch, E. Gröller, Visualizing dynamical systems near critical points, in *Proceedings of the Spring Conference on Computer Graphics and Its Applications*, Budmerice, 1998, pp. 175–184
19. E.N. Lorenz, A study of the predictability if a 28-variable atmospheric model. Tellus **17**, 321–333 (1965)
20. M. Otto, T. Germer, H. Theisel, Uncertain topology of 3d vector fields, in *Proceedings of the IEEE Pacific Visualization Symposium*, Hong Kong, 2011, pp. 67–74
21. R. Peikert, M. Roth, The "parallel vectors" operator – a vector field visualization primitive, in *Proceedings of the IEEE Visualization*, San Francisco, 1999, pp. 263–270
22. R. Peikert, F. Sadlo, Topology-guided visualization of constrained vector fields, in *Topology-Based Methods in Visualization*, ed. by H. Hauser, H. Hagen, H. Theisel (Springer, Berlin/New York, 2007), pp. 21–34
23. R. Peikert, F. Sadlo, Visualization methods for vortex rings and vortex breakdown bubbles, in *Proceedings of the Joint Eurographics/IEEE VGTC Conference on Visualization*, Norrköping, 2007, pp. 211–218
24. R. Peikert, F. Sadlo, Flow topology beyond skeletons: visualization of features in recirculating flow, in *Topology-Based Methods in Visualization II*, ed. by H.-C. Hege, K. Polthier, G. Scheuermann (Springer, Berlin, 2009), pp. 145–160
25. A.E. Perry, M.S. Chong, A description of eddying motions and flow patterns using critical-point concepts. Annu. Rev. Fluid Mech. **19**, 125–155 (1987)
26. A. Pobitzer, R. Peikert, R. Fuchs, B. Schindler, A. Kuhn, H. Theisel, K. Matković, H. Hauser, On the way towards topology-based visualization of unsteady flow – the state of the art, in *Eurographics 2010 State of the Art Reports*, Norrköping, 2010, pp. 137–154
27. S.K. Robinson, Coherent motions in the turbulent boundary layer. Annu. Rev. Fluid Mech. **23**, 601–639 (1991)
28. F. Sadlo, R. Peikert, Efficient visualization of Lagrangian coherent structures by filtered amr ridge extraction. IEEE Trans. Vis. Comput. Graph. **13**(6), 1456–1463 (2007)
29. F. Sadlo, R. Peikert, Visualizing Lagrangian coherent structures and comparison to vector field topology, in *Topology-Based Methods in Visualization II*, ed. by H.-C. Hege, K. Polthier, G. Scheuermann (Springer, Berlin, 2009)
30. F. Sadlo, M. Üffinger, T. Ertl, D. Weiskopf, On the finite-time scope for computing Lagrangian coherent structures from Lyapunov exponents, in *Topological Methods in Data Analysis and Visualization II*, ed. by R. Peikert et al. (Springer, Heidelberg/New York, 2012), pp. 269–281
31. F. Sadlo, D. Weiskopf, Time-dependent 2-d vector field topology: an approach inspired by Lagrangian coherent structures. Comput. Graph. Forum **29**(1), 88–100 (2010)

32. A. Sanderson, G. Chen, X. Tricoche, D. Pugmire, S. Kruger, J. Breslau, Analysis of recurrent patterns in toroidal magnetic fields. IEEE Trans. Vis. Comput. Graph. **16**(6), 1431–1440 (2010)
33. A. Sanderson, C.R. Johnson, R.M. Kirby, Display of vector fields using a reaction-diffusion model, in *Proceedings of the IEEE Visualization*, Austin, 2004, pp. 115–122
34. B. Schindler, R. Fuchs, S. Barp, J. Waser, A. Pobitzer, R. Carnecky, K. Matkovic, R. Peikert, Lagrangian coherent structures for design analysis of revolving doors. IEEE Trans. Vis. Comput. Graph. **18**(12), 2159–2168 (2012)
35. D. Schneider, J. Fuhrmann, W. Reich, G. Scheuermann, A variance based ftle-like method for unsteady uncertain vector fields, in *Topological Methods in Data Analysis and Visualization II*, ed. by R. Peikert et al. (Springer, Heidelberg/New York, 2012), pp. 255–268
36. S.C. Shadden, F. Lekien, J.E. Marsden, Definition and properties of Lagrangian coherent structures from finite-time Lyapunov exponents in two-dimensional aperiodic flows. Phys. D: Nonlinear Phenom. **212**, 271–304 (2005)
37. K. Shi, H. Theisel, H. Hauser, T. Weinkauf, K. Matkovic, H.-C. Hege, H.-P. Seidel, Path line attributes – an information visualization approach to analyzing the dynamic behavior of 3D time-dependent flow fields, in *Topology-Based Methods in Visualization II*, ed. by H.-C. Hege, K. Polthier, G. Scheuermann (Springer, Berlin, 2009), pp. 75–88
38. D. Sujudi, R. Haimes, Identification of swirling flow in 3d vector fields, in *Proceedings of the 12th AIAA Computational Fluid Dynamics Conference*, 1995, pp. 95–1715
39. H. Theisel, J. Sahner, T. Weinkauf, H.-C. Hege, H.-P. Seidel, Extraction of parallel vector surfaces in 3D time-dependent fields and application to vortex core line tracking, in *Proceedings of the IEEE Visualization*, Minneapolis, 2005, pp. 631–638
40. H. Theisel, T. Weinkauf, H.-C. Hege, H.-P. Seidel, Saddle connectors – an approach to visualizing the topological skeleton of complex 3d vector fields, in *Proceedings of the IEEE Visualization*, Seattle, 2003, pp. 225–232
41. H. Theisel, T. Weinkauf, H.-C. Hege, H.-P. Seidel, Stream line and path line oriented topology for 2d time-dependent vector fields, in *Proceedings of the IEEE Visualization*, Austin, 2004, pp. 321–328
42. X. Tricoche, C. Garth, A. Sanderson, K. Joy, Visualizing invariant manifolds in area-preserving maps, in *Topological Methods in Data Analysis and Visualization II*, ed. by R. Peikert et al. (Springer, Heidelberg/New York, 2012), pp. 109–124
43. X. Tricoche, M. Hlawitschka, S. Barakat, C. Garth, Beyond topology: a Lagrangian metaphor to visualize the structure of 3d tensor fields, in *New Developments in the Visualization and Processing of Tensor Fields*, ed. by D. Laidlaw, A. Vilanova (Springer, Berlin/New York, 2012)
44. M. Üffinger, F. Sadlo, T. Ertl, A time-dependent vector field topology based on streak surfaces. IEEE Trans. Vis. Comput. Graph. **19**(3), 379–392 (2013)
45. T. Weinkauf, H. Theisel, H.-C. Hege, H.-P. Seidel, Boundary switch connectors for topological visualization of complex 3D vector fields, in *Proceedings of the VisSym*, Konstanz, 2004, pp. 183–192

Part V
Applications

Definition, Extraction, and Validation of Pore Structures in Porous Materials

Ulrike Homberg, Daniel Baum, Alexander Wiebel, Steffen Prohaska, and Hans-Christian Hege

Abstract An intuitive and sparse representation of the void space of porous materials supports the efficient analysis and visualization of interesting qualitative and quantitative parameters of such materials. We introduce definitions of the elements of this void space, here called pore space, based on its distance function, and present methods to extract these elements using the extremal structures of the distance function. The presented methods are implemented by an image-processing pipeline that determines pore centers, pore paths and pore constrictions. These pore space elements build a graph that represents the topology of the pore space in a compact way. The representations we derive from μCT image data of realistic soil specimens enable the computation of many statistical parameters and, thus, provide a basis for further visual analysis and application-specific developments. We introduced parts of our pipeline in previous work. In this chapter, we present additional details and compare our results with the analytic computation of the pore space elements for a sphere packing in order to show the correctness of our graph computation.

1 Introduction

Soil materials of different particle size distributions form pore structures exhibiting different properties that impact the transport processes of particles through the pore space. Kinds of internal erosion like suffusion comprise such transport processes and may weaken the particle structure of the soil and, thus, its stability. To investigate the risk of suffusion, it is important to know which parts of the pore

U. Homberg (✉) • D. Baum • A. Wiebel • S. Prohaska • H.-C. Hege
Zuse Institute Berlin, Takustr. 7, 14195 Berlin, Germany
e-mail: homberg@zib.de; baum@zib.de; wiebel@zib.de; prohaska@zib.de; hege@zib.de

P.-T. Bremer et al. (eds.), *Topological Methods in Data Analysis and Visualization III*, Mathematics and Visualization, DOI 10.1007/978-3-319-04099-8_15,

space can be reached by particles of what size. The sizes as well as the portions of such mobile particles provide information on the stability of the soil.

For the assessment of soil properties, an understanding of the three-dimensional formations and arrangements of its particles and its pore structure may be of great benefit. To gain insight into soil structures, CT scans of realistic and undisturbed soil material can be acquired and analyzed.

An analysis of the pore space and possible transport pathways needs information on pore space elements like pores, constrictions, and paths. While the pores and their connecting paths determine where particles can move, the constrictions define the size of the particles that can move from one pore to another. Thus, a complete representation that preserves the arrangement and the connectivity of these pore space elements enables an efficient analysis of transport paths, blocking constrictions and sizes of potentially mobile particles.

We previously described [10] the pore space elements based on the extremal structures of the distance map of the pore space and presented methods to extract these structures. The proposed methods start from a segmentation of the soil structure [11] and generate a graph that is a geometric embedding into the pore space and compactly represents the topology of the pore space. This approach avoids anabranches that are caused by irregular particles. A hierarchical merge process enables a clustering of pores by the significance of their connecting constrictions.

In this chapter, we present a validation of the extracted pore structures. For this purpose, we first recall parts of our approach with additional details and then present a comparison with the result of an exact Voronoi graph algorithm applied to a sphere packing to measure the quality of our results.

2 Related Work

Existing methods analyze the pore space at different levels of detail. Statistical approaches use simple methods for phase segmentation and to compute parameters and distributions on the pore-solid relations [3, 9, 16, 20]. In these methods, measurements of the captured pore space as a whole are proposed, but a localization and differentiation of pores is not considered.

Other approaches differentiate the pore space elements and investigate the networks they build. One possibility is to use the maximal inscribed spheres map, which is equal to the distance map [12] of the pore space. Sweeney and Martin [23] and Silin and Patzek [22] accomplish the differentiation of pores and constrictions by evaluating and classifying the neighboring spheres. Both works compute a stick-and-ball diagram representing the pore bodies and their connectivity. However, computing the pore volumes based on spheres leads to an underestimation of the pore size.

Skeleton-based methods [6, 14, 15] determine the medial axes of the pore space and use morphological tools to classify the skeleton voxels as belonging to a path or a constriction. Their representations enable the construction of a geometrically embedded graph and the determination of parameters like numbers and sizes of constrictions. These approaches, however, need removal processes to avoid surface-like structures or anabranches caused by isolated particles or irregular particle surfaces.

Some approaches make use of concepts from Morse theory. For example, Ushizima et al. [25] use Reeb graphs to analyze the permeability and maximum flow of gas through porous networks. In contrast to the method we describe, their method depends on the orientation of the data set, that is, on height information in the data set. Gyulassy et al. [8] compute and simplify the Morse-Smale complex of the distance map to extract filament structures in porous solids, which results in a medial axes-related representation. The result depends on the degree of simplification and may produce a superset of the structures we aim to extract.

Another way to analyze pores and constrictions is based on Delaunay/Voronoi partitions. While Reboul et al. [19] uses Delaunay tetrahedrons and their faces to specify the pore bodies and constrictions for sphere packings, Lindow et al. [13] compute the paths from the Voronoi cells of a sphere packing. Glantz and Hilpert [7] proposes an extension to irregular shaped particles in voxel data where the corners of the pore space boundary are tessellated. This results in time-consuming computations and produces many anabranches. Thompson et al. [24] generates a Delaunay tessellation of the particle centers obtained from distance extremes within the particle regions. However, this is not applicable to the realistic soil material of our samples, because the particles have irregular shapes and, thus, multiple distance extremes.

Some authors address the problem of fragmented pores and propose pairwise merge criteria that are based on overlapping spheres [1, 19, 22, 24] or on relations of radii and distances of pores [15]. Such relations allow one to cluster fragmented pores, but elongated pores may stay separated. Also, these pairwise considerations may cluster pores even if there is a significant constriction in the middle. This is, for example, the case when several neighboring pores gradually increase and decrease in size with minor constrictions in between.

3 Determination of the Pore Structure

The methods we present here have been developed to process CT scans of specimens produced from soil aggregates. These scans are acquired at a resolution of 39 μm [2]. Figure 1a shows a volumetric visualization of the dual soil structures of such a CT scan, that is, the particles and pore structure (Fig. 1b). To assess the complex pore space, we follow the intuition of particle transport where a pathway locally runs with maximal distance to the surrounding soil structure and the bottlenecks determine the maximal size of particles that can move along the path. We start with the three-dimensional image domain $I \subset \mathbb{R}^3$, which can be separated into foreground (particle structure) $F \subset I$ and background (pore space) $B = I \setminus F$. For the computation of the pore space elements, we need a particle segmentation (Fig. 1c) where all particles are well separated and each particle has its own identifier [11].

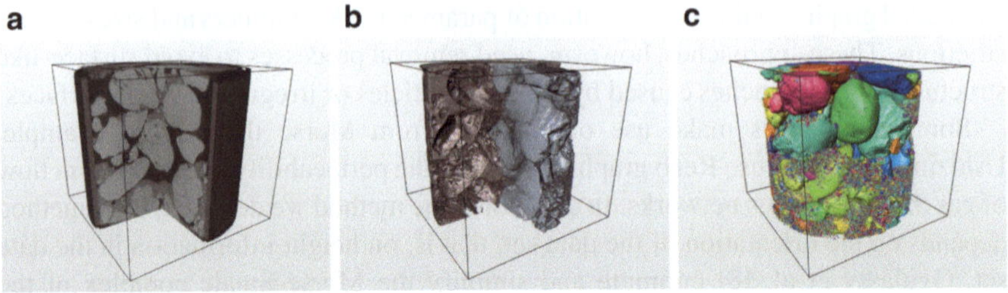

Fig. 1 CT scan of soil material: (**a**) volumetric image data, (**b**) particles and pore space, (**c**) separated particles colored according to their identifier

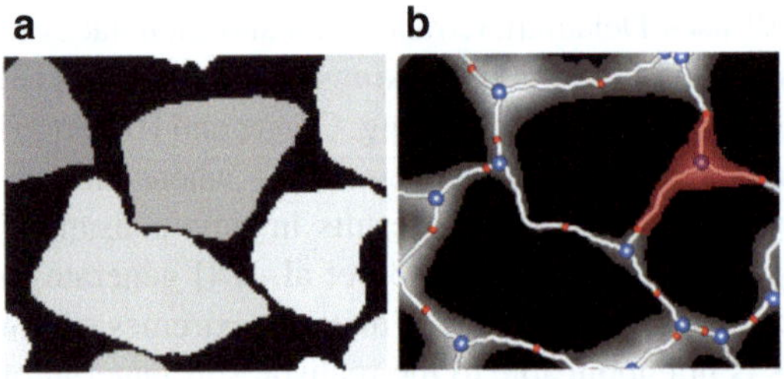

c

Pore Structure	Extremal Structure
pore center	c_{max}
pore body	$S(c_{max}) \cap B$
pore path	$U(c_{S2})$
pore constriction	c_{S2}

Fig. 2 Illustration of the pore space elements (**b**) between particles (**a**): Pore centers (*blue*), paths (*white*), constrictions (*red*), pore bodies (*red region*). The figures are two-dimensional for clarity of the illustration. (**c**) Link between pore structures and extremal structures of the distance map

3.1 *Defining the Elements of the Pore Structure*

Most of the approaches in the literature use methods that encode the distance relations within the pore space. We follow this idea and describe the features of the pore space by the three-dimensional signed distance map [12] $d : I \to \mathbb{R}$ according to the boundaries between the segmented foreground F and background B where each image point is assigned the distance as follows:

$$d(p) = \begin{cases} \min\limits_{q \in F} \|p - q\|, & \text{if } p \in B \\ -\min\limits_{q \in B} \|p - q\|, & \text{if } p \in F \end{cases} \tag{1}$$

To describe the pore structures, we use the extremal structures of the distance map (Fig. 2). These are the critical points c where the gradient of the distance function vanishes, that is $\nabla d(c) = 0$, as well as the integral lines γ of the gradient that allow a grouping into stable and unstable manifolds [4]. The critical points of a 3D function are the maxima c_{max}, the minima c_{min}, and the two types of saddle points (index-1 c_{S1}, index-2 c_{S2}). The integral lines are defined as $\gamma : \mathbb{R} \to \mathbb{R}^3$ whose tangential vectors are parallel to the gradients of d, $\dot{\gamma} = \nabla d(\gamma(t))$ for each $t \in \mathbb{R}$, and connect two critical points where $orig(\gamma) = \lim_{t \to -\infty} \gamma(t)$ is called

Fig. 3 Pore graph computation: (**a**) slice of labeled particles and distance information, (**b**) propagated particle regions, (**c**) pore graph constructed from the boundaries of the propagated regions. The colors of the graph elements correspond to Fig. 2

the origin and $dest(\gamma) = \lim_{t\to\infty} \gamma(t)$ the destination. Each regular point r where $\nabla d(r) \neq 0$, is part of an image of γ and can be assigned to the critical points that are connected by γ. These assignments describe the manifolds where the direction or opposite direction distinguishes between stable or unstable manifolds:

$$\text{stable manifolds:} \qquad S(c) = \{c\} \cup \{r \in d \mid r \in \text{im}\,\gamma,\ \text{dest}(\gamma) = c\} \qquad (2)$$

$$\text{unstable manifolds:} \qquad U(c) = \{c\} \cup \{r \in d \mid r \in \text{im}\,\gamma,\ \text{orig}(\gamma) = c\} \qquad (3)$$

Because the distance values of d are positive within the pore space B, the maxima c_{max} mark those points where the distance is maximal to the surrounding particles. We define these points to be the pore centers. The index-2-saddle points c_{S2} of the 3d distance map and their unstable manifolds $U(c_{S2})$ connect two maxima. They depict the paths from one pore center to another along the maximal distance to the surrounding particles. An index-2-saddle point is the point along a path between two connected maxima having the smallest distance and it corresponds, in terms of pore structure elements, to the pore constriction, which determines the maximal size of particles that can move from one pore to another. Finally, a pore body consists of all points in the pore space B that end in the same maximum when following the steepest ascent. The table in Fig. 2c summarizes these correspondences.

3.2 *Voxel-Based Determination of the Pore Space Skeleton*

The defined pore space elements, that is, pore centers, paths, and constrictions, form a network representing the pore space topology. We previously proposed the following voxel-based processing pipeline to extract these elements [10]. Starting point is the segmentation result where the foreground F covers the whole particle structure and decomposes it into separated particles F_i, which are labeled by their identifier i.

In the processing pipeline illustrated in Fig. 3, we use the correspondence between our distance-based definitions and a Voronoi decomposition. In case of a set of points, a Voronoi cell of a Voronoi point consists of all points whose distance is not greater than their distance to any other Voronoi point. The distances on the facets, edges, and vertices of the Voronoi cells are maximal to the closest two, three, or four Voronoi points, respectively. Looking at the distance function according to the Voronoi points, the Voronoi vertices are located at the maxima, and the Voronoi edges are located at the index-2-saddles and their unstable manifolds. The processing pipeline detects these locations between the particles according to the boundary voxels of the segmented particles instead of according to single Voronoi points.

Decomposition. The three-dimensional signed distance map with regard to the boundaries between foreground and background is computed according to Eq. 1. Distance values within the foreground F are negative while distances within the background B are positive. The labeled particles are used as seeds for a watershed transformation [21] that propagates the particles according to increasing distance. As a result, each voxel of the pore space is assigned the identifier of its nearest particle, such that the image volume I is completely partitioned into the propagated particle regions V_i, which meet at points having equal distances to the particles:

$$V_i = \{F_i\} \cup \left\{ p \in B \mid \min_{q \in F_i} \|p - q\| \leq \min_{s \in F_j} \|p - s\|, \ \forall j \neq i \right\} \tag{4}$$

Skeleton. Next, the boundaries of the particle regions are evaluated in order to determine the pore space skeleton. We identify voxel neighborhoods of size $2 \times 2 \times 2$ that contain labels of three or more particle regions V_i, because these are the neighborhoods that contain points having equal distance to at least three particles. For each such neighborhood, we mark a representative voxel (the bottom, left, front voxel) in the resulting volume. All other voxels of the result are marked as background. With this approach, we remove paths that are only surrounded by two or less particles. These paths can occur in case of neighboring concave particles and build isolated and/or small anabranches. Particles that can move there do not change the soil structure and its stability. Therefore, these paths are negligible.

Pore graph. The identified skeleton may have segments with a thickness of more than one voxel. In order to remove these, we apply thinning [18]. Finally, we construct a graph structure by converting this skeleton into vertices, edges and edge points [17]. The resulting graph represents the pore centers by its vertices and the paths by its edges. The edges additionally have edge points and contain radius information at each point. This allows one to mark the point with the smallest radius on each edge as constriction and provides information about the spatial course of the paths and the distances.

3.3 Merging Unstable Pores

Due to the irregular shape and arrangement of particles, multiple local maxima resp. pore centers may appear, as denoted in Fig. 4. The yellow, orange, and green pore

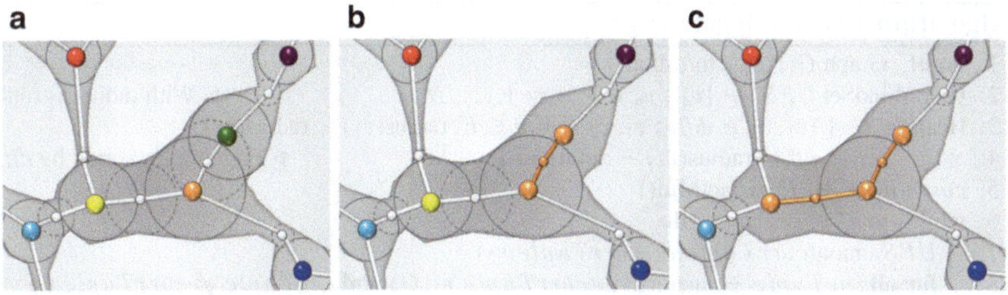

Fig. 4 Hierarchical merge of (**a**) pores (*colored vertices* and *black circles*) separated by their constrictions (*white points* and *dotted circles*): (**b**) the *green* and *orange pores* merge; the *green pore* and the common constriction are assigned the *orange* representative; the constriction between *green* and *purple* is now between the *orange* representative and the *purple* one. (**c**) Merge of the non-overlapping *yellow* and *orange pores*

centers seem to belong to a single pore. This may lead to an under- or overestimation of parameters like pore size and constriction number. Therefore, we merge such pore centers in the following post-processing step.

The method is inspired by topological persistence and simplification [5]. Topological persistence assesses the features of a function by their significance. We use the radius differences of a constriction and their connected pore centers as persistence measure. If the difference is high, the pores are significantly separated by their connecting pore constriction and stay separated. In turn, pore centers that are connected by a constriction having almost the same radius as the pore centers will be merged. The degree of the merge depends on a user-defined threshold.

The algorithm hierarchically merges and updates neighboring pairs of pore centers and their radius differences (Algorithm 1). Each edge, its adjacent vertices and a difference value will be represented by an edge tuple where the difference value is given by the minimum radius difference of the two vertices to the constriction on this edge. The pore center having the larger radius is set to be the representative vertex of the edge tuple. When an edge and a vertex are merged to their representative, the incident edges of the merged vertex will be updated: The edges and the representatives of their vertices build new edge tuples.

Finally, adjacent vertices having the same representative and their connecting edges will be labeled as belonging to the same pore. As a result, merged pore centers, paths and constrictions can be specifically included or excluded from the quantification tasks. The maximal diameter of a pore is then given by the distance information of the representative vertex of the merged pore.

4 Evaluation

The voxel-based pipeline described in Sect. 3 enables us to analyze the pore space of realistic data containing particles of irregular shape. However, the voxel-based methods suffer from inaccuracies due to the discrete nature of the data. The

Algorithm 1 Hierarchical merge

```
input: Graph G(V, E), threshold t
UnionFindSet UFS <- {v_n : v_n ∈ V, n = 1, ..., N}                      ▷ With radius as rank
Heap H <- { (v_1, v_2, e, diff) : v_1, v_2 ∈ V, e ∈ E, radius(v_1) < radius(v_2),
                diff = radius(v_1) - radius(e) }                ▷ Edge tuples sorted by diff
currTuple <- H.extractMin()
while currTuple.diff < t do
    UFS.union(currTuple.v_1, currTuple.v_2)
    for all incTuple <- incidentTo(currTuple.v_1, G) and incTuple != currTuple do
        v_1 <- UFS.find(incTuple.v_1)
        v_2 <- UFS.find(incTuple.v_2)
        e <- incTuple.e
        if radius(v_1) > radius(v_2) then swap(v_1, v_2)
        newDiff <- (radius(v_1) - radius(e))
        newTuple <- (v_1, v_2, e, newDiff)
        H.remove(incTuple)
        H.insert(newTuple)
    end for
    if H.empty() then break
    currTuple <- H.extractMin()
end while
relabel(G, UFS)    ▷ Labels vertices & connecting edges according to the UFS components
```

inaccuracies may even accumulate during processing a pipeline of voxel-based methods. Because we generally apply our pipeline to realistic data, there is no possibility to measure the quality of our pipeline on these data sets, because we do not have any ground truth for the realistic data.

As mentioned above, our pipeline generates a decomposition of the voxel data set such that each voxel is assigned to its nearest segmented particle. This decomposition is related to a Voronoi decomposition. Hence, we call the extraction pipeline Voronoi-like pore space computation.

In order to evaluate the results of our method, we have therefore chosen a sphere packing as comparison basis for two reasons. First, sphere packings are commonly used to simulate porous materials in the area of geotechnics. Second, for sphere packings, it is possible to compute an analytic description of the topology of the distance transform of the spheres by applying an algorithm that computes the Voronoi diagram of spheres (see, for example, Lindow et al. [13]). By considering the result of the analytic algorithm as ground truth, we are able to evaluate the results of our voxel-based Voronoi-like pore graph computation.

4.1 Evaluation Strategy

The input to the analytic Voronoi diagram algorithm is a set of weighted points given by the sphere centers of the sphere packing where the weights are the radii of the spheres. The output of the algorithm is an analytic description of the edge

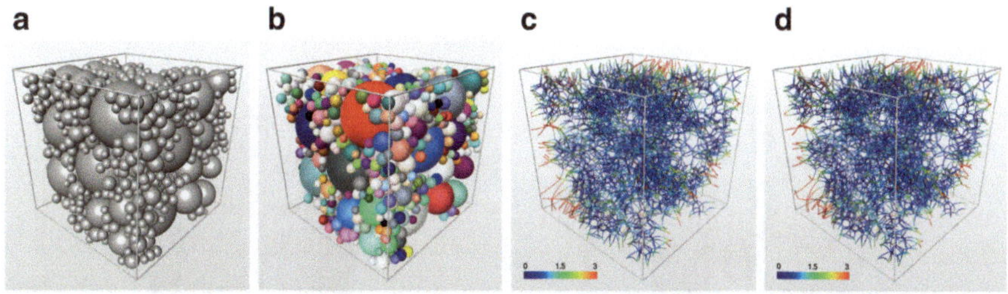

Fig. 5 Comparison data: sphere packing (**a**) and voxelized spheres (**b**), Voronoi graph (**c**) and pore graph (**d**) with radius information

graph of the Voronoi diagram (Fig. 5c). This is given by the Voronoi vertices, that is, the vertices where four Voronoi regions meet, and the Voronoi edges between these vertices. To compare the Voronoi graph with the pore graph resulting from the voxel-based pore space computation, we sample the edges of the Voronoi graph, such that neighboring points on the edges have a maximum distance smaller than a given threshold D_s. For each of these sample points, which we call edge points, as well as for the Voronoi vertices, we also store the minimal distance to any of the neighboring spheres. This distance corresponds to the radius of the maximal ball located at these points without intersecting the spheres.

To compute the pore graph of the sphere packing, we first have to compute a voxel representation of the sphere packing (Fig. 5b). The result is a labeled voxel data set where each voxel is assigned the label of the sphere it belongs or 0 if it belongs to the pore space. The pipeline then computes the pore graph (Fig. 5d) of the sphere packing, consisting of the pore centers, which correspond to the Voronoi vertices, and the pore paths, which correspond to the Voronoi edges. We did not sample the pore paths, because they already have edge points on them with a distance corresponding to the voxel size.

4.2 Graph Matching

Now we have two graphs $G_1 = (V_1, E_1, P_1)$ and $G_2 = (V_2, E_2, P_2)$ representing the Voronoi graph and the pore graph, respectively. Here, V_1 and V_2 are the sets of vertices corresponding to the local maxima of the distance transformation, both in the analytical and the discrete case. E_1 and E_2 are the sets of edges, which are not directly considered in the comparison but only via the sets P_1 and P_2, which represent the edge points. We further define the radius functions $d_i : V_i \cup P_i \to \mathbb{R}$ that assign to each vertex or edge point the minimal distance to any of the neighboring particles or spheres.

To measure the difference of G_1 to G_2, we are interested in two things. (I) How much differ the two graphs geometrically? The geometric distance can be calculated

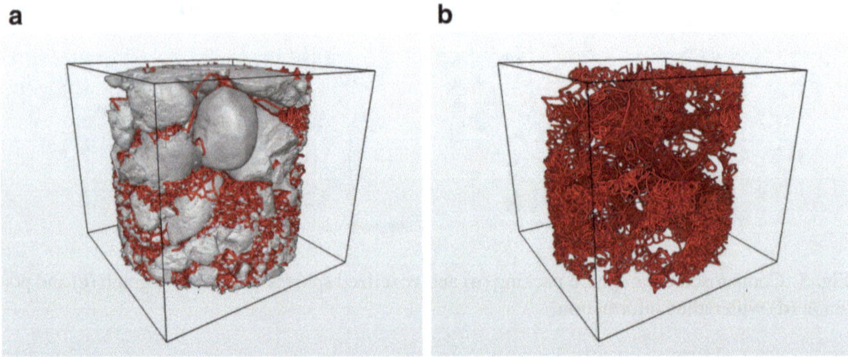

Fig. 6 Pore graph (**a**) with and (**b**) without the particle structure computed from a CT scan of soil

by determining the shortest distance of each vertex in V_1 to any vertex in V_2. Similarly, we need to determine the shortest distance of each point in P_1 to any point in P_2. (II) How much differs the radius information on the graphs? The radius gives the size of the constrictions, which are of particular interest to us. For this, we need matchings $m_{V12} : V_1 \rightarrow V_2$ with $m_{V12}(v) = \text{argmin}_w ||v - w||$ and $m_{P12} : P_1 \rightarrow P_2$ with $m_{P12}(p) = \text{argmin}_q ||p - q||$ where $v \in V_1, w \in V_2$, $p \in P_1, q \in P_2$ and $||\cdot||$ denotes the Euclidean length. Note that the matchings m_{V12} and m_{P12} are, in general, not bijections. This is not a problem, because the matchings are only a means to compute the geometric differences between the graphs. Given the matchings, we can then compute the differences between the radius information of the matched vertices and edge points as $|d_i(v) - d_i(m_{12}(v))|$ and $|d_i(p) - d_i(m_{12}(p))|$, respectively where $|\cdot|$ denotes the absolute value. We use an octree to efficiently compute m_{V12} and m_{P12}.

5 Results and Discussion

In this section, we compare our voxel-based pipeline to the Voronoi graph computation on the basis of a sphere packing. We will not discuss results of the pipeline applied to realistic data. Such results are available in our previous work [10]. (Fig. 6)

The comparison was carried out using a sphere packing of 1,250 spheres of varying radii (1.5–5.2 mm), which were placed in a box of $50 \times 50 \times 50$ mm. For the voxel-based pipeline, we scan-converted the sphere packing into a voxel data set. We chose a sampling rate of 50 voxels per minimal sphere diameter. This resulted in a voxel size of $40 \times 40 \times 40\,\mu$m corresponding to the resolution of our CT scans. The resulting data set had a size of $1{,}250 \times 1{,}250 \times 1{,}250$ voxels. The computation of the Voronoi graph took less than 2 s while the voxel-based pipeline took 52 min. The edges of the Voronoi graph were sampled with a maximal distance $D_s = 0.05$ mm.

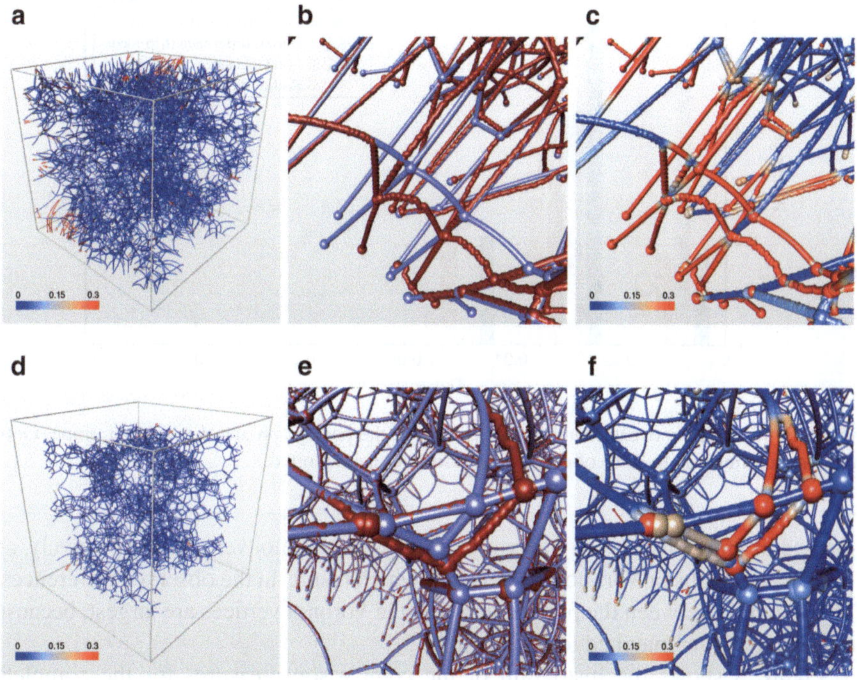

Fig. 7 Differences between pore graph and Voronoi graph. (**a**) Color-coded distances between the matched points and vertices. (**b**) Close-up of the *bottom left* corner in (**a**): pore graph (*red*) and Voronoi graph (*blue*). (**c**) Distances color-coded on the same close-up. (**d**) The graphs are trimmed by 5 mm from each side. (**e**) and (**f**) Close-ups with many pore centers and Voronoi vertices

We applied the matching procedure described in Sect. 4. Figure 7 depicts images showing the vertex and edge point distances between the two graphs. If we consider the complete graphs (Fig. 7a), we can observe that the main differences lie at the border in regions where there is a large empty space (also see Fig. 5a). The reason for this large empty space is that no spheres were placed in this region due to restrictions caused by the border. As a result, we get long pore paths ending at the border (see, for example, the left bottom corner in Fig. 7a which has been zoomed in to in Fig. 7b, c). The further away the pore path is from the nearest spheres, the smaller is the difference in the distances when looking orthogonally around the pore path. Hence, small differences in the distances to the nearest spheres correspond to large differences in the position of the pore path. Thus, small errors in the computation of the distances can result in large errors of the pore path position. For the statistics we want to carry out later, trimming the graph gives us much more realistic data. The graphs in Fig. 7d were trimmed by 5 mm from each side. The major differences that remain between the graphs after trimming appear in regions where many pore centers/Voronoi vertices occur (see, for example, the close-ups in Fig. 7e, f). Here, small errors in the distances to the nearest spheres might result in either pores being

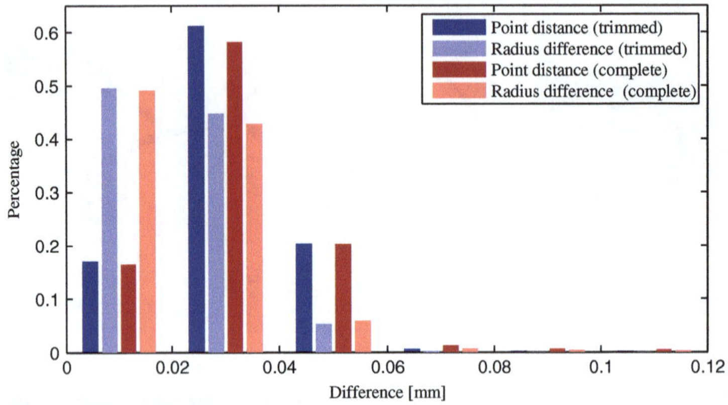

Fig. 8 Differences between matched edge points of the pore graph and the Voronoi graph. Only the bins summing up to 99 % of each of the distribution are plotted

merged or split. Hence, too many pore centers or Voronoi vertices, respectively, as well as pore paths or Voronoi edges occur, which results in the observed differences. The distances between the pore centers and the Voronoi vertices are largest, because they can only get matched to one another.

Figure 8 shows the histogram of the comparison measures for the complete pore graph in red and the pore graph with trimmed boundary region in blue. The distances are plotted in dark and the radius differences in light red and light blue. As mentioned above, we trimmed the pore graph by 5 mm from each side to eliminate boundary effects. This is important for the final analysis of the pore space, because the constrictions on edges near the boundaries exhibit unrealistic radius values due to the restricted growth of the distance map towards the boundary. This, in turn, would lead to biased statistics on the pore constrictions.

It is obvious that the accuracy of the voxel-based computation depends on the resolution of the data set. Taking into account the voxel size, we binned the data points with a bin size of 0.02 mm. This corresponds to half the voxel length. Regarding the distances between the edge points (dark colored bars), this means, that the matched edge points of the first group can be assumed to lie within the same voxel and the points of the second and third group within the direct neighborhood. Furthermore, we assume the radius differences (bright colored bars) within the three first groups to lie on the same or on the very next distance iso-value, whether they share the same position or not.

Accumulating the first three bins shows that 94.9 % of the edge points of the untrimmed and 98.6 % of the trimmed graph match the edge points of the Voronoi graph at least within the direct neighborhood. According to this, the edge points of the first three bins exhibit a radius difference to the Voronoi edge points of less than 0.06 mm. This concerns 97.8 % of the edge points of the untrimmed graph and 99.5 % of the trimmed graph. It is worth noting that regarding the aim of analyzing the transport pathways for particles of certain sizes, the radius information along a path is even more important than the exact position of points along the path way. The maximal error is 0.79 and 0.34 mm in point distance and radius difference of the trimmed pore graph. Such discrepancies mainly occur at the boundary (Fig. 7c) and only make up a very small fraction of the whole set of edge points. Above all, the presented matching procedure is only able to provide an upper bound of the error. Because it matches each edge point from one graph to the other even if there is no direct correspondence, large differences in the distance and the radius comparison measure may be acquired in these cases. In Fig. 8 we can see that the error of 99 % of our edge points is below 0.12 mm which is less than a tenth of the radius of the smallest sphere in the packing. We can assume that the errors of the remaining 1 % of edge points do not invalidate conclusions that can be drawn from derived statistics of the extracted pore structures.

6 Conclusion

We have described a method to extract the topology of the pore space in a porous material and validated the results of our method based on an analytic example. Regarding the given details, the presented description of the method goes beyond what is available in our previous work. Thus, it allows for a straightforward implementation of the pipeline. Comparing our pore graph to the Voronoi graph of the analytic sphere-packing example allowed us to analyze the accuracy and validity of our approach. The analysis has shown that differences between the graphs are mostly in the scale of one voxel and, thus, the accuracy is very high. Together with a simultaneous visual inspection of the two graphs, the high accuracy allowed us to illustrate the validity of our approach. Since the computation of the pore space elements for a sphere packing does not differ from that for a realistic data set, we conclude that our method works also correct for realistic data.

Acknowledgements This work was partly funded by the German Research Foundation (DFG) in the project "Conditions of suffosive erosion phenomena in soil". Special thanks go to Norbert Lindow for providing his implementation of the Voronoi graph algorithm.

References

1. R. Al-Raoush, K. Thompson, C.S. Willson, Comparison of network generation techniques for unconsolidated porous media. Soil Sci. Soc. Am. J. **67**(6), 1687–1700 (2003)
2. R. Binner, U. Homberg, S. Prohaska, U. Kalbe, K.J. Witt, Identification of descriptive parameters of the soil pore structure using experiments and CT data, in *Proceedings of the 5th International Conference Scour and Erosion (ICSE-5)*, San Francisco, 2010, pp. 397–407
3. M.E. Coles, R.D. Hazlett, P. Spanne, W.E. Soll, E.L. Muegge, K.W. Jones, Pore level imaging of fluid transport using synchrotron X-ray microtomography. J. Pet. Sci. Eng. **19**(1–2), 55–63 (1998)

4. H. Edelsbrunner, J. Harer, A. Zomorodian, Hierarchical Morse complexes for piecewise linear 2-manifolds, in *Proceedings of the 17th Annual Symposium on Computational Geometry*, Medford (ACM, 2001), pp. 70–79
5. H. Edelsbrunner, D. Letscher, A. Zomorodian, Topological persistence and simplification. Discret. Comput. Geom. **28**(4), 511–533 (2002)
6. R. Glantz, Porennetzwerke von Erdstoff-Filtern: mathematisch-morphologische Beschreibung kernspintomographischer Aufnahmen. Ph.D. thesis, University Karlsruhe, Karlsruhe, 1997
7. R. Glantz, M. Hilpert, Dual models of pore spaces. Adv. Water Resour. **30**(2), 227–248 (2007)
8. A. Gyulassy, M. Duchaineau, V. Natarajan, V. Pascucci, E. Bringa, A. Higginbotham, B. Hamann, Topologically clean distance fields. IEEE Trans. Vis. Comput. Graph. **13**(6), 1432–1439 (2007)
9. M. Hadwiger, L. Fritz, C. Rezk-Salama, T. Höllt, G. Geier, T. Pabel, Interactive volume exploration for feature detection and quantification in industrial CT data. IEEE Trans. Vis. Comput. Graph. **14**(6), 1507–1514 (2008)
10. U. Homberg, D. Baum, S. Prohaska, U. Kalbe, K.J. Witt, Automatic extraction and analysis of realistic pore structures from μCT data for pore space characterization of graded soil, in *Proceedings of the 6th International Conference Scour and Erosion (ICSE-6)*, Paris, 2012, pp. 66–73
11. U. Homberg, R. Binner, S. Prohaska, V.J. Dercksen, A. Kuß, U. Kalbe, Determining geometric grain structure from X-ray micro-tomograms of gradated soil, in *Internal Erosion*, ed. by K.J. Witt. Schriftenreihe Geotechnik, Bauhaus-Universität Weimar, vol. 21 (2009), pp. 37–52
12. M.W. Jones, J.A. Bærentzen, M. Sramek, 3D distance fields: a survey of techniques and applications. IEEE Trans. Vis. Comput. Graph. **12**(4), 581–599 (2006)
13. N. Lindow, D. Baum, H.C. Hege, Voronoi-based extraction and visualization of molecular paths. IEEE Trans. Vis. Comput. Graph. **17**(12), 2025–2034 (2011)
14. W.B. Lindquist, S.M. Lee, D.A. Coker, K.W. Jones, P. Spanne, Medial axis analysis of void structure in three-dimensional tomographic images of porous media. J. Geophys. Res. **101**(B4), 8297–8310 (1996)
15. W.B. Lindquist, A. Venkatarangan, J. Dunsmuir, T. Wong, Pore and throat size distributions measured from synchrotron X-ray tomographic images of fontainebleau sandstones. J. Geophys. Res. **105**(B9), 21509–21527 (2000)
16. A. Pierret, Y. Capowiez, L. Belzunces, C.J. Moran, 3d reconstruction and quantification of macropores using X-ray computed tomography and image analysis. Geoderma **106**(3–4), 247–271 (2002)
17. S. Prohaska, Skeleton-based visualization of massive voxel objects with network-like architecture. Ph.D. thesis, University of Potsdam, 2007
18. C. Pudney, Distance-ordered homotopic thinning: a skeletonization algorithm for 3d digital images. Comput. Vis. Image Underst. **72**(3), 404–413 (1998)
19. N. Reboul, E. Vincens, B. Cambou, A statistical analysis of void size distribution in a simulated narrowly graded packing of spheres. Granul. Matter **10**(6), 457–468 (2008)
20. F. Rezanezhad, W.L. Quinton, J.S. Price, D. Elrick, T.R. Elliot, R.J. Heck, Examining the effect of pore size distribution and shape on flow through unsaturated peat using 3-d computed tomography. Hydrol. Earth Syst. Sci. **13**(10), 1993–2002 (2009)
21. J.B.T.M. Roerdink, A. Meijster, The watershed transform: definitions, algorithms and parallelization strategies. Fundam. Inform. **41**(1–2), 187–228 (2001)
22. D. Silin, T. Patzek, Pore space morphology analysis using maximal inscribed spheres. Phys. A: Stat. Theor. Phys. **371**(2), 336–360 (2006)
23. S.M. Sweeney, C.L. Martin, Pore size distributions calculated from 3-D images of DEM-simulated powder compacts. Acta Mater. **51**(12), 3635–3649 (2003)
24. K.E. Thompson, C.S. Willson, C.D. White, S. Nyman, J.P. Bhattacharya, A.H. Reed, Application of a new grain-based reconstruction algorithm to microtomography images for quantitative characterization and flow modeling. SPE J. **13**(2), 164–176 (2008)
25. D.M. Ushizima, D. Morozov, G.H. Weber, A.G.C. Bianchi, E.W. Bethel, Augmented topological descriptors of pore networks for material science. IEEE Trans. Vis. Comput. Graph. **18**(12), 2041–2050 (2012)

Visualization of Two-Dimensional Symmetric Positive Definite Tensor Fields Using the Heat Kernel Signature

Valentin Zobel, Jan Reininghaus, and Ingrid Hotz

Abstract We propose a method for visualizing two-dimensional symmetric positive definite tensor fields using the Heat Kernel Signature (HKS). The HKS is derived from the heat kernel and was originally introduced as an isometry invariant shape signature. Each positive definite tensor field defines a Riemannian manifold by considering the tensor field as a Riemannian metric. On this Riemmanian manifold we can apply the definition of the HKS. The resulting scalar quantity is used for the visualization of tensor fields. The HKS is closely related to the Gaussian curvature of the Riemannian manifold and the time parameter of the heat kernel allows a multiscale analysis in a natural way. In this way, the HKS represents field related scale space properties, enabling a level of detail analysis of tensor fields. This makes the HKS an interesting new scalar quantity for tensor fields, which differs significantly from usual tensor invariants like the trace or the determinant. A method for visualization and a numerical realization of the HKS for tensor fields is proposed in this chapter. To validate the approach we apply it to some illustrating simple examples as isolated critical points and to a medical diffusion tensor data set.

V. Zobel (✉)
Zuse Institute Berlin, Berlin, Germany
e-mail: zobel@zib.de

J. Reininghaus
Institute of Science and Technology Austria, Klosterneuburg, Austria
e-mail: jan.reininghaus@ist.ac.at

I. Hotz
German Aerospace Center (DLR), Braunschweig, Germany
e-mail: ingrid.hotz@dlr.de

P.-T. Bremer et al. (eds.), *Topological Methods in Data Analysis and Visualization III*, Mathematics and Visualization, DOI 10.1007/978-3-319-04099-8_16,

1 Introduction

The Heat Kernel Signature (HKS), introduced by Sun et al. in [8], is known to be a powerful shape signature. In [8] it is shown that the HKS is not only an isometric invariant but contains almost all intrinsic information of a surface. Thus it is well suited for detecting similar shaped regions of surfaces. The HKS is derived from the process of heat diffusion and consequenlty equipped with a time parameter. This multiscale property allows to adjust the size of the neighborhood that influences the value of the HKS at a point. Additionally, the HKS is not sensitive to small perturbations of the underlying surface, e.g. a tunnel between small sets of points. Several methods employ the HKS to detect similar shaped surfaces globally, see [1, 4–6]. Also from a visual point of view the HKS characterizes a surface very well, since, for small time values, it is closely related to the Gaussian curvature of the surface. For large time values it can be considered as the curvature on a larger scale. Our idea is to use the HKS for the visualization of tensor fields. Since positive definite tensor fields can be considered as Riemannian metrics, i.e. together with its domain as Riemannian manifolds, the definition of the HKS is still applicable for positive definite tensor fields. Consequently, we obtain a scalable Gaussian curvature of the Riemannian manifold associated with tensor field.

For a better understanding we illustrate the relation between the HKS of a two dimensional surface M and a positive definite tensor field (i.e. the metric tensor field of the surface) in Fig. 1. If g is the metric of the surface M and $f : \mathbb{R}^2 \supset U \mapsto \mathbb{R}^3$ a parametrization of M, i.e. $f(U) = M$, we can compute the pull back of the metric g on U by f, denoted by f^*g. The metric f^*g is a positive definite tensor field on U which is well characterized by the HKS of the surface. In this chapter we propose a method for computing the HKS directly for a positive definite tensor field defined on $U \subset \mathbb{R}^2$, interpreting the tensor field as the metric of a surface. We do not demand that there exists a simple embedding of the surface in some Euclidean space. We restrict ourselves to 2D positive definite tensor fields in this chapter, but the definition of the HKS and the numerical realization presented here is also valid in higher dimensions. However, the computational complexity will be a problem in higher dimensions, see also the remark in Sect. 6.

In Sect. 2 we give a short introduction to the HKS. The application of the HKS to tensor fields is explained in Sect. 3. To compute the HKS we need to compute the eigenvalues of a the Laplacian on a Riemannian manifold (M, g). In case of surfaces the embedding in the Euclidean space can be utilized, whereas, in the case of tensor fields, all computations must be done by using the tensor only. A finite element method to achieve this for tensor fields on a uniform grid is proposed in Sect. 4, alongside with some numerical tests. Results of our method are shown in Sect. 5.

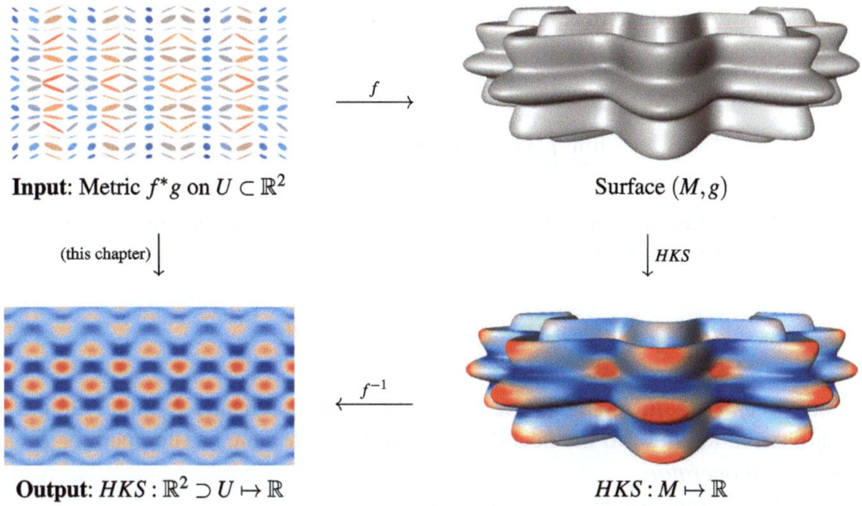

Fig. 1 Commutative diagram illustrating the relation between the HKS of a surface and a positive definite tensor field. Metric of the surface depicted as ellipses (*top left*), the parametrized surface (*top right*), HKS on the surface (*bottom right*) and the HKS on U (*bottom right*)

2 Heat Kernel Signature

The Heat Kernel Signature (HKS) is typically used for the comparison of surfaces. It is derived from the heat equation and assigns each point of the surface a time dependent function $[0, \infty) \to \mathbb{R}$ which depends only on the metric of the surface. Conversely, all information about the metric are contained in the HKS under quite weak assumptions. For smaller time values the HKS at a point is governed by smaller neighbourhoods, i.e. one can control the portion of surface which should be taken into account. This makes the HKS a powerful tool for the identification of similar shaped parts with different level of detail by comparing the HKS for different time values. However, the HKS is not restricted to surfaces, it is defined for arbitrary Riemannian manifolds. We employ this fact to apply the HKS on positive definite tensor fields. A short introduction to the HKS is given in this section. For details we refer the reader to [8]. A detailed treatment of the heat operator and the heat kernel can be found in [7].

Let (M, g) be a compact, oriented Riemannian manifold and Δ the Laplace-Beltrami operator (also called just Laplacian) on M which is a equivalent to the usual Laplacian in case of flat spaces. Given an initial heat distribution $h(x) = h(0, x) \in C^{\infty}(M)$ on M, considered to be perfectly insulated, the heat distribution $h(t, x) \in C^{\infty}(\mathbb{R}^{+} \times M)$ at time t is governed by the *heat equation*

$$(\partial_t + \Delta)h(t, x) = 0 \ .$$

One can show that there exists a function $k(t,x,y) \in C^{\infty}(\mathbb{R}^{+} \times M \times M)$ satisfying

$$(\partial_t + \Delta_x)k(t,x,y) = 0 \ ,$$

$$\lim_{t \to 0} \int k(t,x,y)h(y)\,dy = h(x) \ ,$$

where Δ_x denotes the Laplacian acting in the x variable. The function $k(t,x,y)$ is called *heat kernel*. Let now H_t be the integral operator defined by

$$H_t h(x) = \int_M k(t,x,y)h(y)\,dy \ ,$$

then $h(t,x) = H_t h(x)$ satisfies the heat equation. Consequently H_t takes an initial heat distribution $h(x)$ to the heat distribution $h(t,x)$ at time t. The operator H_t is called *heat operator*.

The heat kernel can be computed by the formula

$$k(t,x,y) = \sum_i e^{-\lambda_i t}\phi_i(x)\phi_i(y) \ , \tag{1}$$

where λ_i and ϕ_i are the eigenvalues and eigenfunctions of Δ. Since Δ is invariant under isometries, Eq. (1) shows that this is also true for the heat kernel. Moreover, the metric can be computed from the heat kernel by the formula

$$\lim_{t \to \infty} t \log k(t,x,y) = -\frac{1}{4}d^2(x,y) \ ,$$

where $d(x,y)$ denotes the geodesic distance between two points $x, y \in M$. Thus, for a given manifold M, the information contained by the heat kernel and the metric are equivalent. Another important property of the heat kernel is its multi-scale property. For the heat kernel t plays the role of a spatial scale of influence, i.e. $k(t,x,\cdot)$ depends mainly on small neighborhoods of x for small t, whereas $k(t,x,\cdot)$ is influenced by larger neighborhoods of x for larger t.

The HKS is defined in [8] to be the function $HKS \in C^{\infty}(\mathbb{R}^{+} \times M)$ given by

$$HKS(t,x) = k(t,x,x) \ . \tag{2}$$

Since the heat kernel is much more complex than the HKS, one might expect to loose a lot of information when regarding the HKS instead of the heat kernel. But, as shown in [8], the metric can be reconstructed from the HKS under quite weak assumptions. This means that the HKS of a positive definite tensor field contains almost all information of the tensor field itself and is consequently much more informative than usual scalar quantities like the trace or the determinant.

2.1 Relation to Curvature

In order to obtain a more intuitive understanding of the HKS we study its relation to the curvature of the manifold M. For small values of the time parameter t the HKS has the row expansion

$$HKS(t,x) = \frac{1}{4\pi t} \sum_{i=1}^{\infty} u_i(x) t^i \ . \tag{3}$$

The general form of the functions $u_i(x)$ is discussed in [7]. For the two-dimensional manifolds considered in this chapter the first three functions can be written as

$$\begin{aligned} u_0(x) &= 1 \ , \\ u_1(x) &= \frac{1}{3} K(x) \ , \\ u_2(x) &= \frac{1}{45} \left(4K(x)^2 - 3\Delta K(x)\right) \ , \end{aligned}$$

where K is the Gaussian curvature of M. Consequently, for a small value of t, the value of the HKS consists mainly of $\frac{1}{3}K$ plus a constant. The derivation of the stated u_i from the general case can be found in the Appendix.

3 HKS for Tensor Fields

The HKS introduced in Sect. 2 is defined for any compact, oriented Riemannian manifold. Thus the HKS is not restricted to surfaces embedded in $\mathbb{R}^n$. If we have a metric tensor g, i.e. a symmetric positive definite tensor field, defined on a region $U \subset \mathbb{R}^n$, then (U, g) forms a Riemannian manifold. Since there is a Riemannian manifold associated with a positive definite tensor field in this way, we can compute the HKS for any positive definite tensor field. In this section we illustrate the relation of the HKS for surfaces and tensor fields by considering a parametrized surface and the pullback of its metric.

Let $f : \mathbb{R}^2 \supset U \to \mathbb{R}^3$ be a parametrized surface. On the one hand, we can compute the HKS for the surface $f(U)$. On the other hand, we can define a metric g on U by

$$g \ : \ T_u(U) \times T_u(U) \to \mathbb{R} \ , \quad (v, w) \mapsto \langle J(u)v \ , \ J(u)w \rangle \ , \tag{4}$$

where $J(u)$ denotes the Jacobian of f at $u \in U$ and $\langle \cdot \ , \cdot \rangle$ the standard scalar product on $\mathbb{R}^3$. That is, g is the pullback $f^*\langle \cdot \ , \cdot \rangle$ of $\langle \cdot \ , \cdot \rangle$ by f. The components of g are given by $g_{ij} = (J^T J)_{ij}$.

This makes (U, g) a Riemannian manifold which is isometric to $f(U)$ (equipped with the metric induced by $\mathbb{R}^3$) and f the associated isometry. Now we can compute the HKS directly on U by using g as metric. This is equivalent to computing the HKS on the surface $f(U)$ and then pull it back to the parameter space U by f, i.e.

$$HKS_U(t,u) = HKS_{f(U)}(t, f(u)) \ ,$$

where HKS_U and $HKS_{f(U)}$ denote the HKS on U and $f(U)$, respectively. In other words: The diagram in Fig. 1 commutes.

Figure 1 also shows that the HKS of (U, g) (bottom left) is a meaningful visualization of the metric. Thus we are interested in a method for computing the HKS directly for tensor fields, so that no embedded surface with the tensor field as metric tensor needs to be constructed. We propose such a method in Sect. 4.

4 Numerical Realization

To our knowledge, the HKS has only been used for triangulated surfaces, so far. We want to use the HKS for the visualization of two-dimensional symmetric positive definite tensor fields T defined on a rectangular region $U \subset \mathbb{R}^2$. Thus we need a method to compute the HKS of T or, more precisely, of the Riemannian manifold (U, T) associated with T. A finite element method for solving this problem is proposed in this section. Moreover, we discuss the boundary conditions and check the correctness of our results numerically.

From Eq. (1) follows that we can compute the heat kernel signature by the formula

$$HKS(t,x) = \sum_i e^{-\lambda_i t} \phi_i(x)\phi_i(x) \ ,$$

where λ_i and ϕ_i are the eigenvalues and eigenfunctions of the Laplacian Δ on (U, T). Thus we need a suitable discretization of Δ. Our first idea was to adapt the Laplacian from the framework of discrete exterior calculus, see [3], which is closely related to the cotangent Laplacian and widely used for triangulated surfaces. However, this discretization makes intensive use of edge lengths, whereas triangulating the domain U and computing edge lengths by the metric g results in triangles which might not even satisfy the triangle inequality. Thus there seems to be no easy modification of this approach. Instead we propose a finite element method to compute the eigenvalues of the Laplacian.

According to Sect. 3 we can think of T as the metric of a surface in local coordinates. In this case the Laplacian is given by

$$\Delta f = \frac{1}{\sqrt{|T|}} \operatorname{div}\left(\sqrt{|T|}\, T^{-1} \nabla f\right)$$

for any function $f \in C^\infty(M)$, where div and ∇ denote the divergence and the gradient on U, respectively. Hence we have to solve the eigenvalue equation

$$\frac{1}{\sqrt{|T|}} \operatorname{div}\left(\sqrt{|T|}\, T^{-1} \nabla \phi\right) = \lambda \phi \ ,$$

or equivalently

$$\operatorname{div}\left(\sqrt{|T|}\, T^{-1} \nabla \phi\right) = \lambda \sqrt{|T|}\, \phi \ .$$

The weak formulation of this problem is given by

$$\int_U \operatorname{div}\left(\sqrt{|T|}\, T^{-1} \nabla \phi\right) \psi \, dx = \lambda \int_U \sqrt{|T|}\, \phi \psi \, dx$$

while this equation must hold for every smooth function ψ. We can rewrite the left hand side to

$$\begin{aligned}
&\int_U \operatorname{div}\left(\sqrt{|T|}\, T^{-1} \nabla \phi\right) \psi \, dx \\
&= \int_U \operatorname{div}\left(\sqrt{|T|}\, T^{-1} (\nabla \phi) \psi\right) dx - \int_U \sqrt{|T|}\, T^{-1} \nabla \phi \cdot \nabla \psi \, dx \\
&= \int_{\partial U} \left(\sqrt{|T|}\, T^{-1} (\nabla \phi) \psi\right) \cdot n \, dx - \int_U \sqrt{|T|}\, T^{-1} \nabla \phi \cdot \nabla \psi \, dx \ ,
\end{aligned}$$

where n denotes the outward pointing normal of the boundary. If we apply Neumann boundary conditions, i.e. $\nabla \phi \cdot n = 0$, the first term vanishes. Finally, we have to solve the equation

$$\int_U \sqrt{|T|}\, T^{-1} \nabla \phi \cdot \nabla \psi \, dx = -\lambda \int_U \sqrt{|T|}\, \phi \psi \, dx \ .$$

Choosing basis functions h_i the stiffness matrix L and the mass matrix M are given by

$$\begin{aligned}
L_{ij} &= \int_U \sqrt{|T|}\, T^{-1} \nabla h_i \cdot \nabla h_j \, dx \ , \\
M_{ij} &= -\int_U \sqrt{|T|}\, h_i h_j \, dx \ ,
\end{aligned}$$

and we solve the generalized eigenvalue equation

$$L v = \lambda M v \ .$$

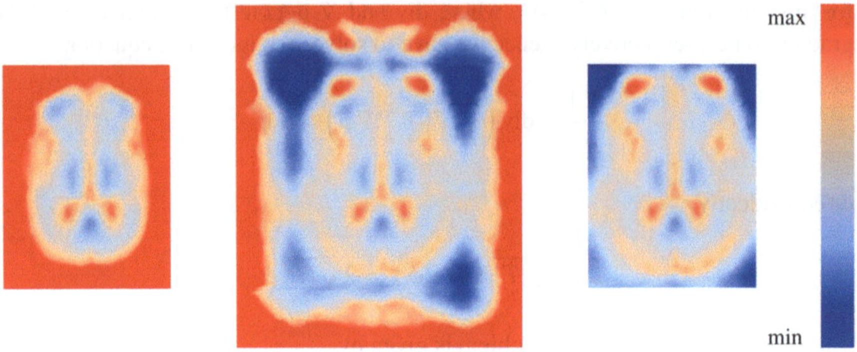

Fig. 2 The result on the *left* is strongly influenced by the boundary. This effect can be reduced significantly by reflecting a portion of the tensor field on the boundary (*middle*) and cropping the result (*right*)

In our examples the tensor fields are given on regular grids and we use bilinear basis functions h_k.

Usual boundary conditions like Dirichlet or Neumann boundary conditions influence the HKS significantly. In particular for large time values the influence is not limited to the immediate vicinity of the boundary. Neumann boundary conditions cause the HKS to have higher values close to the boundary; their physical meaning is that the heat is perfectly insulated. Dirichlet boundary conditions cause the HKS to have a fixed value at the boundary. To overcome this problem we reflect a part of the field at the boundary. Now we can use Neumann boundary conditions for the sake of simplicity and obtain a significantly reduced influence of the boundary, see Fig. 2. The physical meaning of these reflected boundaries is that the heat at the boundary can diffuse outwards in the same way than inwards.

4.1 Numerical Verification

We check the correctness of the FEM described above experimentally by comparing the HKS for a surface with the pull back of its metric, i.e. the commutativity of Fig. 1 is reflected by our verification. Consider a bumpy torus parametrized by

$$f(u, v) = \begin{pmatrix} cos(u)\,(r_1(u) + r_2(v)cos(v)) \\ sin(u)\,(r_1(u) + r_2(v)cos(v)) \\ r_2(v)sin(v) \end{pmatrix} ,$$

where the radii are modulated, i.e. major radius and minor radius are given by $r_1(u) = 3 + \frac{1}{2}cos(10u)$ and $r_2(v) = 1 + \frac{1}{5}cos(8v)$, respectively. This results in the surface that is also used in Fig. 1. We compute the metric of the bumpy torus

Table 1 Relative difference between HKS for tensors and surfaces

Resolution	50^2	100^2	200^2	400^2
Relative difference	0.38369	0.07599	0.00920	0.00075

on $[0, 2\pi]^2$ by formula (4) and sample the resulting tensor field with four different resolutions of 50^2, 100^2, 200^2 and 400^2 points. For this four datasets we compute the HKS with the FEM described above. The results are compared with the HKS for the bumpy torus given as a triangulated surface with 400^2 points, while the HKS is computed by the standard FEM Laplacian for triangulated surfaces, see e.g. [9]. If we denote the HKS for the tensor field by HKS_T and for surfaces by HKS_S, the relative difference for $t = 1$ is given by

$$\frac{\|HKS_T(1,\cdot) - HKS_S(1,\cdot)\|_{L_2}}{\|HKS_S(1,\cdot)\|_{L_2}} = \frac{\left(\int_{[0,2\pi]^2} (HKS_T(1,x) - HKS_S(1,x))^2\, dx\right)^{\frac{1}{2}}}{\left(\int_{[0,2\pi]^2} (HKS_S(1,x))^2\, dx\right)^{\frac{1}{2}}} .$$

See Table 1 for the relative difference between HKS_T for different resolutions and HKS_S. It is obvious that HKS_T approaches HKS_S quickly for increasing resolutions.

5 Results

We show several results of our method in this section. In the following we investigate the significance of the HKS and the meaning of Gaussian curvature in the general tensor context. To get a more intuitive understanding, we analyze the influence of the eigenvalues and eigenvectors on the HKS, and consider synthetic tensor fields with constant eigenvectors and eigenvalues, respectively. As central structural components of tensor fields we also consider isolated degenerate points in our analysis. As real world example we apply the method to a diffusion tensor data set of the brain. During the whole section we use the colormap shown in Fig. 2, which ranges from the minimum to the maximum over all results in one figure, unless otherwise stated.

In Fig. 3 we analyze easy examples of diagonal tensor fields, i.e. $T_{12} = 0$. These fields serve as examples of tensor fields with variable eigenvalues but constant eigenvectors. For the tensor field T^1, where component T^1_{11} is a Gaussian function depending on u_1 and T^1_{22} is constant, the HKS is constant. The field T^2 is very similar to T^1, the only difference is that T^2_{11} depends on u_2. In this case the HKS is not constant anymore. To understand this we consider the formula (3) for $t = 1$, i.e.

$$HKS(1,x) = \frac{1}{4\pi}\left(1 + \frac{1}{3}K(x) + \frac{1}{45}\left(4K(x)^2 - 3\Delta K(x)\right) + \ldots\right) , \quad (5)$$

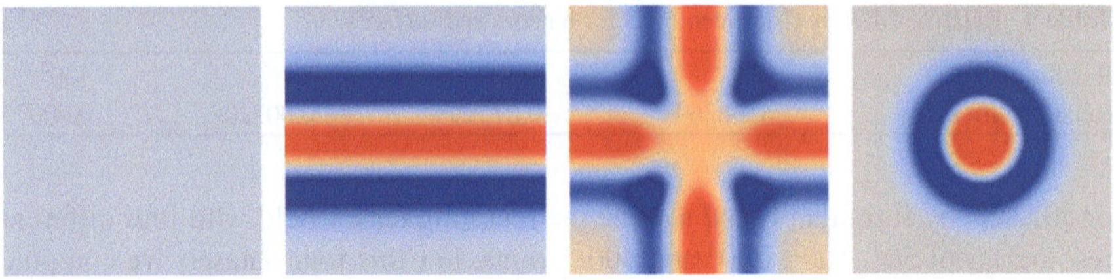

$T^1 = \begin{pmatrix} f(u_1) & \\ & 1 \end{pmatrix}$ $T^2 = \begin{pmatrix} f(u_2) & \\ & 1 \end{pmatrix}$ $T^3 = \begin{pmatrix} f(u_2) & \\ & f(u_1) \end{pmatrix}$ $T^4 = \begin{pmatrix} f(|u|) & \\ & f(|u|) \end{pmatrix}$

Fig. 3 HKS of diagonal tensor fields for small t. The function f is given by $f(x) = 1 + 10e^{-x^2}$ and the tensor fields are defined for $u \in [-5, 5]^2$

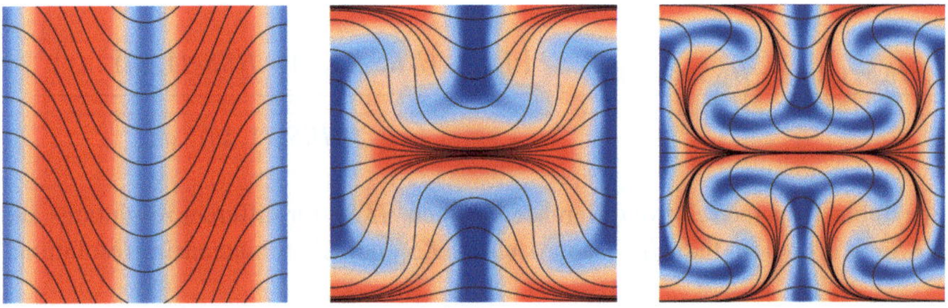

$\begin{pmatrix} \cos\left(\frac{3}{8}\pi \sin(2\pi x)\right) \\ \sin\left(\frac{3}{8}\pi \sin(2\pi x)\right) \end{pmatrix}$ $\begin{pmatrix} \cos\left(\frac{\pi}{2} \sin(2\pi x)\sin(2\pi y)\right) \\ \sin\left(\frac{\pi}{2} \sin(2\pi x)\sin(2\pi y)\right) \end{pmatrix}$ $\begin{pmatrix} \cos\left(\pi \sin(2\pi x)\sin(2\pi y)\right) \\ \sin\left(\pi \sin(2\pi x)\sin(2\pi y)\right) \end{pmatrix}$

Fig. 4 Tensor fields defined on $[0, 1]^2$ by constant eigenvalues and an analytic major eigenvector field (see caption). The HKS is shown for small t and the major eigenvector field is visualized by some integral lines. The dependency of the HKS on the eigenvectors demonstrates a significant difference to other scalar quantities like trace, determinant and anisotropy. The colormap ranges from the minimum to the maximum of the single images

while the Gaussian curvature for diagonal T is given by

$$K = -\frac{1}{2\sqrt{T_{11}T_{22}}} \left(\partial_{u_1} \frac{\partial_{u_1} T_{22}}{\sqrt{T_{11}T_{22}}} - \partial_{u_2} \frac{\partial_{u_2} T_{11}}{\sqrt{T_{11}T_{22}}} \right) . \tag{6}$$

Consequently, if $\partial_{u_1} T_{22} = 0$ and $\partial_{u_2} T_{11} = 0$ we have $K = 0$ and thus $HKS(1, x)$ is constant for T^1. The tensor field T_3 has components T^3_{11} and T^3_{22} depending on u_2 and u_1, respectively, consequently both diagonal components have influence on the HKS. The field T^4 satisfies $T^4_{11} = T^4_{22}$ where T^4_{11} and T^4_{22} depend radially on u. As expected, the HKS depends also radially on u.

The HKS for tensor fields with constant eigenvalues and variable eigenvectors are shown in Fig. 4. These fields are defined by choosing fixed eigenvalues and a variable major eigenvector field, which is visualized by some integral lines. We observe that the HKS is also influenced by the eigenvectors and has high values in compressing regions and low values in expanding regions. This shows also that the HKS and other scalar quantities like trace, determinant and anisotropy, which depend only on the eigenvalues.

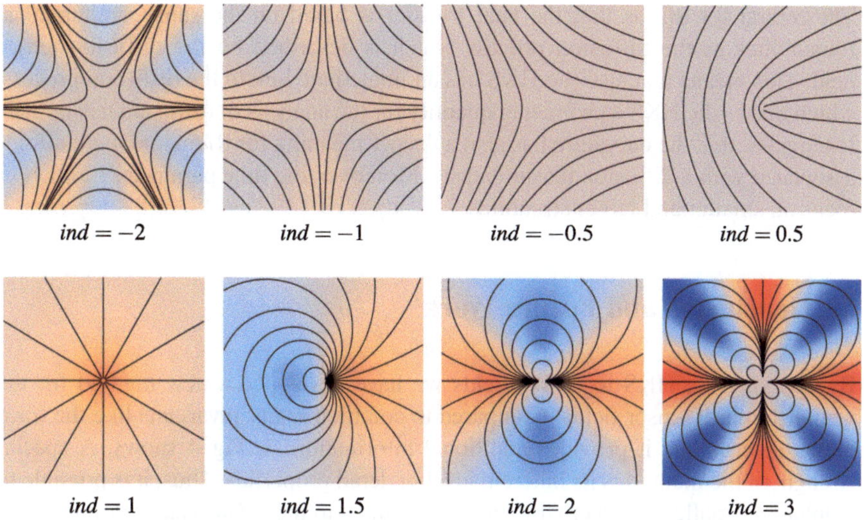

Fig. 5 HKS of degenerate points with different index ind for small t. The tensor field is defined on $u \in [-1, 1]^2$ and the eigenvalues are given by $10 + 3|u|$ and $10 - 3|u|$

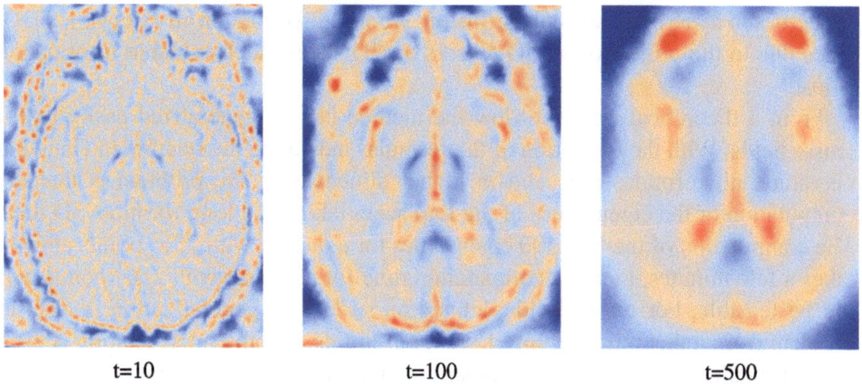

Fig. 6 HKS of a brain dataset for different t. The colormap ranges from the minimum to the maximum of the single images (Brain dataset courtesy of Gordon Kindlmann at the Scientific Computing and Imaging Institute, University of Utah, and Andrew Alexander, W.M. Keck Laboratory for Functional Brain Imaging and Behavior, University of Wisconsin-Madison)

In Fig. 5 we consider degenerate points of tensor fields with indices between -2 and 3. The index ind is defined to be the number of rotations of the eigenvector fields along a curve enclosing the degenerate point (and no other degenerate points). For a more formal definition see [2]. The results show that the HKS also hints at topological features like degenerate points, although there seems to be no obvious way to derive the tensor field topology from the HKS.

As a last example Fig. 6 shows a diffusion tensor dataset of a brain. Again, the HKS is evaluated for different timesteps. Although the extraction of a slice might cut valuable information the structure of the brain becomes obvious by the HKS.

The reason for this is that the heat transfer is based on the same mathematical foundations as the diffusion process which is described by the diffusion tensor field. This means that also the HKS has an immediate link to the diffusion process, which makes the HKS a promising quantity for the analysis of diffusion tensor data. It suggests, that the curvature has also a deeper meaning for Riemannian manifolds associated with a diffusion tensor dataset. Moreover, the time parameter t allows to focus on smaller and larger structures.

6 Conclusion and Future Work

By applying the HKS to tensor fields we have developed a new method for the visualization of tensor fields. Compared to common scalar invariants like the trace or the determinant it provides additional information, as Fig. 4 shows. A special strength of the method is its inherent level of detail properties. Thus, it is possible to emphasize smaller or larger structures. In contrast to naive Gaussian smoothing the scaling is directly driven by the tensor data itself. For diffusion tensor data the results are very promising. For the future we plan on further investigating the significance of the HKS for further applications. It might be of interest to compare the scaling properties to ideas of anisotropic diffusion. We will also work on an extension to 3D, since due to the projection of 3D tensors on 2D slices much valuable information is lost.

From a theoretical point of view the method can be generalized easily to 3D tensor fields. With the exception of the formulas indicating the relation to Gaussian curvature, all formulas are valid in higher dimensions. The problem is that the computation of the eigenvalues of the Laplacian takes very long for most 3D data. The computation of the first 500 eigenvalues for a dataset with 256^2 points already takes a few minutes, thus the computation time for a dataset with 256^3 points will not be feasible. For tensor fields defined on surfaces a generalization is also no problem from a theoretical point of view, but in this case the interpretation is even more difficult. The HKS of the standard metric on the surface results in the usual HKS for surfaces, i.e. the HKS is influenced not only by the tensor field but also by the surface itself.

Acknowledgements This research is partially supported by the TOPOSYS project FP7-ICT-318493-STREP.

Appendix

The row expansion of the HKS for small values of t is given by

$$HKS(t,x) = \frac{1}{4\pi t} \sum_{i=1}^{\infty} u_i(x) t^i \ .$$

In the general case the functions u_i are given by

$$\begin{aligned}
u_0(x) &= 1 \ , \\
u_1(x) &= \frac{1}{6} R(x) \ , \\
u_2(x) &= \frac{1}{360} \left(2 R_{ijkl} R^{ijkl}(x) + 2 R_{jk} R^{jk}(x) + 5 R^2(x) - 12 \Delta R(x)\right) \ ,
\end{aligned}$$

where R_{ijkl} is the Riemann curvature Tensor, $R_{jk} = R^i_{\ jik}$ the Ricci tensor and $R = R^j_j$ the Ricci scalar or scalar curvature. For surfaces these tensors can be written in terms of the Gaussian curvature by

$$\begin{aligned}
R_{ijkl} &= K(g_{ik} g_{jl} - g_{il} g_{jk}) \ , \\
R_{jk} &= K g_{jk} \ , \\
R &= 2K \ .
\end{aligned}$$

Thus we find

$$\begin{aligned}
R_{ijkl} R^{ijkl} &= K(g_{ik} g_{jl} - g_{il} g_{jk}) K(g^{ik} g^{jl} - g^{il} g^{jk}) \\
&= K^2 \left(g_{ik} g_{jl} g^{ik} g^{jl} - g_{ik} g_{jl} g^{il} g^{jk} - g_{il} g_{jk} g^{ik} g^{jl} + g_{il} g_{jk} g^{il} g^{jk}\right) \\
&= K^2 \left(\delta^i_i \delta^j_j - \delta^l_k \delta^k_l - \delta^l_k \delta^k_l + \delta^i_i \delta^j_j\right) \\
&= K^2 (4 - 2 - 2 + 4) = 4K^2 \ ,
\end{aligned}$$

$$R_{jk} R^{jk} = K^2 g_{jk} g^{jk} = K^2 \delta^j_j = 2K^2 \ .$$

And consequently the functions u_i can be expressed in terms of K by

$$\begin{aligned}
u_0(x) &= 1 \ , \\
u_1(x) &= \frac{1}{3} K(x) \ , \\
u_2(x) &= \frac{1}{45} \left(4K(x)^2 - 3\Delta K(x)\right) \ .
\end{aligned}$$

References

1. M. Bronstein, I. Kokkinos, Scale-invariant heat kernel signatures for non-rigid shape recognition, in *IEEE Conference on Computer Vision and Pattern Recognition (CVPR), 2010*, San Francisco (IEEE, 2010), pp. 1704–1711
2. T. Delmarcelle, L. Hesselink, The topology of symmetric, second-order tensor fields, in *Proceedings of the Conference on Visualization'94*, Washington, DC (IEEE, 1994), pp. 140–147
3. M. Desbrun, E. Kanso, Y. Tong, Discrete differential forms for computational modeling, in *SIGGRAPH '06: ACM SIGGRAPH 2006 Courses* (ACM, New York, 2006), pp. 39–54. doi:http://doi.acm.org/10.1145/1185657.1185665
4. T. Dey, K. Li, C. Luo, P. Ranjan, I. Safa, Y. Wang, Persistent heat signature for pose-oblivious matching of incomplete models, in *Computer Graphics Forum*, vol. 29 (Wiley Online Library, 2010), pp. 1545–1554
5. M. Ovsjanikov, A. Bronstein, M. Bronstein, L. Guibas, Shape google: a computer vision approach to isometry invariant shape retrieval, in *IEEE 12th International Conference on Computer Vision Workshops (ICCV Workshops), 2009*, Kyoto (IEEE, 2009), pp. 320–327
6. D. Raviv, M. Bronstein, A. Bronstein, R. Kimmel, Volumetric heat kernel signatures, in *Proceedings of the ACM Workshop on 3D Object Retrieval*, Firenze (ACM, 2010), pp. 39–44
7. S. Rosenberg, *The Laplacian on a Riemannian Manifold: An Introduction to Analysis on Manifolds* (Cambridge University Press, Cambridge, 1997)
8. J. Sun, M. Ovsjanikov, L. Guibas, A concise and provably informative multi-scale signature based on heat diffusion, in *Proceedings of Eurographics Symposium on Geometry Processing (SGP)*, Berlin, 2009
9. H. Zhang, O. van Kaick, R. Dyer, Spectral mesh processing, in *Computer Graphics Forum* (Wiley, 2010)

Topological Features in Glyph-Based Corotation Visualization

Sohail Shafii, Harald Obermaier, Bernd Hamann, and Kenneth I. Joy

Abstract This chapter introduces a novel method for vortex detection in flow fields based on the corotation of line segments and glyph rendering. The corotation measure is defined as a point-symmetric scalar function on a sphere, suitable for direct representation in the form of a three-dimensional glyph. Appropriate placement of these glyphs in the domain of a flow field makes it possible to depict vortical features present in the flow. We demonstrate how topological analysis of this novel glyph-based representation of vortex features can reveal vortex characteristics that lie beyond the capabilities of visualization techniques that consider vortex direction and magnitude information only.

1 Introduction

The extraction and visualization of vortical features, such as vortex cores, has a long and successful history in fields such as aerodynamics. As new vortex definitions emerge, their visualizations are able to convey more characteristics associated with vortices. While the visualization of individual vortex cores by means of core line or hull extraction is successful in illustrating vortex direction and extent, it is limited in its capability to visualize more complex interactions between vortical features. We take advantage of a relatively new vortex descriptor, based on the concept of local corotation [1], to create a glyph-based visualization. This shape-based representation is an encoding of a spherical function that denotes strength of local corotation for arbitrary directions in space, and allows for the examination of vortices by means of a topological analysis. Automatic extraction and visual

S. Shafii (✉) • H. Obermaier • B. Hamann • K.I. Joy
Department of Computer Science, Institute for Data Analysis and Visualization, University of California, Davis, CA 95616-8562, USA
e-mail: ssshafii@ucdavis.edu; hobermaier@ucdavis.edu; bhamann@ucdavis.edu; kijoy@ucdavis.edu

P.-T. Bremer et al. (eds.), *Topological Methods in Data Analysis and Visualization III*, Mathematics and Visualization, DOI 10.1007/978-3-319-04099-8_17,

analysis of maxima and corresponding topological regions in this representation provides insight into possible splitting or merging behavior.

This chapter is structured as follows. We first provide a summary of related work (Sect. 2) and analyze residual vorticity as a measure of local corotation (Sect. 3). Topological analysis of the created glyphs is presented in Sect. 4. We conclude by visualizing and analyzing two data sets.

2 Related Work

The work presented in this chapter spans the areas of glyph rendering and vortex visualization with a focus on analysis of topology. In the following we give a brief summary of related literature in these fields.

2.1 *Vortex Extraction*

Volumetric vortex features in flow fields may be extracted by a number of different vortex classifiers, such as the Q-criterion [2], the Δ-criterion [3–5] or the λ_2-criterion [6]. Kolář [7] introduced a residual vorticity method that not only removes the effects of shear by using the triple decomposition of the velocity-gradient tensor, but is also applicable to compressible flow fields. A similar paper by Kolář et al. [1] derived a simplification of this method using the *corotation of line segments*, which is the method that we discuss in this chapter. Since region-based methods do not readily create global vortical features, many have developed line-based techniques as an alternative. Examples are based on the parallel vectors operator [8], eigenvector analysis [9, 10], pressure-based predictor corrector schemes [11–13], ridge extraction [14, 15], or variations thereof [16–19]. Note that isosurface and medial-axis based representations allow for the visualization of complex vortex behaviors such as splitting. For a summary of various vortex detection methods, we refer to Post et al. [20] and Jiang et al. [21].

2.2 *Vortex Visualization*

The visualization of vortical features is mainly based on isosurfaces, line-like features from predictor-corrector, and skeleton or ridge extraction techniques [12]. Others [22, 23] use (flow) surface or voxel visualization techniques to verify and illustrate vortex cores. Topological analysis and visualization of vortex structures in combination with volume rendering was performed by Tricoche et al. [24].

2.3 *Glyph Visualization*

Glyphs, such as superquadrics [25], are frequently used to encode relevant quantities in vector [26, 27] and tensor fields [28]. Especially in medical visualization, topological properties of tensor glyphs are helpful to identify salient features in the data, such as fiber crossings in DW-MRI data [29,30]. These glyphs can be rendered as discrete meshes or, using modern graphics hardware, be ray-traced [31, 32]. A survey of gylph-based visualization techniques is discussed by Borgo et al. [33]. Our work introduces spherical glyphs as a novel approach to corotation-based vortex visualization. Topological features of these glyphs are evaluated and classified automatically to provide the means for a robust analysis of rotational properties and topological characteristics in vortical features. While previous vortex extraction approaches indicate the strength and directional components of vortices, they do not indicate all possible vortex directions that may exist at a point in a data set. Our glyphs provide a visual representation of vortex axes besides the dominant one that exists at a vortex, as that may indicate branching or merging behavior.

3 Corotation on a Sphere

The extraction of vortex cores relies on the availability of robust mathematical measures of flow rotation. In this work we investigate topological features of glyphs that are derived from a novel rotation measure known as *local corotation* or *residual vorticity*, as described in the following sections.

3.1 *Physical Interpretation*

Given a point $p \in \mathbb{R}^3$, the classic vorticity vector has two main properties. First, its direction corresponds to the normal of the plane along which two arbitrary line segments exhibit the maximal average angular velocity in the flow field [7]. Second, its magnitude represents twice the angular speed of the average rotation of these line segments. Residual vorticity [1], on the other hand, makes use of a similar physical interpretation with one significant difference: The residual vorticity vector is normal to the plane with maximal local corotation of line segments in a plane at p, for all possible planes at p. Instead of maximizing an average rotation speed, the residual vorticity technique maximizes a minimal common rotation speed. The difference of these two concepts is depicted in Fig. 1. Note that this alternative notion of rotation reduces the effect of strain and shear components on the computation of rotation directions.

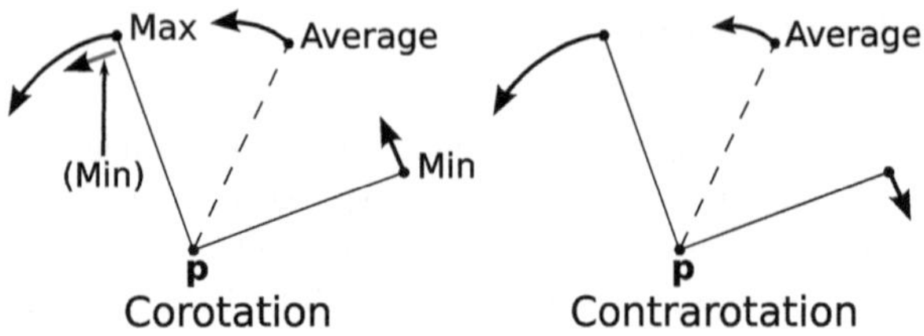

Fig. 1 An illustration of corotation and contrarotation of two line segments at a point p, based on Fig. 1 from Kolář et al. [1]. "Max" is the maximal rotation speed, while "average" and "min" are average and minimum rotational speeds in the corotation example, respectively. On the *left* we have corotation, where residual vorticity is the least absolute value angular velocity of the line segments shown (min). Average is just ordinary vorticity for both cases. Vorticity is still proportional to the average of all rotation speeds in the contrarotation example, which does not correspond to physical rotation in the plane

3.2 *Local Corotation*

The fact that residual vorticity is evaluated by maximizing local corotation over all possible plane orientations in 3D space makes it possible to analyze rotation in directions that do not correspond to the orientation with the globally maximal corotation. In the following we establish the mathematical background by describing how local corotation is defined for arbitrary orientations in 3D space.

Given a position $p \in \mathbb{R}^3$ in a velocity field $f : \mathbb{R}^3 \rightarrow \mathbb{R}^3$ with local gradient ∇f and an arbitrary orientation $v(\alpha, \beta) = (cos(\alpha) \cdot sin(\beta), sin(\alpha) \cdot sin(\beta), cos(\beta))$, quasiplanar residual vorticity as a measure of local corotation [1] is defined as:

$$\omega_{res}(\alpha, \beta) = sign(\omega(\alpha, \beta))(|\omega(\alpha, \beta)| - |s_D(\alpha, \beta)|), \tag{1}$$

where ω and S_D are the two-dimensional vorticity and deviatoric strain on the plane with normal v. Additionally, if $|\omega| \leqslant |s_D|$ then $\omega_{res} = 0$. Two-dimensional vorticity and deviatoric strain are computed as follows.

The quasiplanar effects of the velocity gradient tensor on an arbitrary plane with normal v is obtained by projecting the three-dimensional velocity gradient tensor into a two-dimensional coordinate frame orthogonal to v:

$$\nabla f^* = \begin{pmatrix} cos(\alpha) \cdot cos(\beta) & -sin(\alpha) \\ sin(\alpha) \cdot \cos(\beta) & cos(\alpha) \\ -sin(\beta) & 0 \end{pmatrix}^T \nabla f \begin{pmatrix} cos(\alpha) \cdot cos(\beta) & -sin(\alpha) \\ sin(\alpha) \cdot \cos(\beta) & cos(\alpha) \\ -sin(\beta) & 0 \end{pmatrix}. \tag{2}$$

With this two-dimensional velocity gradient tensor, quasiplanar vorticity and deviatoric strain are defined as:

$$\omega = \frac{\nabla f_{21}^* - \nabla f_{12}^*}{2} \quad \text{and} \quad S_D = \frac{\sqrt{(\nabla f_{11}^* - \nabla f_{22}^*)^2 + (\nabla f_{12}^* + \nabla f_{21}^*)^2}}{2} \tag{3}$$

As a result, local corotation in the form of residual vorticity can be defined for all positions v on a unit-sphere, with corotation scalar magnitude representing the minimal common angular rotation of line elements on a corresponding plane. In the following we visualize and analyze the spherical function given by ω_{res}.

4 Corotation Visualization and Topological Features

The fact that ω_{res} is capable of not only revealing the direction with maximal rotation, as commonly used for vortex core extraction, but the amount of rotation present in other directions in space makes it a prime candidate for detailed rotation analysis. We now discuss how we visualize the corotation function using spherical glyphs, and explain the various topological properties of these shapes.

4.1 Glyph Creation

Since positions on a unit sphere encode all possible orientations in 3D space, the complete domain of ω_{res} can be visualized by modifying such a spherical representation. Note that the spherical function ω_{res} may in theory be approximated as a higher-order tensor, allowing the application of existing glyph generation techniques, such as the one specified by Schultz and others [29, 30]. Such an approximation, however, is outside the scope of this chapter. The spherical meshes used in this work are created using icosahedron subdivision, as the resulting triangles of this mesh have equal areas and are suitable for easy level-of-detail control as shown by Schultz and Kindlmann [34]. Unlike a triangulation based on spherical coordinates, these triangles do not become distorted around the poles of the sphere.

A depiction of a spherical glyph is shown in Fig. 2, where corotation values are indicated by offsetting vertices of a unit sphere along the normal according to the local magnitude of ω_{res}. For this purpose, we first convert the Cartesian coordinates of each mesh vertex (relative to the center of the sphere) into a spherical coordinate representation, obtaining azimuthal and polar angles α and β, respectively. Next, we compute the mesh offset as the corotation scalar magnitude for the angle pair (α, β), and normalize the offset based on the range of corotation values over the entire sphere. We choose this type of local offset normalization over a global data-set-wide normalization, since exceptionally strong vortical features in flow simulations tend to lead to an unbalanced scaling. The offset value used per angle pair is the absolute value of ω_{res}, as we are interested in highest values of corotation magnitude that exist on the sphere.

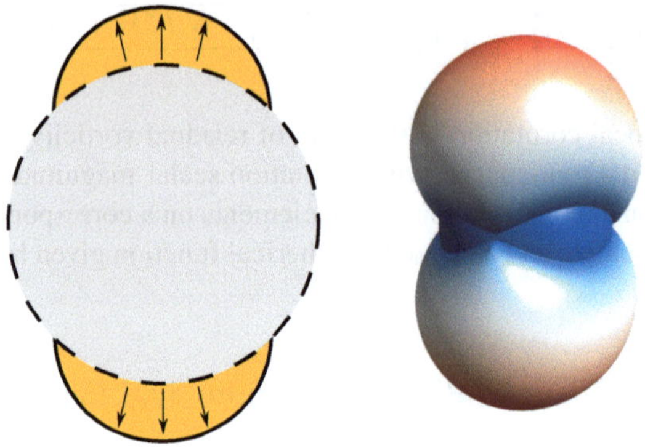

Fig. 2 Example of a spherical glyph, where its vertices are offset at positions where the corotation scalar magnitude is larger than zero. The *arrows* in the figure indicate where these non-zero values exist, while the base sphere (which corresponds to zero) is drawn with a *dashed line*. The colormap of the glyph varies from *blue* (weak corotation) to *red* (strong corotation)

4.2 Glyph and Function Properties

The scalar residual vorticity function and the resulting glyph representation have many interesting characteristics, some of which influence the applicability of topological methods as described in the next section. The ω_{res} function that defines the glyphs is specified in the domain $[0, \pi]$ for the angles α and β, and is periodic beyond that domain. As a result, the spherical glyph is point-symmetric with respect to its center, meaning that each extremum exists twice.

The variable behavior of the gradient magnitude of residual vorticity can often make the glyph-based representation difficult to predict. As a consequence, the shape of maximal regions in a glyph can vary strongly, and can include nearly flat regions with small gradient values, and peaks with extreme gradients. In addition, the use of absolute value and sign operators in the computation of residual vorticity introduces discontinuous derivatives. This can be observed in the form of C^0 continuous regions in glyph shapes. Large regions of the function may be zero, for directions where the field is not rotating, or deviatoric strain exceeds quasiplanar vorticity.

In general, i.e., when the flow field is not irrotational, ω_{res} has at least one maximum pointing along the direction of maximal rotation (two opposing maxima on the glyph representation). There are, however, multiple cases when additional local maxima can occur. The analysis of glyph topology with respect to the number of maxima present and shape characteristics of the corresponding topological regions can aid in understanding vortical features in the data set as described in the following section.

4.3 Topological Analysis of Individual Glyphs

The analysis of the topological properties of the glyph has the potential to indicate the existence of one or multiple related vortex directions. For instance, a glyph with two (opposing) peaks with two associated topological regions indicates that there is a vortex that runs through the center of the glyph, with the axis of the vortex being parallel to those peaks. Peaks that are associated with larger and less sharp topological regions indicate multiple rotation directions with a less distinct vortex core direction. If a glyph has additional maxima, then there exist additional vortex directions at those positions of the glyph, especially if the additional peaks are large relative to other peaks that exist. In the following we discuss methods for topological analysis of these glyphs and provide details about a sample implementation.

As mentioned in the previous section, the ω_{res} function and therefore the topology of the glyph shape may contain multiple local maxima. This property, coupled with the lack of C^1-continuity, precludes us from finding the maxima analytically. Furthermore, it is difficult to perform gradient ascent on this function because of the existence of sharp ridges (due to a discontinuous gradient), and because an ω_{res} function for a given ∇f tensor may have multiple maxima. In the latter case, it is challenging to find an appropriate seed point for gradient ascent that gives us the location of the global maximum. An alternative to finding a maximum explicitly could be based on ray-tracing, which is often used to represent the shape of a glyph. While it is possible to ray-trace the glyph in order to examine the topology of the glyph in image space, one will require a high-resolution image to make this analysis work and the analysis would be view-dependent. For these reasons, we employ an approximation of topological structures by directly examining the glyph meshes. One possible solution is to find local ω_{res} maxima within a neighborhood of the mesh. Unfortunately, it is difficult to perform such a maxima search with fixed neighborhood sizes due to the possible existence of high frequency features in the mesh.

We employ a watershed approach [35] instead, which effectively propagates the labels of maxima to other mesh vertices. We first sort all of the vertices of the mesh by function value, and identify the first vertex as a labeled maximum. As we move along the sorted list, we effectively move downwards in corotation value and identify vertices or "nodes" which are adjacent to labeled regions and which propagate those labels, or vertices where new maxima come into existence. We also track the merging behavior of regions with different labels by creating saddle points, which are assigned the label of the maximum with the largest ω_{res} value. We identify regions with nearly identical function values, or "plateaus," and treat these regions as single nodes during the labeling process. If there are multiple maxima connected via a large plateau, the plateau receives the label of the largest maximum and effectively connects it to the other maxima. After the watershed algorithm is completed, we create watershed graphs, which are similar to the "split trees" defined by Carr et al. [36]. We can filter the maxima of this tree that result from sampling artifacts, and color resulting descending manifolds of a glyph. An example of a glyph, and its corresponding unfiltered and filtered trees are shown in Fig. 3.

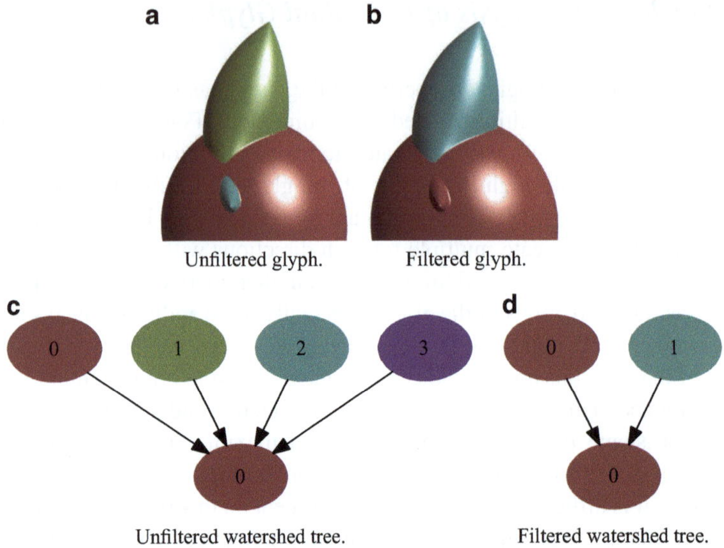

Fig. 3 Example of a glyph (**a**) along with its normal watershed tree (**c**), and the corresponding "filtered" representations (**b**), (**d**). The unfiltered versions contains small peaks, which include the diminutive, *teal-colored* maximum close to the center of the glyph in (**a**) and other (hidden from view) maxima. All maxima are connected to a large plateau (base icosahedron sphere). In the filtered versions, we removed small branches of the tree corresponding to these spurious maxima and subsequently recompute the topological regions

4.4 Topological Analysis of Multiple Glyphs

If multiple glyphs are positioned along a grid of a flow field, one can observe how their topological regions change as the field is traversed. In order to detect these changes, we first identify canonical glyph examples of various topologies that exist in the test data sets used in this chapter. In Fig. 4, we show these glyphs, and discuss how it is possible to have glyphs with well-defined or ambiguous directions for two or four peaks. Since the corotation function is symmetric, values beyond the domains $[0, \pi]$ for α and β are repeated in the graphs for this figure.

In Fig. 4a, c, e, g, we show visualizations and plots of large topological regions for two and four peaks, respectively. Each peak in these cases indicates multiple directions with strong rotations in large areas of the sphere. This is in stark contrast to Fig. 4b, d, f, h, where the pronounced peaks indicate unambiguous vortex directions for the two and four peak cases, respectively. It is notable how this glyph representation allows for the analysis of rotation in multiple directions, a feature that is not possible in classic direction and magnitude-based vortex visualizations.

Glyphs are likely to have two peaks that point along the forward and backward directions of the vortex core's axis at certain positions along a vortex core line, as seen in Fig. 4a, b. In other cases, glyphs might have four peaks (Fig. 4c, d), indicating

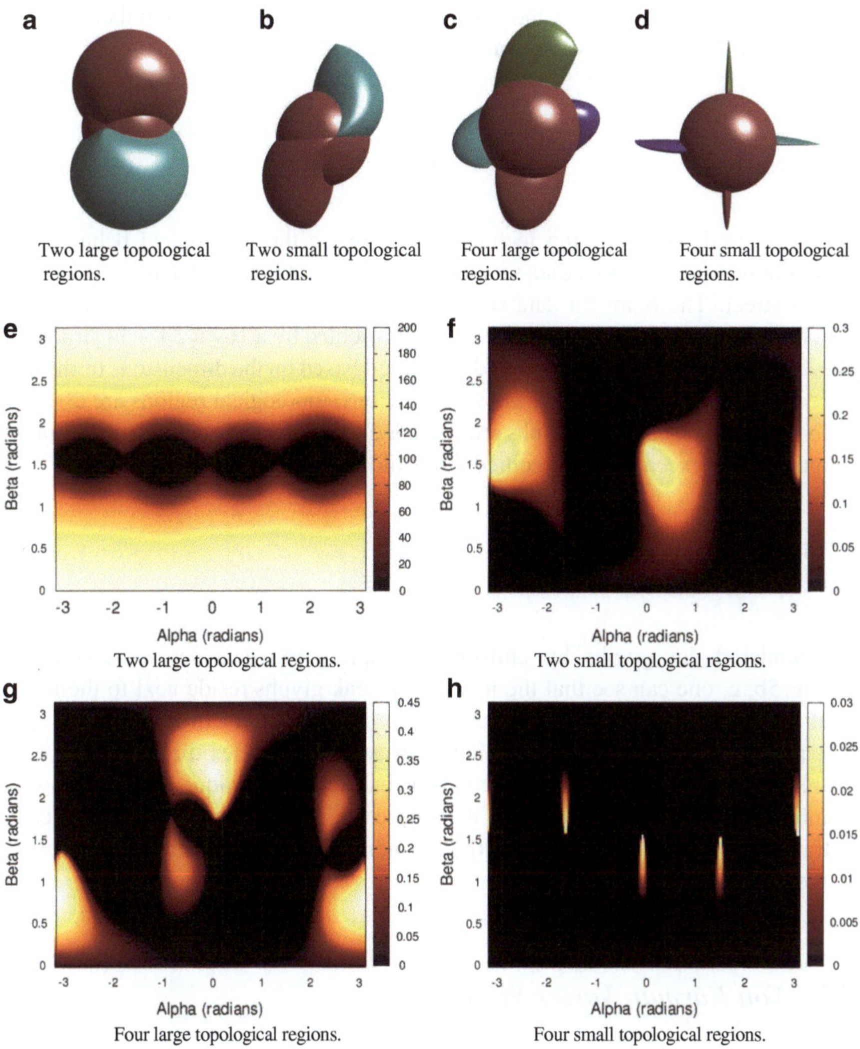

Fig. 4 Visualizations of various types of glyph topologies that we have observed, as well as their corresponding graphs (where the range of scalar corotation values is indicated by a color gradient of *black* to *yellow*). The corotation function is periodic, which explains why the visualizations and graphs portray symmetry. One can see that there may exist multiple directions with strong rotation close to the maximal corotation values in (**a**), (**e**), while in (**b**), (**f**) the maximum point is unique. Similarly, we observe four large topological regions in (**c**), (**g**), while we observe sharp, unambiguous peaks in (**d**), (**h**)

multiple orientations with maximal corotation. If the number of topological regions along a vortex increases from two to four, more vortex directions will start to appear. If two of the glyph's four peaks begin to wane, then the remaining two peaks will indicate dominating vortex directions. Additionally, the widening or narrowing of individual topological regions indicate an increasing or decreasing number of vortex directions in the region of each peak, respectively.

5 Results

We have used our rendering technique to observe the topological behaviors of glyphs in our data sets. Our data sets include the "Blunt Fin" [37] and a von Kármán vortex street. The blunt fin data set is represented by a $40 \times 32 \times 32$ structured curvilinear grid, while the vortex street is represented by a $167 \times 34 \times 34$ structured curvilinear grid. The glyphs rendered are scaled based on the dimensions of the data set cell that they reside in. In these visualizations, topological regions are assigned individual colors. Sampled glyphs are excluded assuming their individual corotation ranges did not exceed a pre-specified threshold value.

5.1 Blunt Fin

We rendered our glyphs by uniformly sampling the data set as seen Fig. 5. In Fig. 5b, c, one can see that the many four-peak glyphs reside next to their two-peak counterparts. In Fig. 5d, e, we observe more instances of four-peak glyphs. The presence of this type of glyph indicates additional vortex directions that exist at this position. A standard vortex core line extraction technique only portrays the dominant vortex axis per point. In the future we hope to employ a sophisticated extraction technique to portray splitting or merging behaviors that could be related to these additional vortex directions.

5.2 Von Kármán Vortex Street

We visualized glyphs in the Von Kármán Vortex Street data set by uniformly sampling the grid as seen in Fig. 6. We observed glyphs near five stable vortex regions parallel to the y-axis of the data set, and the glyphs' peaks are aligned with the y-axis as we expected. Most of the glyphs' topologies are similar to their neighbors, and they mostly point in the same direction. This indicates a lack of additional vortex directions per glyph, a behavior that was present in the Blunt Fin data set.

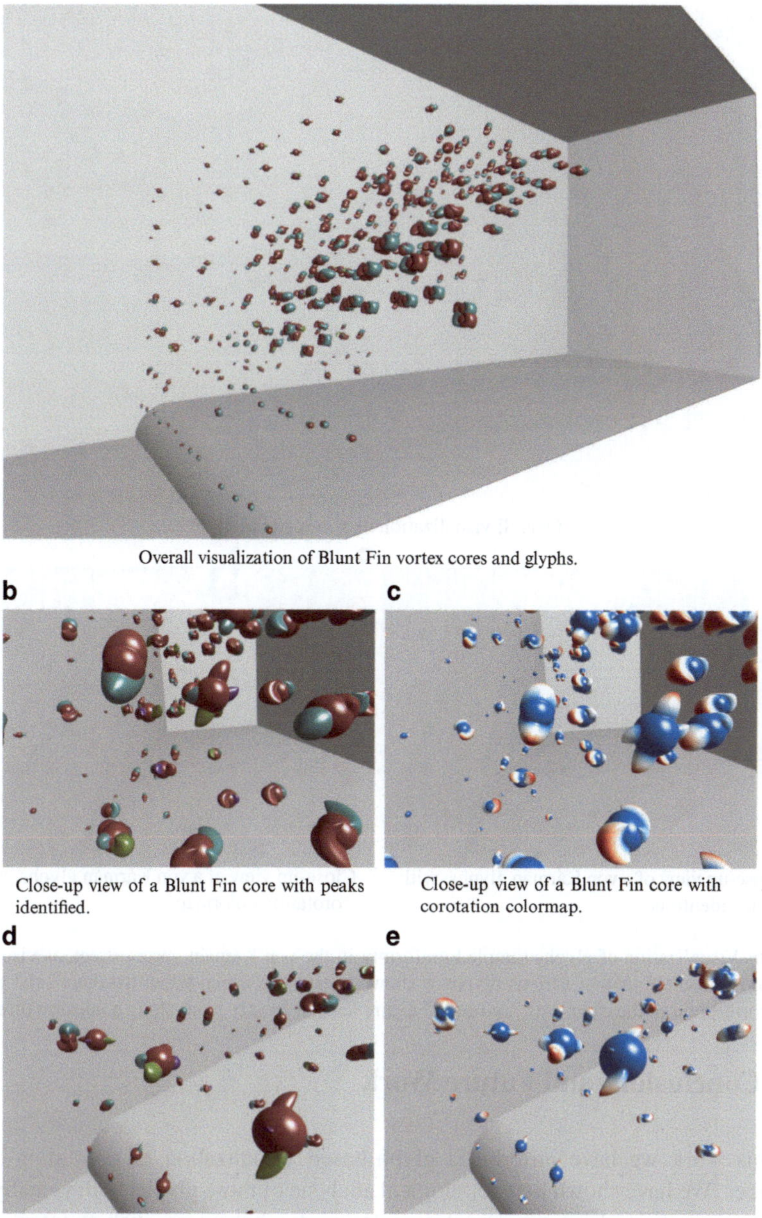

a Overall visualization of Blunt Fin vortex cores and glyphs.

b Close-up view of a Blunt Fin core with peaks identified.

c Close-up view of a Blunt Fin core with corotation colormap.

d Close-up view of a Blunt Fin core with peaks identified.

e Close-up view of a Blunt Fin core with corotation colormap.

Fig. 5 Visualization of glyphs sampled uniformly in the Blunt Fin data set. The vortices in this data set are parallel with the blunt fin boundary and are parallel with the longest dimension of the data set. A visualization of most of the glyphs is shown in (**a**). The first close-up visualization is shown in (**b**), (**c**), where there exist many glyphs that possess four peaks. Figure (**d**), (**e**) show more four-peak glyphs

a

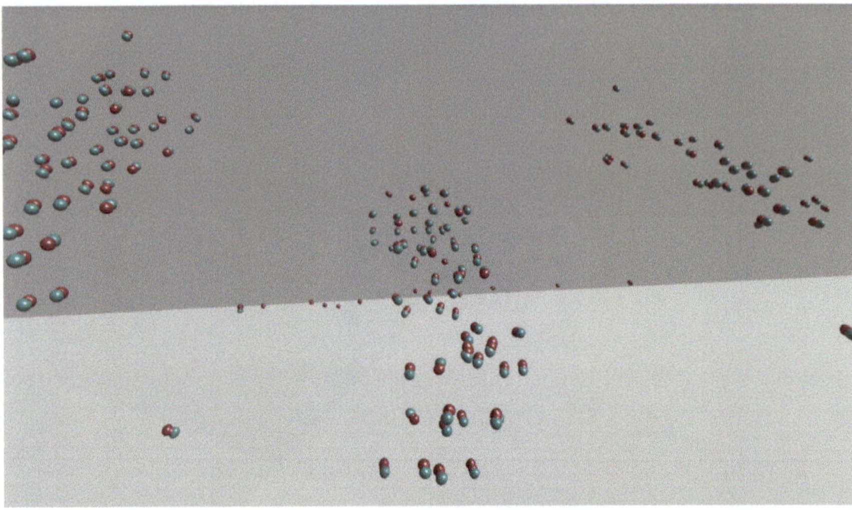

Overall visualization of von Kármán glyphs.

b

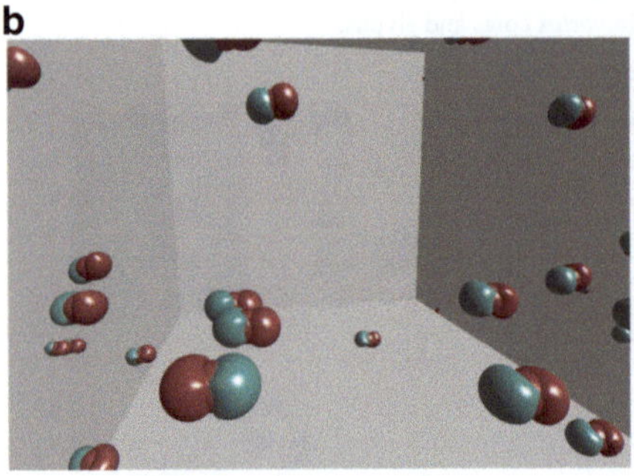

Close-up view of a von Kármán glyphs, with peaks identified.

c

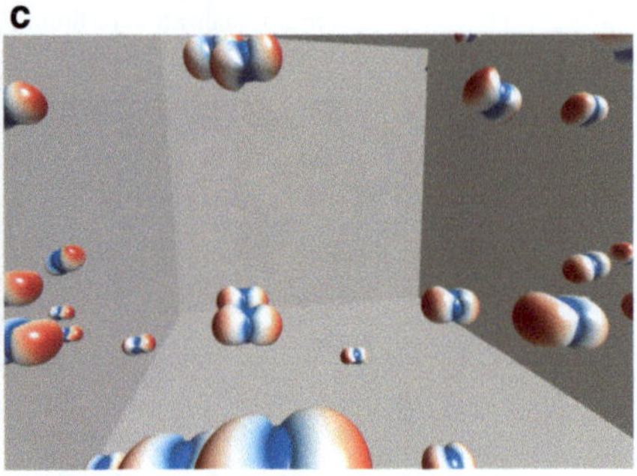

Close-up view of a von Kármán glyphs, with corotation colormap.

Fig. 6 Visualization of glyphs sampled uniformly in the von Kármán vortex street, as visualized in (**a**). A close-up view of various glyphs is shown in (**b**), (**c**), which show that there exist glyphs that correspond to the canonical "two-peak" examples that mostly point along a uniform direction

6 Conclusions and Future Work

In this work we have introduced glyph-based visualization of corotation-based vortices. We have shown how topological analysis of these glyphs both visually and in an automatic fashion can reveal complex vortex behaviors, such as interactions between rotational components present in a single core. In the future we hope to represent the corotation function using multiple levels of detail in order to sample the function more effectively, and investigate how one can extract glyphs in time-dependent flows as the current method applies to discrete time steps. Furthermore, we hope to visually represent vortex core splits and merges in conjunction with the glyphs to investigate a possible relationship between the two.

Acknowledgements This work was partially supported by the Materials Design Institute, funded by the UC Davis/LANL Research Collaboration (LANL Agreement No. 75782-001-09). It was also supported by the NSF under contracts IIS 0916289 and IIS 1018097, the Office of Advanced Scientific Computing Research, Office of Science, of the US DOE under Contract No. DE-FC02-06ER25780 through the SciDAC programs VACET, and contract DE-FC02-12ER26072, SDAV Institute. We thank Simon Stegmaier for making available his software [16].

References

1. V. Kolář, P. Moses, J. Sistek, Local corotation of line segments and vortex identification, in *Proceedings of the Seventeenth Australasian Fluid Mechanics Conference*, Auckland, ed. by G. Mallinson, J. Cater (2010), pp. 251–254
2. J.C.R. Hunt, A. Wray, P. Moin, Eddies, stream, and convergence zones in turbulent flows. Center for Turbulence Research Report CTR-S88, 1988, pp. 193–208
3. U. Dallmann, Topological structures of three-dimensional vortex flow separation, in *16th American Institute of Aeronautics and Astronautics, Fluid and Plasma Dynamics Conference*, Danvers, 1983
4. H. Vollmers, H. Kreplin, H. Meier, Separation and vortical-type flow around a prolate spheroid-evaluation of relevant parameters, in *Proceedings of the AGARD Symposium on Aerodynamics of Vortical Type Flows in Three Dimensions*, Rotterdam, 1983, pp. 14-1–14-14
5. M. Chong, A. Perry, B. Cantwell, A general classification of three-dimensional flow fields. Phys. Fluids **2**, 765–777 (1990)
6. J. Jeong, F. Hussain, On the identification of a vortex. J. Fluid Mech. **285**, 69–94 (1995)
7. V. Kolář, Vortex identification: new requirements and limitations. Intern. J. Heat Fluid Flow **28**(4), 638–652 (2007)
8. M. Roth, R. Peikert, A higher-order method for finding vortex core lines, in *Proceedings of the Conference on Visualization'98*, Minneapolis (IEEE, 1998), pp. 143–150
9. D. Sujudi, R. Haimes, Identification of swirling flow in 3D vector fields, in *AIAA 12th Computational Fluid Dynamics Conference Paper*, San Diego, 1995, pp. 95–1715.
10. D. Kenwright, R. Haimes, Automatic vortex core detection. IEEE Comput. Graph. Appl. **18**, 70–74 (1998)
11. B. Singer, D. Banks, A predictor-corrector scheme for vortex identification. Technical report, TR-94–11, 1994
12. D. Banks, B. Singer, Vortex tubes in turbulent flows: identification, representation, reconstruction, in *Proceedings of the Conference on Visualization'94*, Washington, DC (IEEE, 1994), pp. 132–139
13. D.C. Banks, B.A. Singer, A predictor-corrector technique for visualizing unsteady flow. IEEE Trans. Vis. Comput. Graph. **1**, 151–163 (1995)
14. J. Sahner, T. Weinkauf, H. Hege, Galilean invariant extraction and iconic representation of vortex core lines, in *IEEE VGTC Symposium on Visualization*, Leeds, 2005, pp. 151–160
15. J. Sahner, T. Weinkauf, N. Teuber, H.C. Hege, Vortex and strain skeletons in eulerian and lagrangian frames. IEEE Trans. Vis. Comput. Graph. **13**(5), 980–990 (2007)
16. S. Stegmaier, U. Rist, T. Ertl, Opening the can of worms: an exploration tool for vortical flows, in *IEEE Visualization Conference*, Minneapolis, 2005, pp. 463–470
17. T. Schafhitzel, D. Weiskopf, T. Ertl, Interactive investigation and visualization of 3D vortex structures, in *Electronic Proceedings of 12th International Symposium on Flow Visualization*, Gottingen, Sept 2006
18. T. Schafhitzel, J. Vollrath, J. Gois, D. Weiskopf, A. Castelo, T. Ertl, Topology-preserving $\lambda 2$-based vortex core line detection for flow visualization, in *Computer Graphics Forum*, vol. 27 (Wiley Online Library, 2008), pp. 1023–1030

19. K. Baysal, T. Schafhitzel, T. Ertl, U. Rist, Extraction and visualization of flow features, in *Imaging Measurement Methods for Flow Analysis*. Volume 106 of Notes on Numerical Fluid Mechanics and Multidisciplinary Design, ed. by W. Nitsche, C. Dobriloff (Springer, Berlin/Heidelberg, 2009), pp. 305–314
20. F.H. Post, B. Vrolijk, H. Hauser, R.S. Laramee, H. Doleisch, The state of the art in flow visualisation: feature extraction and tracking. Comput. Graph. Forum **22**, 775–792 (2003)
21. M. Jiang, R. Machiraju, D. Thompson, Detection and visualization of vortices, in *The Visualization Handbook*, ed. by C.D. Hansen, C.R. Johnson (Elsevier, Amsterdam, 2005), pp. 295–309
22. C. Garth, X. Tricoche, T. Salzbrunn, T. Bobach, G. Scheuermann, Surface techniques for vortex visualization, in *VisSym*, Konstanz, 2004, pp. 155–164
23. M. Jankun-Kelly, M. Jiang, D. Thompson, R. Machiraju, Vortex visualization for practical engineering applications. IEEE Trans. Vis. Comput. Graph. **12**(5), 957–964 (2006)
24. X. Tricoche, C. Garth, G. Kindlmann, E. Deines, G. Scheuermann, M. Ruetten, C. Hansen, Visualization of intricate flow structures for vortex breakdown analysis, in *Proceedings of the Conference on Visualization, VIS'04*, Washington, DC (IEEE, 2004), pp. 187–194
25. C.D. Shaw, D.S. Ebert, J.M. Kukla, A. Zwa, I. Soboroff, D.A. Roberts, Data visualization using automatic perceptually motivated shapes, in *SPIE Conference on Visual Data Exploration and Analysis*, San Jose, 1998, pp. 208–213
26. F.H. Post, F.J. Post, T.V. Walsum, D. Silver, Iconic techniques for feature visualization, in *Proceedings of the 6th Conference on Visualization, VIS'95*, Washington, DC (IEEE, 1995), pp. 288–295
27. A. Wiebel, S. Koch, G. Scheuermann, Glyphs for non-linear vector field singularities, in *Topological Methods in Data Analysis and Visualization II*, ed. by R. Peikert, H. Hauser, H. Carr, R. Fuchs. Mathematics and Visualization (Springer, Berlin/Heidelberg, 2012), pp. 177–190
28. G. Kindlmann, Superquadric tensor glyphs, in *Proceedings of the Sixth Joint Eurographics – IEEE TCVG Conference on Visualization, VISSYM'04*, Aire-la-Ville (Eurographics Association, 2004), pp. 147–154
29. T. Schultz, C.F. Westin, G. Kindlmann, Multi-diffusion-tensor fitting via spherical deconvolution: a unifying framework, in *Proceedings of the 13th International Conference on Medical Image Computing and Computer-Assisted Intervention: Part I, MICCAI'10*, Beijing (Springer, Berlin/Heidelberg, 2010), pp. 674–681
30. T. Schultz, Towards resolving fiber crossings with higher order tensor inpainting, in *New Developments in the Visualization and Processing of Tensor Fields*, ed. by D.H. Laidlaw, A. Vilanova. Mathematics and Visualization (Springer, Berlin/Heidelberg, 2012), pp. 253–265
31. T. Peeters, V. Prckovska, M. van Almsick, A. Vilanova, B. ter Haar Romeny, Fast and sleek glyph rendering for interactive hardi data exploration, in *Visualization Symposium, PacificVis'09*, Beijing (IEEE, Apr 2009), pp. 153–160
32. M. van Almsick, T.H. Peeters, V. Prčkovska, A. Vilanova, B. ter Haar Romeny, GPU-based ray-casting of spherical functions applied to high angular resolution diffusion imaging. IEEE Trans. Visual. Comput. Graph. **17**(5), 612–625 (2011)
33. R. Borgo, J. Kehrer, D.H.S. Chung, E. Maguire, R.S. Laramee, H. Hauser, M. Ward, M. Chen, Glyph-based visualization: foundations, design guidelines, techniques and applications, in *Eurographics State of the Art Reports* (EG STARs, Eurographics Association, May 2013), pp. 39–63. http://diglib.eg.org/EG/DL/conf/EG2013/stars/039-063.pdf
34. T. Schultz, G. Kindlmann, A maximum enhancing higher-order tensor glyph. Comput. Graph. Forum **29**(3), 1143–1152 (2010)
35. S. Beucher, F. Meyer, The morphological approach to segmentation: the watershed transformation (Mathematical morphology in image processing). Opt. Eng. **34**, 433–481 (1993)
36. H. Carr, J. Snoeyink, U. Axen, Computing contour trees in all dimensions. Comput. Geom. Theory Appl. **24**(2), 75–94 (2003)
37. C. Hung, P. Buning, Simulation of blunt-fin-induced shock-wave and turbulent boundary-layer interaction. J. Fluid Mech. **154**(1), 163–185 (1985)

Index

P.-T. Bremer et al. (eds.), *Topological Methods in Data Analysis and Visualization III*, Mathematics and Visualization, DOI 10.1007/978-3-319-04099-8,

MIX
Papier aus verantwortungsvollen Quellen
Paper from responsible sources
FSC® C105338

If you have any concerns about our products,
you can contact us on
ProductSafety@springernature.com

In case Publisher is established outside the EU,
the EU authorized representative is:
Springer Nature Customer Service Center GmbH
Europaplatz 3, 69115 Heidelberg, Germany

Printed by Libri Plureos GmbH
in Hamburg, Germany